Starr, Evers, and Starr's
Biology: Concepts and Applications

EIGHTH EDITION

Michael Windelspecht
Appalachian State University

Amy Fenster
Virginia Western Community College and Hollins University

John D. Jackson
North Hennepin Community College

Jane B. Taylor
Northern Virginia Community College

Edited by

Sylvester Allred
Northern Arizona University

BROOKS/COLE
CENGAGE Learning

Australia • Brazil • Japan • Korea • Mexico • Singapore • Spain • United Kingdom • United States

ISBN-13: 978-0-8400-4816-5
ISBN-10: 0-8400-4816-5

Brooks/Cole
20 Davis Drive
Belmont, CA 94002-3098
USA

Cengage Learning is a leading provider of customized learning solutions with office locations around the globe, including Singapore, the United Kingdom, Australia, Mexico, Brazil, and Japan. Locate your local office at:
www.cengage.com/global

Cengage Learning products are represented in Canada by Nelson Education, Ltd.

To learn more about Brooks/Cole, visit
www.cengage.com/brookscole

Purchase any of our products at your local college store or at our preferred online store
www.cengagebrain.com

Printed in the United States of America
1 2 3 4 5 6 7 14 13 12 11 10

TABLE OF CONTENTS

Credits and Sources

Chapter 4
p.46: Lisa Starr

Chapter 5
p.57: PDB files from NYU Scientific Visualization Lab
p.59: Raychel Ciemma

Chapter 15
p.148: Courtesy of © Genelex Corp.

Chapter 18
p.176-177: Raychel Ciemma and Precision Graphics, Inc.

Chapter 19
p.185: Lisa Starr

Chapter 20
p.193: © Dr. David Phillips/ Visual Unlimited
p.194: (a) © Sinclair Stammers/ Photo Researchers, Inc.; (c) London School of Hygiene & Tropical Medicine/ Photo Researchers, Inc.; (e) Micrograph Steven L'Hernaults

Chapter 22
p.218: Left, Garry T. Cole, University of Texas, Austin/BPS; Right, © Eye of Science/Photo Researchers, Inc.; Art, After T. Rost, et al., *Botany*, Wiley, 1979.

Chapter 23
p.226: (c) © Andrew Syred/SPL/Photo Researchers, Inc.
p.228: Palay/Beaubois

Chapter 24
p.237: Gary Head
p.239: from E. Solomon, L. Berg, and D.W. Martin, *Biology*, Seventh Edition, Thomson Brooks/Cole

Chapter 25
p.248: Raychel Ciemma
p.250: Art, Lisa Starr; Photograph, © Andrew Syred/Photo Researchers, Inc.
p.252: (1-6) Robert and Linda Mitchell Photography
p.252: (7-13) Art, Raychel Ciemma
p.252: (14) © Carolina Biological Supply Company
p.252: (15) Mike Clayton, University of Wisconsin, Botany Department
p.254: (1-11) After Salisbury and Ross, *Plant Physiology*, Fourth Edition, Wadsworth

Chapter 26
p.254: (1-5, 5-6) Michael Clayton, University of Wisconsin, Botany Department
p.254: (1-6 and 10) Omikron/Photo Researchers, Inc.
p.255: (7-9) H.A. Core, W.A. Cote, and A.C. Day, *Wood Structure and Identification*, Second Edition, University Press, 1979.

Chapter 26
p.262: (9-12) Micrograph Chuck Brown

Chapter 27
p.269: Raychel Ciemma
p.270: (left) ©John McAnulty/CORBIS; (right) © Robert Essel NYC/CORBIS
p.272: Raychel Ciemma

Chapter 28
p.287: Robert Demarest

Chapter 29
p.293: Kevin Somerville
p.298: Robert Demarest
p.299: Robert Demarest
p.302: © Colin Chumbley/Science Source/Photo Researchers, Inc.

Chapter 30
p.309: Robert Demarest
p.312: Robert Demarest

Chapter 32
p.327: Raychel Ciemma
p.331: Raychel Ciemma

Chapter 33
p.340: Kevin Somerville
p.341: Raychel Ciemma
p.345: left, © Biophoto Associates/Photo Researchers, Inc.

Chapter 35
p.364: Kevin Somerville

Chapter 37
p.385 (top): Robert Demarest
p.385 (bottom): Precision Graphics, Inc.

Chapter 40
p.425: Preface, Inc. and Precision Graphics, Inc.

PREFACE

Tell me and I will forget, show me and I might remember, involve me and I will understand.

—Chinese Proverb

This insightful proverb outlines three levels of learning, each successively more effective than the preceding one. The writer of the proverb understood that we learn most efficiently when we involve ourselves in the material to be learned. This workbook is like a tutor; when used properly, it increases the efficiency of your study periods. The workbook's interactive exercises actively involve you in the most important terms and central ideas of your text. Specific tasks ask you to recall key concepts and terms and apply them to life; they test your understanding of the facts and indicate items to reexamine or clarify. Your performance on these tasks will help you estimate your performance on in-class test of similar material. Most important, however, this biology workbook and text together help you make informed decisions about matters that affect not only your own well being, but also that of your environment. In the years to come, human survival on Earth will demand that individual, national, and global decisions be based on an informed biological background.

HOW TO USE THIS STUDENT WORKBOOK

Each chapter of this workbook begins with a title and a brief introduction regarding the content of the chapter. This is followed by a set of bulleted focal points. The purpose of the focal points is to direct you to key figures, tables, or material in the chapter. For easy reference to an answer or definition, each question and term in this workbook is accompanied by the appropriate text page(s), and appears in the form: [p.352], for example. The Interactive Exercises begin with a list of Selected Words (other than boldfaced terms) chosen by the authors as those that are most likely to enhance your understanding of the material. In the text chapters, these words appear in italics, quotation marks, or normal type. Next, a list of Boldfaced Terms from the text appears. These terms are essential to your understanding of each text section. Space is provided by each term for you to formulate a definition in your own words. After the terms is a series of different types of exercises that may include completion, short answer, true/false, matching, choice, dichotomous choice, identification, problems, labeling, sequence, concept maps, and complete the table.

A Self Quiz immediately follows the Interactive Exercises. This quiz is composed primarily of multiple-choice questions. Any wrong answers in the Self Quiz indicate portions of the text you need to reexamine. A series of Chapter Objectives/Review Questions follows. These are tasks that you should be able to accomplish if you have understood the assigned reading in the text. Following this section is a Chapter Summary. This is a fill-in-the-blank version of the Key Concepts material at the beginning of the chapter.

Each workbook chapter concludes with the Integrating and Applying Key Concepts section. It invites you to try your hand at applying major text concepts to situations in which a single answer does not necessarily work—and so non are provided in the chapter answer section. Your text generally will provide enough clues to get you started on an answer, but this part is intended to stimulate thought and provoke lively group discussions.

A person's mind, once stretched by a new idea, can never return to its original dimension.

—Oliver Wendell Holmes

STRUCTURE OF THIS STUDENT WORKBOOK

The following outline shows how each chapter in this student workbook is organized.

Chapter Number —————⟶

32

Chapter Title —————⟶

STRUCTURAL SUPPORT AND MOVEMENT

Introduction —————⟶

This includes a brief overview to the subject matter of the chapter. This chapter examines how muscles and bones interact in the process of movement. It begins with an introduction of animals skeletons in general and leads to the more complex. The human skeleton is details with diagram and description. A diagram of the knee is given as well as aliments that can occur to the skeleton. Next the anatomical features of joints with what they can and shouldn't do. The chapter introduces the basic principles of how skeletons and muscles function, as well as some of the diseases that are common with these two systems.

Focal Points —————⟶

- Figure 32.2 [p.540] illustrates how cells are attached to each other to form coherent tissues.
- Figure 32.4 [p.541] shows various types of epithelium.
- Figure 32.5 and 32.6 [pp.542-543] illustrate connective tissues.
- Figure 32.8 [p.544] has images of muscle tissues.
- Figure 32.9 [p.545] shows a typical neuron.
- Figure 32.11 and 32.12 [pp.546-547] describe anatomical terms and outline the major human organ systems.
- Figure 32.13 [p.568] diagrams human skin structure.

Interactive Exercises —————⟶

The Interactive Exercises are divided into numbered sections by titles of main headings and page references. Each section begins with a list of author-selected words that appear in the chapter. This is followed by a list of important boldfaced, page-referenced terms from each section of the chapter. Each section ends with interactive exercises that vary in type and require constant interaction with the important chapter information

Self-Test —————⟶

This is a set of questions designed to provide a quick assessment of how well you understand the information from the chapter.

Chapter Objectives / Review Questions —————⟶

This section provides a list of concepts that you need to master before proceeding to the next chapter. Page numbers from the text are provided should you be unable to answer the questions.

Integrating and Applying Key Concepts —————⟶

These represent "big-picture" or "real-life" applications of the concepts presented in the chapter.

Answers to Interactive Exercises and Self-Test —————⟶

Answers for all the Interactive Exercises and the Self-Test can be found at the end of the student workbook. The answers are arranged by chapter number and main heading

1
INVITATION TO BIOLOGY

INTRODUCTION

This first chapter of the text serves as the introduction to many of the main concepts that will be explored throughout the course. Biology is the science that studies life. There are many amazing life forms on our planet and even though many have been discovered, many more await being found and they are most likely going to be small sized as the biggest life forms are hard to miss. It is almost like a game of hide and seek that biologist participate in while conducting their searches. The next chapter in our study of biology begins with the basic building blocks of life – atoms and eventually the journey ends with an interesting chapter on the behavioral ecology of animals. Open your books and your minds to the exciting study of LIFE on EARTH.

FOCAL POINTS
- Figure 1.1 [p.3] is a peek inside a tropical rainforest.
- Figure 1.3 [pp.4-5] illustrates the levels of organization of life.
- Figure 1.4 [p.5] illustrates the flow of energy and the cycling of materials in a world of life.
- Figure 1.6 [pp.8-9] representatives of the diverse life forms on Earth.
- Figure 1.7 [p.10] Linnaean classification of five species related at different levels.
- Figure 1.8 [p.11] the six kingdoms of life.
- Section 1.6 and 1.7 [pp.12-15] explain how scientists explore nature using the scientific method.
- Section 1.8 discusses asking useful questions
- Section 1.9 discusses the philosophy of science

INTERACTIVE EXERCISES
Note: In the answer sections of this book, a specific molecule is most often indicated by its abbreviation. For example, adenosine triphosphate is ATP.

1.1 THE SECRET LIFE OF EARTH [PP. 2-3]

Selected Terms
extinct [p.3], life [p.3] biology [p.3]

Boldfaced Terms
The boldfaced terms are particularly important. Write a definition for each term in your Terms without looking at the text. Next, compare your definition with that given in the chapter or in the text glossary. If your definition seems inadequate, allow some time to pass and repeat this procedure until you can define each term quickly.

biology [p.3] _____

_____

1.2 THE SCIENCE OF NATURE [PP.4-5]

Selected Terms
In addition to the boldfaced terms, the text features other important terms essential to understanding the assigned material. "Selected Terms" is a list of these terms, which appear in the text in italics, in quotation marks, and occasionally in normal type.
Life [p.4], nonliving [p.4]

Boldfaced Terms
[p.4] emergent properties _____

_____

[p.4] atoms _____

_____

[p.4] molecules _____

_____

[p.5] cell _____

_____

[p.5] organism _____

_____

[p.5] tissue _____

_____

[p.5] organ _____

_____

[p.5] organ system_____

_____

[p.5] population_____

_____

[p.5] community _____

_____

[p.5] ecosystem _____

_____

[p.5] biosphere_____

_____

Matching

Match each of the following terms with its correct description. [pp.4-5]

1. ____ atom
2. ____ organ system
3. ____ organ
4. ____ cell
5. ____ organism
6. ____ community
7. ____ biosphere
8. ____ molecule
9. ____ tissue
10. ____ population
11. ____ ecosystem

a. structural unit of two or more tissues interacting in some task
b. all of the regions of the Earth's water, crust, and atmosphere that hold organisms.
c. smallest unit that can live and reproduce on its own or as part of a multicelled organism.
d. organs interacting physically and/or chemically in some task
e. two or more joined atoms of the same or different elements
f. all populations of all species occupying a specified area
g. smallest unit of an element that still retains the element's properties.
h. a community that is interacting with its physical environment
i. individual made of different types of cells
k. a unit of cells interacting in some task
l. group of individuals of the same species in a given area

Concept Map

In the concept map below, provide the missing terms associated with the numbers in parentheses. [pp.4-5]

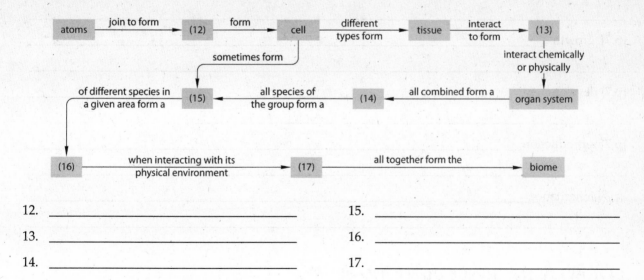

12. _____   15. _____

13. _____   16. _____

14. _____   17. _____

1.3 HOW LIVING THINGS ARE ALIKE [PP.6-7]

Selected Terms
decomposers [p.6], carbon dioxide [p.6]

Boldfaced Terms
[p.6] energy _____

_____

[p.6] nutrient _____

_____

[p.6] producers _____

_____

[p.6] photosynthesis _____

_____

[p.6] consumers _____

_____

[p.7] homeostasis _____

_____

[p.7] DNA _____

_____

[p.7]. growth _____

_____

[p.7] development _____

_____

[p.7] reproduction _____

_____

[p.7] inheritance _____

_____

1.4 HOW LIVING THINGS DIFFER [PP.8-9]

Selected Terms
variation [p.8], single-celled [p.8], decomposers [p.6], *internal* environment [pp.6-7], eukaryotic [p.8], adaptive [p.10], *artificial* selection [p.10], *natural* selection [p.10], "selective agents" [p.10]

Boldfaced Terms

[p.8] biodiversity _____

_____

[p.8] nucleus _____

_____

[p.8] bacteria _____

_____

[p.8] archaeans _____

_____

[p.8] eukaryote _____

_____

[p.8] protists_____

_____

[p.8] fungi _____

_____

[p.8] plants_____

_____

[p.8] animals _____

_____

Matching

Match each of the following definitions with the correct term. [pp.6-7]

1. ____ The capacity to do work.

2. ____ A gas in air

3. ____ A nucleic acid that is the signature molecule of life.

4. ____ Organisms that have the capacity to make their own food.

5. ____ Mechanism by which parents transmit DNA to offspring

6. ____ Transmission of genetic information from parents to offspring.

7. ____ Transformation of cells into an adult organism

8. ____ Maintenance of internal conditions within a range suitable for life

9. ____ Organisms that must get their energy from other organisms.

10. ____ An atom or molecule that has an essential role in growth and survival

a. consumers
b. homeostasis
c. producers
d. development
e. inheritance
f. energy
g. nutrient
h. carbon dioxide
i. DNA
j. reproduction

Choice

For each of the following, choose from one of the following answers. [pp.6-7]

a. energy
b. nutrient
c. both a and b

11. ____ photosynthesis captures this for use by the producers

12. ____ cycled between producers and consumers

13. ____ flows through the world of life in one direction

14. ____ decomposers return this to the environment for the producers

15. ____ consumers must get this from the producers

Matching

Match each of the following kingdom or domain to its correct description. [pp.8-9]

16. _____ protists
17. _____ animals
18. _____ fungi
19. _____ bacteria
20. _____ plants
21. _____ archaeans

a. the simplest of the eukaryotic organisms
b. multicelled, eukaryotic decomposers
c. these organisms are the most numerous on the planet
d. the prokaryotes that are most closely related to the eukaryotes
e. photosynthetic, multicelled eukaryotes
f. multicelled, eukaryotic, consumers

1.5 ORGANIZING INFORMATION ABOUT SPECIES [PP.10-11]

Selected Terms

Linnaeus [p.10], family [p.10], order [p.10], class [p.10], phylum [p.10], kingdom [p.10], domain [p.10], morphological trait [p.10], biochemistry [p.11], biological species concept [p.11]

Boldfaced Terms

[p.10] species _____

_____

[p.10] taxonomy _____

_____

[p.10] genus _____

_____

[p.10] specific epithet _____

_____

[p.10] taxon _____

_____

Matching

Match the following [pp.10-11]

1. _____ a group of species that share a unique set of traits
2. _____ a type of organism
3. _____ a grouping of organisms
4. _____ second part of a species name
5. _____ the science of naming and classifying species

a. species
b. specific epithet
c. taxonomy
d. genus
e. taxon

Matching
Match the following about the three domains of life with their characteristics [p.10]

6. ____ single cell, no nucleus, most ancient lineage
7. ____ single cell, no nucleus, closer to eukaryotes evolutionary
8. ____ cells with nucleus, single and muticelled species

a. Eukarya
b. Bacteria
c. Archaea

Fill in the blank
For each of the following fill in the blank using the numbers 1, 3, 4, or 6. Some numbers may be used more than once [p.11]

9. There is/are _____ domain(s) of life.

10. There is/are _____ kingdom(s) of life.

11. There is/are _____ kingdom(s) of Archaea

12. There is/are _____ kingdom(s) of Bacteria

13. There is/are _____ kingdom(s) of eukaryotes

1.6 THE NATURE OF SCIENCE [PP.12-13]

Selected Terms
Kriticos [p.12]

Boldfaced Terms
[p.12] critical thinking _____

_____

[p.12] science _____

_____

[p.12] hypothesis_____

_____

[p.12] inductive reasoning _____

_____

[p.12] prediction _____

_____

[p.12] deductive reasoning _____

_____

[p.13] model_____

_____

[p.13] experiment _____
_____

[p.13] variables _____
_____

[p.13] independent variable _____
_____

[p.13] dependent variable _____
_____

[p.13] experimental group _____
_____

[p.13] control group _____
_____

[p.13] data _____
_____

[p.13] scientific method _____
_____

Matching

Match each of the following terms with its correct definition [pp.12-13]

1. ____ dependent variable [p.13]
2. ____ independent variable [p.13]
3. ____ critical thinking [p.12]
4. ____ model [p.13]
5. ____ science [p.12]
6. ____ prediction [p.12]
7. ____ hypothesis [p.12]
8. ____ variables [p. 13]
9. ____ control group [p.13]
10. ____ experimental group [p.13]
11. ____ data [p.13]
12. ____ scientific method [p.13]
13. ____ inductive reasoning [p.12]
14. ____ deductive reasoning [p.12]

a. defined or controlled by the person doing the experiment
b. an analogous system for testing a hypothesis
c. characteristics or events that can differ among individuals or over time
d. a statement of a condition that should exist if a hypothesis is valid.
e. a testable answer to a question
f. the judging of information before accepting it
g. the systematic study of nature
h. an observed result that is supposed to be influenced by the independent variable
i. the group in n experiment that has one variable changed
j. a type of reasoning that arrives at a conclusion based on one's observations
k. forming, testing, and evaluating hypotheses
l. a type of reasoning that uses a hypothesis to make a prediction
m. the group in an experiment that has no charge in a variable
n. test results

Concept Map

Provide the missing word(s) for each number in parentheses in the following concept map of scientific inquiry. [p.12]

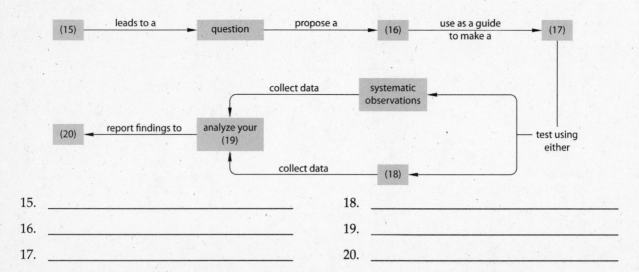

15. _____
16. _____
17. _____

18. _____
19. _____
20. _____

1.7 EXAMPLES OF BIOLOGY EXPERIMENTS [PP.14-15]

Selected Terms
Olestra [p.14], peacock butterfly [p.14],

1.8 ASKING USEFUL QUESTIONS [PP.16-17]

Boldfaced Terms
[p. 16] sampling error_____

_____

[p.17] probability _____

_____

[p.17] statistically significant _____

_____

1.9 PHILOSOPHY OF SCIENCE [PP.18-19]

Selected Terms
atomic theory [p.18], big bang [p.18], cell theory [p.18], evolution [p.18], global warming [p.18], plate tectonics [p.18], supernatural [p.19]

Boldfaced Terms
[p.18] scientific theory _____

_____

[p.18] law of nature _____

_____

Matching

Review the experiment regarding peacock butterfly and birds [pp.14-15]. For each of the following, choose the appropriate component of the scientific method that is best described by the statement. Some answers may be used more than once.

   a.   question
   b.   prediction
   c.   hypothesis
   d.   experimental test
   e.   analysis of results

1. _____ If the sounds deter predatory birds, then individuals who can't make the sound will be more likely to be eaten by predatory birds.

2. _____ All of the butterflies with unmodified wing spots survived.

3. _____ Some butterflies have the sound-making part of the wing removed.

4. _____ Why does the peacock butterfly flick its wings?

5. _____ The hissing and clicking sounds produced by the butterfly rubbing its wings deter predatory birds.

6. _____ If brilliant wing spots deter birds, then birds with no spots will be more likely to be eaten by predatory birds.

7. _____ Opening of the wings exposes spots that resemble owl eyes, thus scaring off predators.

8. _____ Some butterflies had both their spots painted black and the sound-producing portions of the wings removed.

9. _____ Most butterflies with neither spots nor sound structures survived.

Matching

Review the experiment regarding Olestra [p.14]. For each of the following, choose the appropriate component of the scientific method that is best described by the statement. Some answers may be used more than once.

   a.   prediction
   b.   hypothesis
   c.   experimental test
   d.   conclusion

10. _____ Olestra causes intestinal cramps.

11. _____ Individuals who eat potato chips with Olestra are and without Olestra have the same chance of suffering from intestinal cramps.

12. _____ Individuals who eat potato chips with Olestra will be more likely to suffer from intestinal cramps than those individuals who eat potato chips without Olestra.

13. _____ Control groups eats regular potato chips; Experimental groups eats potato chips with Olestra.

Fill in the Blank

Fill in the blanks of the following questions using the terms below.

sampling error, probability, statistically significant, law of nature, scientific theory

14. When a hypothesis meets the criteria of multiple tests over many years and has helped make other predictions, it is considered a(n) _____.

15. A phenomenon that has been observed to occur in every circumstance without fail, but for which we currently do not have a complete scientific explanation is called a(n) _____.

16. The difference between results obtained from a subset and the results obtained from the whole is called a(n) _____.

17. The chance that a particular outcome of an event will occur is called a(n) _____.

18. This term refers to a result that is unlikely to occur by chance. It is called _____.

SELF QUIZ

1. All populations of all species occupying a specific area is the definition of a _____. [p.4]
   a. population
   b. community
   c. ecosystem
   d. biome
   e. domain

2. Which of the following is not a domain of life? [pp.8-9]
   a. archaean
   b. eukarya
   c. protist
   d. bacteria

3. In an ecosystem, energy _____ the ecosystem while nutrients _____ the ecosystem.[p.6]
   a. flows through; cycles within
   b. flows through; flows through
   c. cycles within; cycles within
   d. cycles within; flows through

4. Which of the following terms describes the transformation of a single cell into an adult organism? [p.7]
   a. reproduction
   b. inheritance
   c. development
   d. homeostasis

5. Which of the following are involved with the initial capture of energy by an ecosystem? [p.6]
   a. decomposers
   b. consumers
   c. producers
   d. all of the above

6. In a scientific investigation, what step typically follows after the hypothesis is created? [p.14]
   a. analysis of results
   b. report to the scientific community
   c. experimental testing
   d. forming a prediction

7. A(n) _____ group is a set of individuals that have a certain characteristic or receive a certain treatment.[p.13]
   a. experimental
   b. control
   c. model
   d. statistically significant

8. The ability of life to maintain an internal environment despite changes in the external environment is called _____. [p.7]
   a. inheritance
   b. homeostasis
   c. reproduction
   d. development

9. The signature molecule of life is _____. [p.7]
   a. protein
   b. RNA
   c. DNA
   d. the cell

10. The difference between results obtained from a subset, and the results obtained from the whole is called _____. [p.16]
    a. the sampling error
    b. a scientific theory
    c. a law of nature
    d. a probability

CHAPTER OBJECTIVES/REVIEW QUESTIONS

This section lists general and detailed chapter objectives that can be used as review questions. You can make maximum use of these items by writing answers on a separate sheet of paper. Where blanks are provided, fill in the answers. Use the indicated page numbers or the glossary to check the accuracy of your answers.

1. Understand the definition of biology. [p.3]

2. Understand the levels of organization in nature. [pp.4-5]

3. Understand the concept of an emergent property. [p.4]

4. Understand the definitions of *atom, molecule, cell, tissue, organ, organ system, organism, population, ecosystem, biosphere*. [pp.4-5]

5. Define the role of producers, consumers and decomposers in an ecosystem. [p.6]

6. Illustrate the difference between nutrient and energy flow in an ecosystem. [p.6]

7. Understand the importance of photosynthesis. [p.6]

8. Understand the importance of homeostasis. [p.7]

9. Understand the differences between growth and development. [p.7]

10. Explain the difference between inheritance, reproduction and development. [p.7]

11. Explain the difference between prokaryotes and eukaryotes. [p.8]

12. Be able to define *biodiversity*. [p.8]

13. Distinguish between prokaryotes, protists, fungi, plants and animals. [pp.8-9]

14. Understand the definitions of *taxonomy, species, genus, specific epithet,* and *taxon*. [p.10]

15. List the three domains of life. [p.11]

16. List the six kingdoms of life. [p.11]

17. Understand the importance of critical thinking to the study of science. [p.12]

18. Explain the difference between hypothesis and prediction. [p.12]

19. Explain the difference between inductive and deductive reasoning. [p.12]

20. Understand the concept of model with respect to a scientific inquiry. [p.13]

21. Explain the difference between independent and dependent variable. [p. 13]

22. Explain the difference between a control and an experimental group. [p.13]

23. Describe the components of the scientific method. [p.13]

24. Understand the concept of sampling error. [p.16]

25. Explain how the terms probability and statistically significant are used in science. [p.17]

26. Explain the difference between scientific theory and law of nature. [p.18]

CHAPTER SUMMARY

This section serves as a review of the key concepts presented at the beginning of the chapter. After completing the chapter you should be able to fill in the blanks without assistance from the textbook.

We study the world of life at different levels of (1) _____, which extend from atoms and molecules to the biosphere. The quality known as "life" emerges at the level of (2) _____.

The world of life shows (3) _____, because all organisms are alike in key aspects. They consist of one or more (4) _____, which stay alive through ongoing inputs of energy and raw materials. These sense and respond to changes in their external and internal (5) _____. Their cells contain (6) _____, a type of molecule that offspring inherit from parents and that encodes information necessary for growth, survival, and (7) _____.

The world of life also shows great diversity. Many millions of kinds of organisms, or species, have appeared and disappeared over time. Each (8) _____ is unique in at least some trait – in some aspect of its body form or behavior. Taxonomy is a system of naming and classifying (9) _____. The second part of a species scientific name is the specific (10) _____.

Biologists make systematic (11) _____, form hypotheses, make predictions and tests in the laboratory and in the field. They report their work so that others may repeat their work and check their reasoning.

INTEGRATING AND APPLYING KEY CONCEPTS

1. What would happen if all of the decomposers were removed from an ecosystem?

2. What sorts of topics do scientists usually regard as untestable by applications of the scientific method?

3. Based upon your knowledge of the scientific method and critical thinking, what is the problem with calling religious beliefs "creationist theory"?

2
LIFE'S CHEMICAL BASIS

INTRODUCTION

All living organisms are based upon chemical reactions. This chapter introduces you to the principles of chemistry, namely atoms, elements and chemical bonds. It is important that you develop an understanding of these basic chemical principles early in the course as many of the later topics will build on these concepts. Chemistry is the basis of life.

FOCAL POINTS

- The basic characteristics of an atom. [p.24]
- Uses of isotopes and radioactive decay. [p.25]
- The energy levels of an atom. [p.26]
- Importance of ionic, covalent and hydrogen bonds. [pp.27-29]
- Properties of water. [pp.30-31]
- Figure 2.14 [p.32] and the pH scale.

INTERACTIVE EXERCISES

2.1 MERCURY RISING [p.23]
2.2 START WITH ATOMS [p.24-25]

Selected Terms
PET [p.25]

Boldfaced Terms
[p.24] atoms _____

_____

[p.24] protons _____

_____

[p.24] neutrons _____

_____

[p.24] electrons _____

_____

[p.24] charge _____

_____

[p.24] nucleus _____

_____

[p.24] atomic number _____

_____

[p.24] elements _____

_____

[p.24] periodic table _____

_____

[p.25] isotopes _____

_____

[p.25] mass number _____

_____

[p.25] radioisotopes _____

_____

[p.25] radioactive decay _____

_____

[p.25] tracer _____

_____

Matching

Match each of the following terms to its correct description.

1. ____ atoms
2. ____ protons
3. ____ PET
4. ____ neutrons
5. ____ electrons
6. ____ atomic number
7. ____ radioactive decay
8. ____ elements
9. ____ isotopes
10. ____ radioisotopes
11. ____ tracers
12. ____ periodic table of the elements
13. ____ mass number

a. subatomic particles in the nucleus that carry no charge
b. particles that are the building blocks of all substances
c. an arrangement of the known elements based on their chemical properties
d. the number of protons and neutrons in the nucleus of a single atom
e. occurs when an element spontaneously emits energy or subatomic particles when its nucleus breaks down
f. an isotope that undergoes radioactive decay
g. the number of protons in an atom
h. atoms of a given element that differ in the number of neutrons
i. a procedure that is used to form images of tissues in the body
j. positively charged subatomic particles in the nucleus
k. negatively charged subatomic particles found outside the nucleus
l. a molecule with a detectable substance, such as a radioisotope, attached
m. pure substances containing atoms with the same number of protons

2.3 WHY ELECTRONS MATTER [pp.26-27]
2.4 WHY ATOMS INTERACT? [pp.28-29]

Selected Terms
"shells" [p.26], orbital [p.26], *single, double, triple* covalent bond [p.28], *nonpolar* covalent bond [p.28], structural formula [p.28], *polar* covalent bond [p.29]

Boldfaced Terms
[p.26] shell model _____

_____

[p.27] ion _____

_____

[p.27] electronegativity _____

_____

[p.27] chemical bond _____

_____

[p.27] molecule _____

_____

[p.27] compounds _____

_____

[p.27] mixture _____

_____

[p.28] ionic bond _____

_____

[p.28] covalent bond _____

_____

[p.29] polarity _____

_____

[p.29] hydrogen bond _____

_____

Matching

For each definition below, choose the most appropriate term from the list provided. [pp.24-29]

1. _____ molecules that contain two or more elements in a constant proportion

2. _____ two or more substances that do not form chemical bonds and whose proportions may vary

3. _____ when two or more atoms of the same or different elements form a chemical bond

4. _____ an atom that carries an electrical charge

5. _____ an attractive force between two atoms

a. chemical bond
b. molecule
c. compound
d. ion
e. mixture

Identification

Identify each of the following elements based on the number of electrons. Recall that the number of electrons equals the number of protons.

⊙ electron

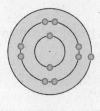

6 _____   7 _____

8 _____   9 _____

10 _____

6. _____   9. _____

7. _____   10. _____

8. _____

Choice

For each of the following, choose the most appropriate type of bond from the list provided. Some answers may be used more than once.

     a.   covalent bonds [pp.28-29]
     b.   hydrogen bonds [p.29]
     c.   ionic bond [p.28]
     d.   all of these

11. ____ these are found in biological molecules

12. ____ occurs when one atom loses it electrons to another atom

13. ____ formed by the sharing of electrons

14. ____ atoms in this bond stay together due to a strong electrical attraction

15. ____ may be either polar or nonpolar

16. ____ does not form molecules from atoms

17. ____ may be single, double, or triple

18. ____ formed between the polar regions of two molecules

2.5 WATER'S LIFE-GIVING PROPERTIES [pp.30-31]
2.6 ACIDS AND BASES [pp.32-33]

Selected Terms

hydrogen bonding [p.30], alkaline [p.30], acid rain [p.33], hyperventilation [p.33], fossil fuel [p.33], *tetany* [p.33]

Boldfaced Terms

[p.30] solvent _____

_____

[p.30] solutes _____

_____

[p.30] salt _____

_____

[p.30] hydrophilic _____

_____

[p.30] hydrophobic _____

_____

[p.31] cohesion _____

_____

[p.31] evaporation _____

_____

[p.31] temperature _____

_____

[p.32] concentration _____

_____

[p.32] pH _____

_____

[p.32] acids _____

_____

[p.32] bases _____

_____

[p.33] buffer _____

_____

Concept Map

The concept map below refers to the properties of water [pp.30-31]. Fill in the missing term or terms for each numbered space.

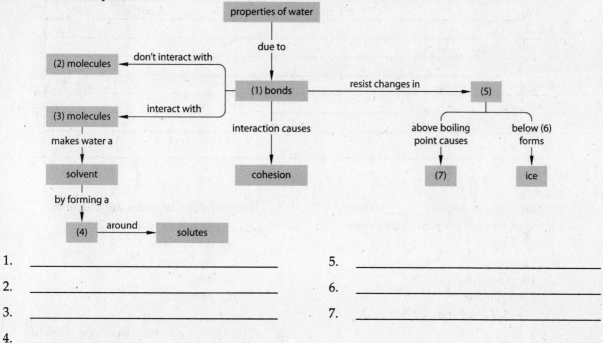

1. _____

2. _____

3. _____

4. _____

5. _____

6. _____

7. _____

Matching

Match each of the following terms to the correct description. [pp.30-33]

8. ____ concentration

9. ____ pH scale

10. ____ acids

11. ____ buffer system

12. ____ bases

13. ____ cohesion

14. ____ tetany

15. ____ alkalosis

16. ____ solute

17. ____ solvent

a. a substance that accepts hydrogen ions from a solution

b. a substance, usually a fluid that can dissolve other substances

c. what a substance becomes when it is dissolved

d. the amount of a particular solute dissolved in a given volume of fluid

e. a measure of hydrogen ion concentration in a solution

f. a potentially lethal rise in blood pH

g. substances that donate hydrogen ions in a solution

h. condition in which water molecules resist separating from each other

i. a period of prolonged muscle spasms

j. chemicals that keep the pH of a solution stable

Complete the Table
Use the information in Figure 2.16 [p.32] to complete the following table.

Substance	pH value	Acid/Base
18. seawater		
19. cola		
20. toothpaste		
21. black coffee		
22. gastric fluid		
23. beer		

SELF QUIZ

1. Each element has a unique _____, which equals the number of protons in its atoms. [p.24]
   a. isotope
   b. atomic number
   c. mass number
   d. number of orbitals

2. A molecule is _____. [p.27]
   a. substance that resists changes in pH
   b. a less stable form of an atom
   c. an element with a variable number of electrons
   d. a combination of two or more atoms

3. Lithium has an atomic number of 3 and an atomic mass of 7. How many neutrons are located in its nucleus? [p.26]
   a. 3
   b. 4
   c. 7
   d. can't be determined

4. A _____ substance is non-polar and repelled by water. [p.30]
   a. radioactive
   b. hydrophilic
   c. covalent
   d. hydrophobic

5. The properties of water are due to the _____ bonds in water. [p.29]
   a. non-polar
   b. hydrogen
   c. ionic
   d. covalent

6. A transfer of electrons from one atom to another forms a _____. [p.27]
   a. hydrogen bond
   b. isotope
   c. acid
   d. ion

7. In a covalent bond, electrons are _____ between two atoms. [p.28]
   a. shared
   b. transferred
   c. destroyed
   d. combined

8. The property of water that allows one water molecule to pull on another is called _____. [p.31]
   a. evaporation
   b. hydrophobic interactions
   c. buffering
   d. cohesion

9. A solution with a pH of 3.0 is _____ times more acidic than a pH of 5.0. [p.32]
    a. 2
    b. 20
    c. 100
    d. 200

10. In a solution, the substance that donates hydrogen ions is called the _____. [p.32]
    a. buffer
    b. ion
    c. acid
    d. salt

CHAPTER OBJECTIVES/REVIEW QUESTIONS

1. Understand the location and charge of each of the subatomic particles. [p.24]

2. Understand how atomic number and mass number relate to an element. [pp.24-25]

3. Define the terms *isotope* and *radioisotope*. [p.25]]

4. Understand the importance of electrons to an atom. [pp.26-27]

5. Understand the concept of a chemical bond. [p.26]

6. Distinguish between a molecule, compound, and mixture. [p.27]

7. Distinguish between an ionic, covalent, and hydrogen bond. [pp.28-29]

8. Understand polarity and be able to define the terms hydrophilic and hydrophobic. [pp.29-30]

9. Understand the importance of hydrogen bonds to the properties of water. [pp.30-31]

10. Be able to list the properties of water. [pp.30-31]

11. Distinguish between a solvent and solute. [p.30]

12. Understand the importance of cohesion and how it occurs. [p.31]

13. Be able to interpret the pH scale. [p.32]

14. Distinguish between an acid and base. [p.32]

15. Understand the importance of a buffer. [p.33]

CHAPTER SUMMARY

(1) _____ are fundamental units of all matter. Protons, electrons and (2) _____ are their building blocks. Elements are pure substances, each consisting entirely of atoms that have the same number of (3) _____. (4) _____ are atoms of the same element that have different numbers of neutrons.

Whether one atom will bond with others depends on the number and arrangement of its (5) _____.

Atoms of many elements interact by acquiring, (6) _____, and giving up electrons. (7) _____, covalent and (8) _____ bonds are the main interactions between atoms in biological molecules.

Life originated in (9) _____ and is adapted to its properties. Its temperature-stabilizing effects, (10) _____, and capacity to act as a solvent for so many other substances are the properties that make life possible on Earth.

Life is responsive to changes in the amounts of hydrogen (11) _____ and other substances dissolved in water.

INTEGRATING AND APPLYING KEY CONCEPTS

1. Explain what would happen if water were a nonpolar molecule. In your explanation, include the implications for the properties of water including temperature-stabilization, cohesion, action as a solvent, and evaporation.

2. Scientists have focused the search for life on other planets around the search for water. Why is liquid water important to life as we know it? Do you think that it is possible that life could be based on other molecules? Explain your answer.

3. Global climate change is based not only on carbon dioxide, but also water's temperature stabilizing effects. How would this property of water serve to increase the world's temperature over time?

3

MOLECULES OF LIFE

INTRODUCTION

Often students are confused as to why introductory biology courses start off with a discussion of chemistry. As you will soon learn, the study of living organisms is the study of how they process chemicals to conduct the processes of life. In this chapter you will be introduced to the principles of organic chemistry, which focuses on the element carbon. The chapter will introduce the importance organic molecules: carbohydrates, lipids, proteins and nucleic acids.

FOCAL POINTS

- Figure 3.1 [p.37] illustrates trans fats.
- Figure 3.2 [p.38] illustrates modeling an organic compound.
- Figure 3.3 [p.39] illustrates the major functional groups of organic molecules.
- Figure 3.4 [p.39] illustrates the difference between condensation and hydrolysis.
- Figure 3.6 [p.41] illustrates the structures of cellulose, starch, and glycogen.
- Figure 3.7 [p.41] illustrates the structure of chitin.
- Figures 3.8 & 3.9 [p.42] illustrate fatty acids and lipids.
- Figure 3.10 [p.42] illustrates phospholipids.
- Figure 3.13 [p.44] illustrates the formation of a polypeptide.
- Figure 3.14 [p.45] illustrates the levels of protein structure.
- Figure 3.17a&b [p.47] illustrate nucleotides and DNA.

INTERACTIVE EXERCISES

3.1 FEAR OF FRYING? [p.37]

Selected Terms:
hydrogenated [p.37], trans fats [p.37]

3.2 MOLECULES OF LIFE – FROM STRUCTURE TO FUNCTION [pp.38-39]

Selected Terms:
methane group [p.38], *hydroxyl* group [p.38], *carbonyl* group [p.38], *carboxyl* group [p.38], *phosphate* group [p.38], amine group [p.38], sulfhydryl [p.38], alcohol [p.39],

Boldfaced Terms:
[p.38] organic _____

_____

[p.38] hydrocarbon _____

_____

[p.38] functional group_____

_____

[p.39] monomers _____

_____

[p.39] polymers _____

_____

[p.39] metabolism _____

_____

[p.39] enzymes_____

_____

[p.39] condensation_____

_____

[p.39] hydrolysis _____

_____

Labeling

Study the structure of the organic molecules below. Using the information presented in Figure 3.3 of the text [p.38], identify each of the circled functional groups.

1. _____   5. _____

2. _____   6. _____

3. _____   7. _____

4. _____

Fill in the blank

8. _____ is a process in which an enzyme covalently bonds two molecules together.

9. _____ is a process that is opposite of condensation, large molecules broken into smaller molecules.

10. An organic molecule that consists of hydrogen and carbon is referred to as a _____ molecule.

11. A cluster of atoms that are covalently bonded to a carbon atom of an organic molecule are referred to as _____ groups.

12. _____ compounds consists primarily of carbon and hydrogen atoms.

13. Molecules that consist of multiple monomers are called _____.

14. _____ refers to activities by which cells acquire and use energy as they make and break apart organic molecules.

Identification

15. Identify the metabolic reactions by writing *condensation* or *hydrolysis* in the space provided. [p.39]

a _____   b _____

3.3 CARBOHYDRATES – THE MOST ABUNDANT ONES [pp.40-41]

Selected Terms
*Mono*saccharides [p.40], "saccharide" [p.40], *oligo*saccharide [p.40], *oligo-* [p.40], *di*saccharides [p.40], "complex" carbohydrates [p.40], *poly*saccharides [p.40], starch [p.41], cellulose [p.41], glycogen [p.41], chitin [p.41]

Boldfaced Terms
[p.40] carbohydrates_____

_____

Complete the Table

Complete the table below by entering the name and class of each carbohydrate described in terms of its function. For class, choose from *monosaccharide, oligosaccharide,* or *polysaccharide.* [pp.40-41]

Carbohydrate	Class	Function
1.		Sugar present in milk
2.		Strengthens external skeletons and other hard body parts of some animals and fungi.
3.		Used by cells as an energy source
4.		The carbohydrate found in plant cell walls
5.		The plant storage carbohydrate for the products of photosynthesis
6.		Table sugar
7. The 5 carbon sugar found in DNA		Found in DNA
8. The 5 carbon sugar found in DNA		Found in DNA
9.		Long-term carbohydrates storage in animals; stored in the liver and muscles

3.4 LIPIDS [pp.42-43]

Selected Terms

"essential fatty acids" [p.42], *neutral* fats [p.42], *saturated* fats [p.42], *unsaturated* fats [p.42], glycerol [p.42], *cis* fatty acids [p.42], *trans* fatty acids [p.42], estrogen [p.43], testosterone [p.43]

Boldfaced Terms

[p.42] lipids _____

_____

[p.42] fatty acids _____

_____

[p.42] fats _____

_____

[p.42] triglycerides _____

_____

[p.43] phospholipid _____

_____

[p.43] wax _____

_____

[p.43] steroid _____

_____

Choice

For each of the following statements, choose the most appropriate class of lipids from the list below.

   a.   steroid [p.43]
   b.   phospholipids [p.43]
   c.   fats [p.42]
   d.   waxes [p.43]

1.   ____   the lipid in the cell membrane

2.   ____   firm, water-repellent lipids found in the plant cuticle

3.   ____   cholesterol belongs to this class

4.   ____   may be either saturated or unsaturated

5.   ____   vitamin D is manufactured using this class

6.   ____   the major component of adipose tissue

7.   ____   has a ring structure and no fatty acid tails

8.   ____   triglycerides are the most abundant type of these

9.   ____   contains both polar and nonpolar regions

10.   ____   if saturated, these may be either cis or trans

11.   ____   steroid hormones are made from these

12.   ____   the richest, most abundant source of energy in vertebrates

3.5 PROTEINS – DIVERSITY IN STRUCTURE AND FUNCTION [pp.44-45]
3.6 WHY IS PROTEIN STRUCTURE SO IMPORTANT? [p.46]

Selected Terms

R group [p.44], domain [p.44], helices [p.44], *primary* structure [p.45], *secondary* structure [p.45], *tertiary* structure [p.45], *quaternary* structure [p.45], *glyco*protein [p.43], *fibrous* proteins [p.45], keratin [p.45], hemoglobin [p.45], globin [p.45], lipoprotein [p.45], LDL [p.45], HDL [p.45], actin [p.45], myosin [p.45]

Boldfaced Terms

[p.44] protein_____

_____

[p.44] amino acid _____

_____

[p.44] peptide bond_____

_____

[p.44] polypeptide _____

_____

[p.46] denature _____

_____

[p.46] prion _____

_____

Matching
Match each term with its description.

1. ____ fibrous protein [p.45]

2. ____ denature [p.46]

3. ____ globin [p.45]

4. ____ glycoproteins [p.45]

5. ____ quaternary structure [p.45]

6. ____ tertiary structure [p.45]

7. ____ domain [p.44]

8. ____ secondary structure [p.45]

9. ____ protein [p.44]

10. ____ primary structure [p.45]

11. ____ polypeptide chain [p.44]

12. ____ peptide bond [p.44]

13. ____ amino acid [p.44]

a. several amino acids linked together

b. two or more polypeptide chains bound together

c. an organic compound containing one or more chains of amino acids

d. a protein with an oligosaccharide attached

e. a type of protein associated with oxygen transport

f. the unraveling of a proteins secondary, tertiary and quaternary structure

g. an organic compound containing an amino and carboxyl group

h. contribute to the structure of cells and tissues

i. the structure of a protein that is based on hydrogen bonds

j. a linear, unique sequence of amino acids

k. the linkage between two amino acids

l. the level at which a protein becomes a working molecule

m. a part of a protein that is organized as a structurally stable unit

3.7 NUCLEIC ACIDS [p.47]

Selected Terms
double helix [p.47]

Boldfaced Terms
[p.47] nucleotide _____

_____

[p.47] ATP _____

_____

[p.47] nucleic acids _____

_____

[p.47] RNA _____

_____

[p.47] DNA _____

_____

Concept Map

Provide the missing term or terms in the numbered spaces of the concept map below. [p.47]

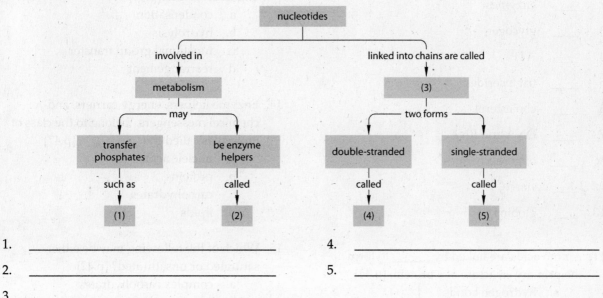

1. _____   4. _____

2. _____   5. _____

3. _____

SELF QUIZ

Choice

For each of the molecules in the list below, choose the appropriate class of organic compounds from the list.

    a.   carbohydrates [pp.40-41]
    b.   proteins [pp.44-46]
    c.   lipids [pp.42-43]
    d.   nucleic acids [p.47]

1. ____ glycoproteins

2. ____ enzymes

3. ____ glycogen

4. ____ ATP

5. ____ triglycerides

6. ____ cholesterol

7. ____ DNA and RNA

8. ____ oligosaccharides

9. ____ starch

10. ____ globins

11. Amino acids are linked by _____ to form the primary structure of a protein. [p.44]
    a.   hydrogen bonds
    b.   peptide bonds
    c.   ionic bonds
    d.   disulfide bonds

12. Primary, secondary and tertiary structure are characteristic of the _____. [p.45]
    a.   carbohydrates
    b.   lipids
    c.   proteins
    d.   nucleic acids

13. Monosaccharides are joined by a process called _____ to form the oligosaccharides.[p.40]
    a.   condensation
    b.   hydrolysis
    c.   functional group transfer
    d.   rearrangement

14. Enzyme helpers, energy carriers, and chemical messengers, belong to the class of molecules called the _____. [p.47]
    a.   nucleic acids
    b.   proteins
    c.   carbohydrates
    d.   lipids

15. Which of the following may be either saturated or unsaturated? [p.42]
    a.   complex carbohydrates
    b.   polypeptides
    c.   nucleotides
    d.   triglycerides

CHAPTER OBJECTIVES/REVIEW QUESTIONS

1. Understand the definition of an organic compound. [p.37]

2. Be able to identify the major functional groups. [p.38]

3. Understand the process of condensation and hydrolysis. [p.39]

4. Understand the role of carbohydrates in a cell. [pp.40-41]

5. Be able to define the terms monosaccharide, oligosaccharide, and polysaccharide and give an example of each. [pp.40-41]

6. Understand the use of the complex carbohydrates or polysaccharides. [p.41]

7. Be able to define the role of the lipids in the cell. [pp.42-43]

8. Understand how the terms saturated, unsaturated, cis and trans relate to fatty acids. [p.42]

9. Understand the role of phospholipids, cholesterol and waxes in living organisms. [p.43]

10. Give examples of how proteins are used by a cell. [pp.44-45]

11. Understand how amino acids are used as building blocks of proteins. [p.44]

12. Understand the four levels of protein structure. [p.45]

13. Understand the how prions can cause diseases. [p.46]

14. Understand the role of nucleic acids and nucleotides in the cell. [p.47]

15. Understand the importance of DNA and RNA [p.47]

CHAPTER SUMMARY

We define cells partly by their ability to build complex carbohydrates and lipids, (1) _____ and nucleic acids. The main building blocks are simple sugars, (2) _____, amino acids and nucleotides. All of these organic compounds ha e a backbone of (3) _____ atoms with (4) _____ groups attached.

(5) _____ are the most abundant biological molecules. Simple sugars function as transportable forms of (6) _____ or as quick energy sources. The (7) _____ carbohydrates are structural materials or energy reservoirs.

Complex lipids function as energy (8) _____, structural materials of cell (9) _____, signaling molecules, and waterproofing or lubricating substances.

Structurally and functionally, (10) _____ are the most diverse molecules of life. They include (11) _____, structural materials, signaling molecules, and transporters.

Nucleotides have major metabolic roles and are building blocks of (12) _____ acids. Two kinds of nucleic acids, (13) _____ and RNA, interact as the cell's system of storing, retrieving and translating information about building (14) _____.

INTEGRATING AND APPLYING KEY CONCEPTS

1. All organic molecules can be used for energy. What is the common link between all organic compounds that makes this possible?

2. Humans can obtain energy from many different food sources. Do you think that this is an advantage or disadvantage in terms of long-term survival and human evolution?

3. We know that the ways in which atoms bond effect molecular shapes. Do you think that the shape of organelles and cells is influenced in the same way by organic molecules? Explain your answer.

4

CELL STRUCTURE

INTRODUCTION

The cell represents the fundamental unit of life and in this chapter you will begin to explore some of the workings of this miniature factory. It is important that you develop an understanding of the major components of the cell early in the course, as you will keep returning to the material in this chapter in later chapters.

FOCAL POINTS

- Figure 4.3 [p.52] illustrates the organization of a bacterial cell and two eukaryotic cells (plant and animal).
- Figure 4.5 [p.54] illustrates the light and electron microscopes.
- Figure 4.8 [pp.56-57] illustrates the plasma membrane and how proteins interact with the membrane.
- Figure 4.11 [p.58] illustrates the general structure of a prokaryotic cell.
- Table 4.2 [p.60] provides the function of the major components of a eukaryotic cell.
- Figure 4.14 [p.61] illustrates the cell nucleus.
- Figure 4.16 [pp.62-63] outlines the operation of the endomembrane system.
- Figure 4.17 [p.64] illustrates the mitochondrion.
- Figure 4.18 [p.65] illustrates a chloroplast.
- Figure 4.26 [p.70] illustrates the structure of plant and animal eukaryotic cells.
- Table 4.3 [p.72] gives a summary of the typical components of cells of various life forms.

INTERACTIVE EXERCISES

4.1 Food for Thought [p.51]

Selected Terms
E. coli 0157:H7 [p.51]

4.2 WHAT, EXACTLY, IS A CELL? [pp.52-53]
4.3 SPYING ON CELLS [pp.54-55]

Selected Terms
eukaryotic cell [p.52], prokaryotic cells [p.52], Robert Brown [p.53], "cellae" [p.53], Matthias Schleiden [p.53], Theodor Schwann [p.53], Rudolf Virchow [p.53], Anton van Leeuwenhoek [p.53], *light microscopes* [p.54], phase-contrast microscopes [p.54], micrographs [p.54], *fluorescence microscope* [p.54], *electron microscopes* [p.54], transmission electron microscopes [p.55], scanning electron microscopes [p.55], various metric system measurements [p.55]

Boldfaced Terms
[p.52] cell _____

_____

[p.52] plasma membrane _____

_____

[p.52] cytoplasm _____

_____

[p.52] organelle _____

_____

[p.52] nucleus _____

_____

[p.52] surface-to-volume ratio _____

_____

[p.53] cell theory _____

_____

Matching

Match each of the following terms with its correct description. [p.52]

1. ____ organelles
2. ____ cell
3. ____ cell theory
4. ____ eukaryotic cell
5. ____ cytoplasm
6. ____ prokaryotic cell
7. ____ surface-to-volume ratio
8. ____ nucleus
9. ____ plasma membrane

a. the smallest unit that has the properties of life
b. the cell's outer membrane
c. refers to size limits with cells
d. a cell with internal functional compartments and a nucleus
e. semifluid mixture of water, ions, sugars and proteins
f. smaller, simpler cells lacking a nucleus
g. location of the DNA in a eukaryotic cell
h. structures that carry out special metabolic functions inside of a cell
i. four generalizations about cells make this

Short Answer

10. Explain how the concept of "surface-to-volume ratio" influences the size and shape of cells. [pp.52-53]

_____

_____

_____

_____

11. List the four generalizations that constitute the cell theory. [p.53]

_____

_____

_____

_____

Choice

For each of the following statements, choose one of the following types of microscopes. Some answers may be used more than once. [pp.54-55]

     a. light microscope
     b. fluorescent microscope
     c. scanning electron microscope
     d. transmission electron microscope
     e. phase-contrast microscope

12. ____ visible light serves as the light source

13. ____ the specimen is coated first with a metal

14. ____ a cell or molecule emits light that is detected by the microscope

15. ____ the electrons detect surface structures

16. ____ electrons pass through the thin sample

17. ____ the stained areas that are darkest show up with this microscope

4.4 MEMBRANE STRUCTURE AND FUNCTION [pp.56-57]
4.5 INTRODUCING BACTERIA AND ARCHAEANS [pp.58-59]

Selected Terms

hydrophobic [p.56], hydrophilic [p.56], enzyme [p.56], phospholipid [p.56], *extracellular fluid [p.57]*, prokaryote [p.58], capsule [p.58],

Boldfaced Terms

[p.56] lipid bilayer _____

_____

[p.56] fluid mosaic _____

_____

[p.56] adhesion proteins_____

_____

[p.56] recognition proteins_____

_____

[p.57] receptor proteins _____

_____

[p.57] transport proteins _____

_____

[p.58] ribosome _____

_____

[p.58] plasmid _____

_____

[p.58] nucleoid _____

_____

[p.58] flagella _____

_____

[p.58] pili _____

_____

[p.58] cell wall _____

_____

[p.59] biofilms _____

_____

Matching
Match each term with its description. [pp.56-57]

1. ____ fluid mosaic model
2. ____ phospholipid
3. ____ adhesion proteins
4. ____ recognition proteins
5. ____ transport proteins
6. ____ receptor proteins
7. ____ "fluid"
8. ____ "mosaic"

a. describes the mixed nature of the components of the cell membrane
b. function as identity tags for a cell type, individual, or species
c. describes the movement of the phospholipids in the bilayer
d. trigger changes in cell activities
e. may either actively or passively move compounds
f. the description of the organization of cell membranes
g. helps cells stick to one another to form tissues
h. the major component of the cell membrane

Labeling
Label each of the indicated parts of the prokaryotic cell. [p.58]

9. _____   12. _____

10. _____   13. _____

11. _____   14. _____

Fill in the Blank [pp.58-59]

Ribosomes are sites where (15) _____ are assembled. (16) _____ are small circles

of DNA that carry a few genes that can provide advantages to bacteria. A long, slender cellular structure

used for movement is called a (17) _____. A durable cell (18) _____ surrounds the plasma

membrane of most bacteria and archaea. A communal living arrangement in which single-celled

organisms live in a shaped mass of slime is called a (19) _____.

4.6 INTRODUCING EUKARYOTIC CELLS [p.60]
4.7 THE NUCLEUS [p.61]
4.8 THE ENDOMEMBRANE SYSTEM [pp.62-63]

Selected Terms
Eu- [p.60], *karyon* [p.60], *cyto-* [p.60], nucleus [p.61], *rough* ER [p.62], *smooth* ER [p.62]

Boldfaced Terms
[p.61] nuclear envelope _____

_____

[p.61] nucleoplasm _____

_____

[p.61] nucleolus _____

_____

[p.62] endomembrane system _____

_____

[p.62] endoplasmic reticulum (ER)_____

_____

[p.62] vesicles _____

_____

[p.63] vacuole _____

_____

[p.63] central vacuole _____

_____

[p.63] lysosomes _____

_____

[p.63] peroxisomes _____

_____

[p.63] Golgi body _____

_____

Matching

Match each of the following cellular structures to its correct description. [p.60]

1. ____ cytoskeleton
2. ____ ribosomes
3. ____ centriole
4. ____ chloroplast
5. ____ vesicles
6. ____ Golgi body
7. ____ endoplasmic reticulum
8. ____ nucleus

a. protects and controls access to the DNA
b. contributes to cell shape, internal organization and movement
c. routing and modifying new polypeptide chains, synthesizes lipids
d. assembles polypeptide chains
e. modifies new polypeptide chains, sorting, shipping proteins and lipids
f. anchor for cytoskeleton
g. assembles sugars in plants using sunlight
h. transporting, storing, or digesting substances in a cell

Complete the Table

9. Complete the table about the eukaryotic nucleus by entering the name of each of the described nuclear components. [p.61]

Nuclear Component	Description
a.	A construction site where large and small subunits of ribosomes are assembled
b.	Consists of two lipid bilayers folded together into a single membrane
c.	The cells genetic material
d.	Semifluid interior of the nucleus
e.	An organized cluster of membrane proteins in the nuclear membrane

Concept Map

Complete the following concept map on the export of materials using the endomembrane system. [pp.62-63]

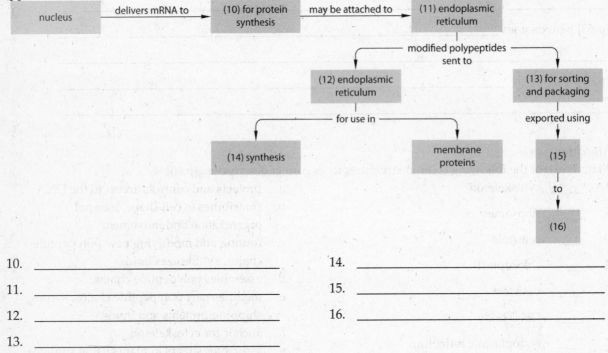

10. _____

11. _____

12. _____

13. _____

14. _____

15. _____

16. _____

4.9 MITOCHONDRIA AND PLASTIDS [pp.64-65]
4.10 THE DYNAMIC CYTOSKELETON [pp.66-67]

Selected Terms

ATP [p.64], aerobic respiration [p.64], endosymbiosis [p.65], grana [p.65], stroma [p.65], photosynthesis [p.65], thylakoid membrane [p.65], chlorophyll [p.65], chromoplasts [p.65], amyloplasts [p.65], tubulin [p.66], actin [p.66], myosin [p.66], lamins [p.66], flagella [p.67], false feet [p.67], 9 + 2 array [p.67], pseudopod [p.67],

Boldfaced Terms

[p.64] mitochondrion _____

_____

[p.65] plastids _____

_____

[p.65] chloroplasts _____

_____

[p.66] cytoskeleton _____

_____

[p.66] microtubules _____

_____

[p.66] microfilaments _____

_____

[p.66] cell cortex _____

_____

[p.66] intermediate filaments _____

_____

[p.66] motor proteins _____

_____

[p.67] cilia _____

_____

[p.67] centriole _____

_____

[p.67] basal body _____

_____

Choice

For the following questions, choose from one of the following answers. [pp.64-65]

    a. mitochondria
    b. chloroplasts
    c. both mitochondria and chloroplasts

1. ____ occur only in photosynthetic eukaryotic cells

2. ____ make far more ATP from the same compounds than prokaryotic cells

3. ____ hydrogen ions released from the breakdown of organic compounds accumulate in the inner compartment by operation of transport systems

4. ____ a muscle cell might have a thousand or more

Cell Structure **43**

5. ____ use sunlight energy to form ATP and other products, which in turn are used in sugar-producing reactions

6. ____ ATP-forming reactions require oxygen

7. ____ two outer membranes surround the stroma, a semifluid interior that bathes an inner membrane

8. ____ resemble bacteria in their size, structure, and biochemistry

9. ____ an energy powerhouse of eukaryotic cells

10. ____ contains the thylakoid membranes

Choice

For each of the following, choose the appropriate component of the cytoskeleton from the list below. [pp.66-67]

    a. microtubules
    b. microfilaments
    c. intermediate filaments

11. ____ made from the protein actin

12. ____ used to separate chromosomes prior to cell division

13. ____ lamins are an example

14. ____ made from the protein tubulin

15. ____ make of the cell cortex

16. ____ the major components of flagella and cilia

17. ____ responsible for the action of pseudopods

4.11 CELL SURFACE SPECIALIZATIONS [pp.68-69]

Selected Terms
pectin [68], middle lamella [p.68], chitin [p.68], epidermis [p.68], collagen [p.68]

Boldfaced Terms
[p.68] extracellular matrix (ECM)_____

_____

[p.68] cuticle_____

_____

[p.68] primary wall_____

_____

[p.68] secondary wall _____

_____

[p.68] lignin _____

_____

[p.69] cell junctions _____

_____

[p.69] tight junction _____

_____

[p.69] adhering junction _____

_____

[p.69] gap junction _____

_____

[p.69] plasmodesmata _____

_____

Matching

Match each of the following terms to its correct description. [pp.68-69]

1. _____ cuticle
2. _____ primary wall
3. _____ tight junctions
4. _____ adhering junctions
5. _____ secondary wall
6. _____ extracellular matrix
7. _____ gap junctions
8. _____ plasmodesmata

a. the cell wall made from lignin
b. a protective covering that prevents water loss in plants
c. seals cells together to prevent movement of water-soluble substances
d. a cell wall made of cellulose
e. connect the cytoplasm of adjacent cells
f. anchor cells and the extracellular matrix together
g. non-living material secreted by cells for support and anchoring
h. connects the primary wall of two plant cells to allow quick movement of substances

4.12 A VISUAL SUMMARY OF EUKARYOTIC CELL COMPONENTS [p.70]

Identification

Identify each lettered organelle in the diagrams. Organelles that are found in both plants and animals may have two letters. [p.70]

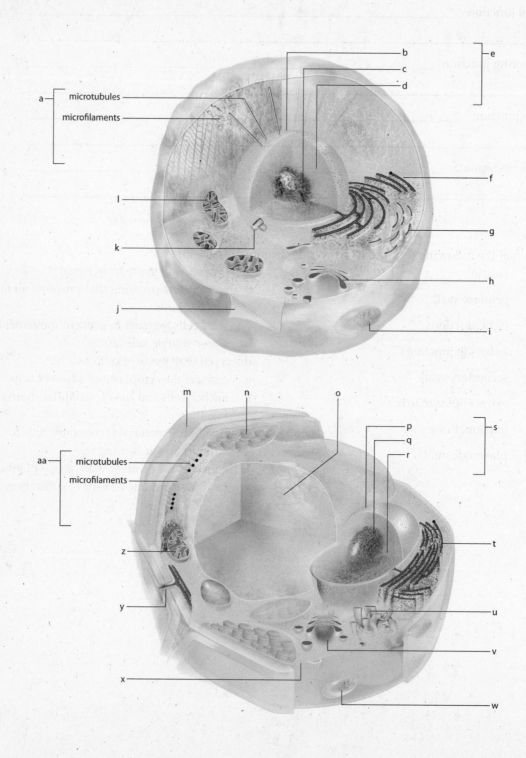

1. ____ central vacuole

2. ____ lysosome

3. ____ nucleus

4. ____ Golgi body

5. ____ chloroplast

6. ____ plasma membrane

7. ____ nucleolus

8. ____ smooth ER

9. ____ mitochondrion

10. ____ rough ER

11. ____ plasmodesma

12. ____ cytoskeleton

13. ____ centrioles

14. ____ cell wall

15. ____ nuclear envelope

16. ____ DNA in nucleoplasm

4.13 THE NATURE OF LIFE

Selected Terms
life [p.71], homeostasis [p.71], hereditary [p.71], adapting [p.71]

SELF QUIZ

1. As a cell increases in size, its volume increases with the ____ of its diameter, and the surface area increases with the _____ of the diameter. [pp.52-53]
   a. square ; square
   b. cube; cube
   c. cube; square
   d. square; cube

2. The major component of a cell membrane is _____. [p.56]
   a. protein
   b. triglycerides
   c. cellulose
   d. phospholipids

3. Which of the following would not be found in a prokaryotic cell? [p.58]
   a. ribosomes
   b. DNA
   c. cytoplasm
   d. cell membrane
   e. nucleus

4. Most cell membrane functions are carried out by _____ embedded in the bilayer or positioned at one of its surfaces. [p.58]
   a. carbohydrates
   b. proteins
   c. phospholipids
   d. cholesterol

5. Which of the following organelles sorts and ships molecules coming into and leaving the cell? [p.60]
   a. Golgi body
   b. plasma membrane
   c. rough endoplasmic reticulum
   d. ribosomes
   e. smooth endoplasmic reticulum

6. What is the site where the small and large subunits of the ribosome are assembled? [p.61]
   a. nucleus
   b. nucleoplasm
   c. Golgi body
   d. nucleolus

7. The _____ is free of ribosomes and makes the lipid molecules for the cell membranes. [p.62]
   a. smooth endoplasmic reticulum
   b. rough endoplasmic reticulum
   c. vesicles
   d. lysosomes

8. Which of the following part of the cytoskeleton is involved in moving chromosomes prior to cell division? [p.66]
   a. microfilaments
   b. microtubules
   c. intermediate filaments
   d. none of the above

9. Which of the following junctions forms a water-tight seal? [p.69]
   a. gap junctions
   b. plasmodesmata
   c. adhering junctions
   d. tight junctions

10. Which of the following is found in both plant and animal cells and produces the majority of ATP used by the cell? [p.64]
    a. chloroplasts
    b. mitochondria
    c. ribosomes
    d. peroxisomes

CHAPTER OBJECTIVES/REVIEW QUESTIONS

1. Understand the basic components of all cells. [p.52]

2. Understand the concept of the surface-to-volume ratio [p.52]

3. Know the four principles of the cell theory. [p.53]

4. Recognize the major types of microscopes. [p.54]

5. Understand the principles of the fluid mosaic model. [p.56-57]

6. Understand the structure of a prokaryotic cell. [pp.58-59]

7. Define the term biofilm and explain its importance to the study of bacteria. [p.59]

8. Understand the function of the major eukaryotic organelles. [p.60]

9. Understand the structure of the nucleus and the role of the major components within the nucleus. [p.61]

10. Know the components of the endomembrane system. [pp.62-63]

11. Be able to trace the movement of a substance through the endomembrane system. [pp.62-63]

12. Distinguish between the mitochondria and chloroplast. [pp.64-65]

13. Give the function of microfilaments, microtubules and intermediate filaments. [pp.66-67]

14. Be able to identify the major types of cell junctions. [p.68-69].

15. Understand the structure of a cell wall and the importance of the extracellular matrix to some organisms. [pp.68-69]

16. Identify the major parts of a eukaryotic plant and animal cell. [p.70]

CHAPTER SUMMARY

Each cell has a (1) _____, a boundary between its interior and exterior environment. The interior consists of organelles contained in the (2) _____.

Microscopic analysis supports four generalizations of the cell theory. Each organism consists of one or more (3) _____and their products, a cell contains hereditary material, which can be passed on, each new cell is descended from a living cell, and the cell is the structural and functional unit of life.

All cell membranes are mostly a lipid (4) _____- two layers of lipids – and a variety of (5) _____. The proteins have diverse tasks, including control over which (6) _____cross the membrane at any given time.

Archaeans and (7) _____are prokaryotic cells, which have few, if any, internal membrane-bound compartments. In general, they are the (8) _____and structurally simplest cells.

Cells of protists, plants, fungi, and animals are (9) _____; they have a (10) _____and other organelles. They differ in internal parts and surface specializations.

Diverse (11) _____filaments reinforce a cell's shape and keep its parts organized. As some filaments lengthen and shorten, they move (12) _____or other structures to new locations.

INTEGRATING AND APPLYING KEY CONCEPTS

1. Cells are typically described as biological factories. Compare a typical eukaryotic cell to a modern factory by relating the function of the organelles to the functions you think are necessary for a factory to run efficiently (i.e. power, waste reduction, etc.).

2. How do you think that the process of evolution has made cell's efficient?

3. In the past, some researchers believed that they had found evidence of "life" on the surface of Mars. Critics stated that since these "cells" were only 1/10th the size of a bacteria on Earth, that they were probably not alive. Using your knowledge of cells, explain the limitations that a cell this size would have.

4. Suppose that you wanted to develop a drug that inhibited the ability of a cell to export proteins. What organelles would you target and why?

5

GROUND RULES OF METABOLISM

INTRODUCTION

Having completed a tour of the cell in chapter 4, it is now time to look at how cell's distinguish themselves from their external environments. In this chapter you will begin to examine the role of enzymes in metabolic reactions and how cell's move compounds needed for metabolic pathways across their membranes. Many of the principles introduced in this chapter will be used in later chapters on energy pathways.

FOCAL POINTS

- Figure 5.3 [p.76] diagrams entropy and energy.
- Figure 5.5 [p.77] diagrams energy from the environment into living organisms.
- Figure 5.8 [p.79] diagrams activation energy.
- Figure 5.9 [p.79] diagrams ATP, the energy currency of the cell.
- Figure 5.11 [p. 80] diagrams how an enzyme enhances the rate of a reaction by lowering its activation energy.
- Figure 5.12 [p.81] shows how enzymes respond to temperature.
- Figure 5.13 [p.81] shows how enzymes respond to pH changes.
- Figure 5.14 [p.82] shows feedback inhibition.
- Figure 5.17 [p.84] diagrams osmosis.
- Figure 5.19 [p.86] illustrates the process of passive transport.
- Figure 5.21 [p.87] illustrates the process of active transport.
- Figure 5.23 [p.88] diagrams the relationship between endocytosis and exocytosis.

INTERACTIVE EXERCISES

5.1 A TOAST TO ALCOHOL DEHYDROGENASE [p.75]

Selected Terms
Alcohol dehydrogenase [p.75]

5.2 ENERGY AND THE WORLD OF LIFE [pp.76-77]
5.3 ENERGY IN THE MOLECULES OF LIFE [pp.78-79]

Selected Terms
therm [p.76], dynam [p.76], consumer [p.77], producer [p.77], kinetic energy [p.77]

Boldfaced Terms
[p.76] energy _____

_____

[p.76] kinetic energy _____

_____

[p.76] first law of thermodynamics _____

_____

[p.76] entropy _____

_____

[p.76] second law of thermodynamics _____

_____

[p.77] potential energy _____

_____

[p.78] reaction _____

_____

[p.78] reactants _____

_____

[p.78] products _____

_____

[p.78] endergonic _____

_____

[p.78] exergonic _____

_____

[p.79] activation energy _____

_____

[p.79] ATP _____

_____

[p.79] phosphorylation _____

_____

[p.79] ATP/ADP cycle _____

_____

Choice

For each of the following statements, choose the most appropriate law of thermodynamics to which the statement applies. [p.76]

      a.   first law of thermodynamics
      b.   second law of thermodynamics

1. _____ States that entropy is increasing in the universe.

2. _____ Energy cannot be created or destroyed.

3. _____ Energy may be converted between forms.

4. _____ An example is the death and decay of an organism.

5. _____ An example is corn plants producing starch molecules

6. _____ An example is oak logs burning in a fireplace

7. _____ Energy disperses spontaneously

Matching

Match each of the following terms to the correct description.

8. _____ consumer [p.77]

9. _____ kinetic energy [p.76]

10. _____ producers [p.77]

11. _____ exergonic reaction [p.78]

12. _____ potential energy [p.77]

13. _____ endergonic reaction [p.78]

14. _____ energy [p.76]

    a. a capacity to do work
    b. the energy of motion
    c. reactions that require an input of energy
    d. capacity to cause change based on where an object is located or how its parts are arranged
    e. reactions that release energy
    f. examples are plants
    g. examples are animals, most fungi, and many protists, bacteria

Concept Map

Provide the terms indicated in the following concept map of the ATP/ADP cycle. [p.79]

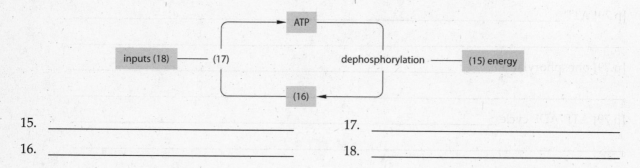

15. _____   17. _____

16. _____   18. _____

5.4 HOW ENZYMES WORK [pp.80-81]

Selected Terms

enzymes [p.80], transition state [p.80], [p.78], temperature, [p.81], pH [p.81], salinity [p.81], free radicals [p.81], catalase [p.81]

Boldfaced Terms

[p.80] catalysis _____

_____

[p.80] substrates _____

_____

[p.80] active sites_____

_____

[p.80] induced-fit model _____

_____

[p.81] cofactors _____

_____

[p.81] coenzymes _____

_____

[p.81] antioxidant _____

_____

Labeling

Chose the correct term for each numbered label in the diagram below. [p. 80]

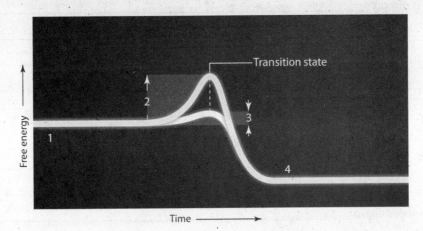

1. ____

2. ____

3. ____

4. ____

a. reactants
b. activation energy without enzyme
c. activation energy with enzyme
d. products

Matching

Match each of the following terms with its correct description.

5. ____ induced-fit-model [p.80]

6. ____ coenzyme [p.81]

7. ____ catalysis [p.80]

8. ____ cofactor [p.81]

9. ____ antioxidants [p.81]

10. ____ active sites [p.81]

11. ____ transition state [p.80]

a. a process where an enzyme makes a reaction run much faster than it would on its own

b. an enzyme that reduces strong oxidizers such as free radicals

c. a situation in which an enzyme's active site is not quite complementary to its substrate

d. the location where substrates bind in an enzyme

e. atoms or molecules (other than proteins) that associate with enzymes and are necessary for their function

f. Vitamin C is an example of this

g. when the substrate's bonds reach a breaking point allowing the reaction to run spontaneously to product

5.5 METABOLISM: ORGANIZED, ENZYME-MEDIATED REACTIONS [pp.82-83]

Selected Terms

metabolism [p.82], cyclic pathways [p.82], oxidized [p.83], reduced [p.83], allo [p.83]

Boldfaced Terms

[p.82] metabolic pathways _____

_____

[p.82] feedback inhibition _____

_____

[p.82] allosteric _____

_____

[p.83] redox reactions _____

_____

[p.83] electron transfer chain _____

_____

Matching

Match each of the following terms to the correct definition. [pp.82-83]

1. ____ metabolism
2. ____ oxidized
3. ____ redox reactions
4. ____ reduced
5. ____ electron transfer chain
6. ____ allo
7. ____ metabolic pathway

a. membrane-bound arrays of enzymes that transfer electrons in a series of steps.
b. refers to activities by which cells acquire and use energy as they build and break down organic molecules
c. the stepwise series of reactions in which an organic compound is built, rearranged, or broken down
d. this means other
e. the combination of oxidation and reduction in a single reaction
f. a condition of a molecule when it accepts electrons from another molecule
g. a condition that occurs to a molecule when its electrons are given to another molecule

5.6 MOVEMENT OF IONS AND MOLECULES [pp.84-85]
5.7 MEMBRANE – CROSSING MECHANISMS [pp.86-87]

Selected Terms

molecule [p.84], ion [p.84], selectively permeable [p.85], tonicity [p.85], transport protein [p.86], extracellular fluid [p.86], passive transporter [p.86], solute [p.86], gradient [p.87], cotransporter [p.87]

Boldfaced Terms

[p.84] concentration _____

_____

[p.84] concentration gradient _____

_____

[p.84] diffusion _____

_____

[p.84] hypotonic _____

_____

[p.84] hypertonic_____

_____

[p.84] isotonic _____

_____

[p.85] osmosis _____

_____

[p.85] tugor _____

_____

[p.85] osmotic potential _____

_____

[p.86] passive transport _____

_____

[p.87] active transport _____

_____

Matching

Match each of the following statements with the correct term. [pp.84-85]

1. ____ the net movement of like molecules or ions down a concentration gradient

2. ____ the diffusion of water across a membrane

3. ____ pressure that a fluid exerts against a structure that contains it

4. ____ the difference in the number per unit volume of molecules or ions between two adjacent areas.

5. ____ the amount of turgor that stops osmosis

a. concentration gradient
b. turgor
c. osmotic pressure
d. osmosis
e. diffusion

Short Answer

6. List five factors, besides electric and pressure gradients, which influence the rate of diffusion across a membrane. [p.84]

_____

_____

_____

_____

_____

7. List two differences between passive and active transport. [pp.86-87]

_____

_____

_____

_____

Choice

For each statement, choose the form of transport that is described by the statement.

     a.   passive transport [p.86]

     b.   active transport [pp.86-87]

8. _____ the movement of molecules down a concentration gradient

9. _____ requires an input of energy

10. _____ an example is facilitated diffusion

11. _____ small, non-polar molecules usually move by this mechanism

12. _____ assists the movement of polar molecules across the membrane.

Labeling

The diagrams below represent stages of the facilitated transport of glucose. Place these diagrams in order by writing the letter of the first step in #13, the second in #14, etc. [p.86]

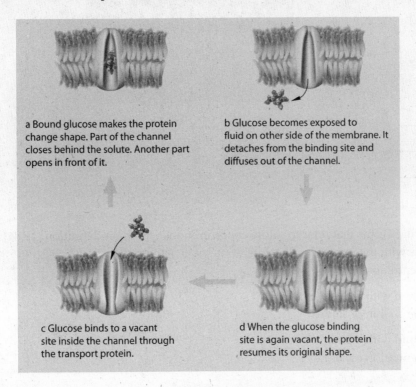

a Bound glucose makes the protein change shape. Part of the channel closes behind the solute. Another part opens in front of it.

b Glucose becomes exposed to fluid on other side of the membrane. It detaches from the binding site and diffuses out of the channel.

c Glucose binds to a vacant site inside the channel through the transport protein.

d When the glucose binding site is again vacant, the protein resumes its original shape.

13. _____

14. _____

15. _____

16. _____

Labeling

The diagrams below represent stages of the active transport of calcium ions. Place these diagrams in order by writing the letter of the first step in #17, the second in #18, etc. [p.87]

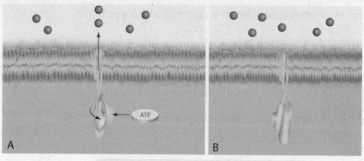

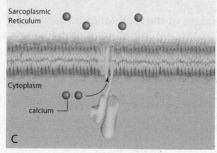

17. _____

18. _____

19. _____

Matching

Match the following Terms that refer to solute concentrations with their definitions [p.84]

20. _____ fluids with the same overall solute concentrations

21. _____ the fluid with the lower overall concentration of solute

22. _____ the fluid with the higher overall concentration of solute

    a. hypotonic
    b. hypertonic
    c. isotonic

Labeling
Indicate whether the fluid in each beaker is *hypotonic, hypertonic,* or *isotonic* in relation to the fluid in the bag. [p.84]

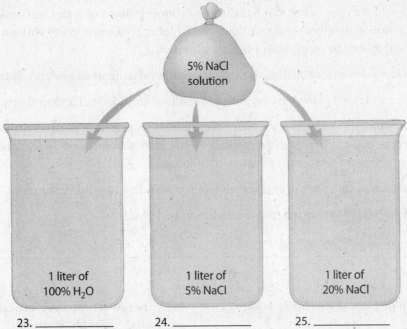

23. _____   24. _____   25. _____

5.8 MEMBRANE TRAFFICKING [pp.88-89]

Selected Terms
phospholipid [p.88], vesicle [p.88], pseudopod [p.88], lysosome [p.88], ER [p.89], Golgi bodies [p.89]

Boldfaced Terms
[p.88] exocytosis _____

_____

[p.88] endocytosis _____

_____

[p.88] phagocytosis _____

_____

Matching
Match each term with the correct definition.

1. ____ exocytosis [p.88]

2. ____ phagocytosis [p.88]

3. ____ endocytosis [p.88]

a. this term means "cell eating"
b. general term for the taking up of substances near a cell's surface
c. general term for the movement of a vesicle to the cell membrane, releasing its contents into the external environment

True-False

If the statement is true, write a T in the blank. If the statement is false, correct it by changing the underlined word(s) and writing the correct word(s) in the answer blank. [pp.88-89]

4. _____ Consider a membrane that has different concentrations of water molecules on each side; the side with fewer solute particles has a higher concentration of water molecules than the fluid on the other side of the membrane.

5. _____ Water moves from an area of high concentration to areas of low concentration.

6. _____ An animal cell placed in a hypertonic solution would swell and perhaps burst.

7. _____ Physiological saline is 0.9% NaCl; red blood cells placed in such a solution will not gain or lose water; therefore, one could correctly state that the fluid in red blood cells is hypertonic in relation to physiological saline.

8. _____ Red blood cells shrivel and shrink when placed in a hypotonic solution.

9. _____ Plant cells placed in a hypotonic solution will swell.

SELF QUIZ

1. Essentially, the first law of thermodynamics states that _____. [p.76]
   a. one form of energy cannot be converted into another
   b. entropy is increasing in the universe
   c. energy cannot be created or destroyed
   d. energy cannot be converted into matter or matter into energy

2. An important principle of the second law of thermodynamics states that _____. [p.76]
   a. energy can be transformed into matter, and because of this we can get something for nothing
   b. energy can be destroyed only during nuclear reactions, such as those that occur inside the sun
   c. if energy is gained by one region of the universe, another place in the universe also must gain energy in order to maintain the balance of nature
   d. matter tends to become increasingly more disorganized

3. "Energy in" reactions are called [p.78]
   a. phosphorylation
   b. biosynthetic
   c. exergonic
   d. endergonic

4. Any substance (for example, NAD+) that accepts electrons is _____. [p.83]
   a. oxidized
   b. a catalyst
   c. reduced
   d. a substrate

5. This term means "a process in which an enzyme makes a reaction run much faster than it would on its own." [p.80]
   a. redox reaction
   b. endergonic
   c. catalysis
   d. exergonic

6. A _____ is a compound that stops other molecules from reacting with oxygen. [p.81]
   a. cofactor
   b. antioxidant
   c. coenzyme
   d. enzyme

7. Allosteric sites are _____. [p.82]
    a. found on enzyme that can change shape during catalysis.
    b. active sites found on cofactors.
    c. regions found on coenzymes that are active sites
    d. regions of an enzyme (other than the active site) where regulatory molecules bind

8. The net movement of molecules or ions in response to a concentration gradient is called _____. [p.84]
    a. diffusion
    b. active transport
    c. osmosis
    d. passive transport

9. Which of the following processes would require an expenditure of energy by the cell? [p.87]
    a. diffusion
    b. osmosis
    c. passive transport
    d. active transport

10. Enzymes function by _____ the activation energy of a reaction. [p.80]
    a. maintaining
    b. reversing
    c. raising
    d. lowering

CHAPTER OBJECTIVES/REVIEW QUESTIONS

1. Understand the two laws of thermodynamics and how they relate to biological systems. [p.76]

2. Understand the difference between producers and consumers with regard to how they obtain their carbon. [p.77]

3. Understand reactants, products, and reactions. [p.78]

4. Be able to diagram the ATP-ADP cycle. [p.79]

5. Recognize the importance of enzymes to biological systems. [pp.80-81]

6. Understand how enzymes act as catalysts. [pp.80-81]

7. Understand how the environmental and allosteric mechanisms control enzyme function. [pp.80-81]

8. Recognize the relationship between coenzymes and cofactors in enzyme function. [p.81]

9. Understand the principles of a redox reaction. [p.83]

10. Understand the process of diffusion and the factors that influence the rate of diffusion across a membrane. [pp.84-85]

11. Be able to predict the tonicity relationship of two environments when given solute concentrations. [p. 84]

12. Define osmosis, turgor, osmotic pressure, and tonicity. [p.85]

13. Distinguish between active and passive transport. [pp.86-87]

14. Be able to distinguish between endocytosis and exocytosis. [p.88]

CHAPTER SUMMARY

(1) _____ tends to disperse spontaneously. Each time energy is transferred, some if it disperses. Organisms maintain their complex (2) _____ only by continuously harvesting energy.

(3) _____ couples metabolic reactions that release usable energy with reactions that require energy input. On their own, those reactions proceed too slowly to sustain life. (4) _____ increase reaction rates. (5) _____ factors influence enzyme activity.

(6) _____ are energy-driven sequences of enzyme-mediated reactions. They concentrate, convert, or dispose of materials in cells. Controls over (7) _____ that govern key steps in these pathways can shift cell activities fast.

(8) _____ gradients drive the directional movements of ions and molecules into and out of cells. Transport (9) _____ raise and lower water and (10) _____ concentrations across the plasma membrane and internal cell membranes. Other mechanisms move larger molecules across the plasma membrane.

INTEGRATING AND APPLYING KEY CONCEPTS

1. A piece of dry ice left sitting on a table at room temperature vaporizes. As the dry ice vaporizes into CO_2 gas, does its entropy increase or decrease? Tell why you answered as you did.

2. Apply your knowledge of isotonic, hypertonic, and hypotonic to explain why salt water can kills grass on your lawn?

6
WHERE IT STARTS — PHOTOSYNTHESIS

INTRODUCTION

The process of photosynthesis is responsible for capturing the energy necessary to sustain life on Earth. While initially it may seem to be a complicated series of chemical processes, in reality it represents a series of logical steps in which the energy in sunlight is first captured and converted to chemical energy, and then this chemical energy is used to manufacture sugar. Pay close attention to the diagrams in this chapter as they will assist you in understanding the basic flows of chemicals and ions in the photosynthetic pathways.

FOCAL POINTS

- Figure 6.2 [p.94] illustrates the concept of an absorption spectrum.
- Figure 6.3 [p.95] illustrates the photosynthetic pigments.
- Figure 6.5 [p.97] illustrates the chloroplasts.
- Figure 6.7 [p.99] demonstrates the light dependent reactions of photosynthesis.
- Figure 6.8 a and b [p.100] diagrams the differences between the cyclic and non-cyclic light-dependent reactions.
- Figure 6.10 [p.101] illustrates the Calvin-Benson (light-independent) cycle.

INTERACTIVE EXERCISES

6.1 GREEN ENERGY [p.93]
6.2 SUNLIGHT AS AN ENERGY SOURCE [pp.94-95]
6.3 EXPLORING THE RAINBOW [p.96]

Selected Terms
photon [p.94], prism [p.94], electromagnetic spectrum [p.94], chloroplast [p.95], phycobilins [p.95], carotenoids [p.95], anthocyanins [p.95], retinal [p.95], UV light [p.95], Englemann [p.96], *Chladophora* [p.96], absorption spectrum [p.96]

Boldfaced Terms
[p.93] autotroph _____

_____

[p.93] photosynthesis _____

_____

[p.93] heterotroph _____

_____

[p.94] wavelength _____

_____

[p.94] pigment _____

_____

[p.94] chlorophyll *a* _____

_____

Matching
Match each of the following terms with the correct definition.

1. ____ carotenoids [p.95]
2. ____ absorption spectrum [p.96]
3. ____ autotroph [p.93]
4. ____ chlorophyll a [p.94]
5. ____ heterotroph [p.93]
6. ____ pigment [p.94]
7. ____ wavelength [p.94]

a. extract energy and carbon from inorganic material in the environment; makes its own food
b. a unit that is used to identify light
c. the main photosynthetic pigment of plants
d. gets energy and carbon by breaking down organic molecules assembled by other organisms
e. first discovered by Engelmann, this is used to identify the wavelengths of light absorbed by a pigment
f. an accessory pigment that extends the range of photosynthesis
g. an organic compound that selectively absorbs light of specific wavelengths

Short Answer
8. In an absorption spectrum, such as the one shown in Figure 6.4c [p.96], what do the peaks represent?

_____

_____

_____

6.4 OVERVIEW OF PHOTOSYNTHESIS [p.97]
6.5 LIGHT-DEPENDENT REACTIONS [pp.98-99]

Selected Terms
matrix [p.97], noncyclic [p.97], cyclic [p.97], NADPH [p.97], NADP+ [p.97], gradient [p.98], transfer chain [p.99]

Boldfaced Terms
[p.97] chloroplast _____

_____

[p.97] stroma _____

_____

[p.97] thylakoid membrane _____

_____

[p.97] photosystems _____

_____

[p.97] light-dependent reactions_____

_____

[p.97] light-independent reactions _____

_____

[p.98] photosystems _____

_____

[p.98] photolysis _____

_____

[p.98] electron transfer phosphorylation_____

_____

Fill in the Chart
Provide the missing information for each numbered item in the table below. [p.97]

Light-Dependent Reaction

Right Side Reactants	Left Side Products
1a.	1b.
2a.	2b.
3a.	3b.
4.	

Fill in the Chart
Provide the missing information for each numbered item in the table below. [p.97]

Light-Independent Reaction

Right Side Reactants	Left Side Products
5a.	5b.
6a.	6b.
7a.	7b.
8.	

True-False

If the statement is true, place a T in the blank. If the statement is false, correct the underlined word to make the statement correct. [p.97]

9. _____ The chloroplast is the plant organelle that specializes in photosynthesis.

10. _____ During the light-independent reactions, light energy is converted to chemical energy.

11. _____ The light-harvesting pigments of photosynthesis are found in the stroma of a chloroplast.

12. _____ The sugars of photosynthesis are manufactured in the thylakoids.

13. _____ Glucose synthesis occurs in the light-independent reactions of photosynthesis.

Labeling

Provide the missing term for each numbered item in the diagram below. [p.99]

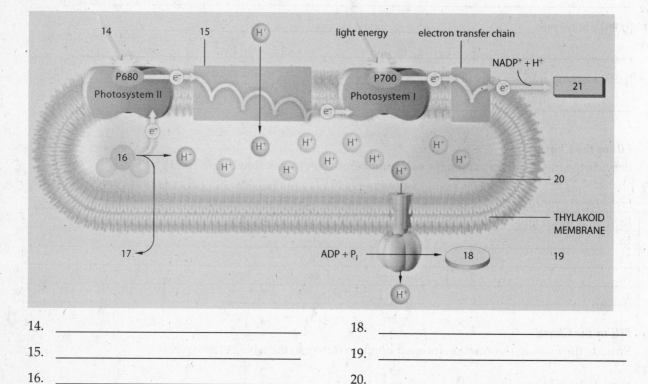

14. _____	18. _____
15. _____	19. _____
16. _____	20. _____
17. _____	21. _____

6.6 ENERGY FLOW IN PHOTOSYNTHESIS [p.100]
6.7 LIGHT-INDEPENDENT REACTIONS: THE SUGAR FACTORY [p.101]

Selected Terms

680-nanometer (P680) [p.100], 700 – nanometer (P700) [p.100], PGAL [p.101], PGA [p.101]

Boldfaced Terms

[p.101] Calvin-Benson cycle_____

_____

[p.101] carbon fixation _____

_____

[p.101] rubisco_____

_____

Choice

Choose the appropriate form of electron flow for each of the following statements. [p.100]

    a. cyclic

    b. non-cyclic

    c. both cyclic and non-cyclic

1. ____ Forms ATP

2. ____ Forms NADPH

3. ____ Lost electrons are replaced by photolysis

4. ____ Electrons proceed through an electron transfer chain

5. ____ Is used when NADPH supplies are plentiful in the cell

6. ____ Oxygen is released as a byproduct.

7. ____ The first to evolve in photosynthetic organisms

Identification

Identify each numbered part of the diagram, using abbreviations where appropriate. Complete the exercise by entering the letter of the correct statement in the parentheses following each label. [p.101]

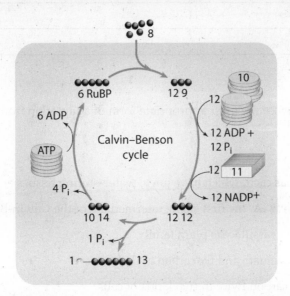

8. _____ ( )   12. _____ ( )

9. _____ ( )   13. _____ ( )

10. _____ ( )   14. _____ ( )

11. _____ ( )

a. Donates hydrogens and electrons to PGA molecules
b. Gets phosphate groups from ATP, priming them for synthesis reactions that generation RuBP
c. Rubisco attaches the carbon atom of CO2 to RuBP, which starts the Calvin-Benson cycle; this compound forms from splitting of an intermediate compound
d. Gets a phosphate group from ATP, plus hydrogen and electrons
e. In the air spaces inside a leaf; diffuses into photosynthetic cells
f. Formed by a combination of two of twelve PGAL molecules
g. Donates phosphate groups to PGA

6.8 ADAPTATIONS: DIFFERENT CARBON-FIXING PATHWAYS [p.102]

Selected Terms
Crassulaceae family [p.102]

Boldfaced Terms
[p.102] stomata _____

_____

[p.102] C3 plants _____

_____

[p.102] photorespiration_____

_____

[p.102] C4 plants _____

_____

[p.102] CAM plants_____

_____

Choice
For each of the following, choose the most appropriate form of adaptation from the list below. [p.102]
        a. C3 plants
        b. C4 plants
        c. CAM plants

1. ____ Succulents such as cacti, which have juicy, water-storing tissues and thick surface layers

2. ____ The three-carbon PGA, the first stable intermediate of the Calvin-Benson cycle

3. ____ Named after the Crassulaceae plant family

4. ____ They open their stomata and fix carbon at night

5. ____ These plants fix carbon twice in two kinds of cells

6. ____ Evolved independently over millions of years in many lineages

7. ____ With CO_2 levels rising, these plants may again have the edge

8. ____ Do not grow well in hot, dry climates without steady irrigation

9. ____ These plants survive prolonged droughts by closing stomata even at night

10. ____ An example is a jade plant

11. ____ CO_2 is fixed by repeated turns of a type of C4 cycle, then it enters the Calvin-Benson cycle the next day

12. ____ An example is the basswood tree

13. ____ These plants lose less water and make more sugar than the C3 plants can when days are dry

14. ____ When oxygen level is high, rubisco uses oxygen instead of CO_2 in an alternate reaction that yields only one molecule of PGA

15. ____ Common in grasses, corn, and other plants that evolved in the tropics

Short Answer

16. Briefly explain the primary cause of global warming. [p.103]

_____

_____

_____

_____

SELF QUIZ

1. Plants need _____ and _____ as the raw materials to carry on photosynthesis. [p.97]
   a. oxygen; water
   b. oxygen; CO2
   c. CO2; H2O
   d. sugar; water
   e. CO2; NADPH

2. Chlorophyll is found _____. [p.97]
   a. on the outer chloroplast membrane
   b. inside the mitochondria
   c. inside the stroma
   d. in the thylakoid membranes
   e. on the membranes of the stroma

3. The cyclic route functions mainly to _____. [p.99]
   a. make NADPH
   b. make PGAL
   c. set up conditions for making ATP
   d. regenerate RuBP
   e. break down CO2

4. During the noncyclic pathway, the origin of the electrons passed to $NADP^+$ is _____. [pp.98-99]
   a. CO2
   b. glucose
   c. sunlight
   d. water
   e. ATP

5. Two products of the light-dependent reactions are required to drive the chemistry of the light-independent reactions. They are _____ and _____. [p.97]
   a. O2; NADPH
   b. CO2; H2O
   c. O2; inorganic phosphate
   d. ATP; NADPH
   e. RuBP; PGA

6. ATP synthases are involved in _____. [p.98]
   a. creating electron transport chains
   b. photosystems
   c. attaching inorganic phosphates to ADP during the cyclic route
   d. the noncyclic pathway
   e. allowing H+ flow to attach inorganic phosphate to ADP

7. The Calvin-Benson cycle uses _____. [p.101]
   a. CO_2
   b. hydrogen and electrons from NADPH
   c. phosphate group transfers from ATP
   d. an enzyme known as rubisco
   e. all of the above

8. In the Calvin-Benson cycle, RuBP is regenerated by _____. [p.101]
   a. rearrangement of PGAL molecules
   b. the combination of two PGAL molecules
   c. PGA receiving phosphate groups from ATPs
   d. glucose entering other reactions that form carbohydrates
   e. rubisco attaching carbon atoms to RuBP

9. C4 plants have an advantage in hot, dry conditions because _____. [p.102]
   a. their leaves are covered with thicker wax layers than those of C3 plants
   b. their stomates open wider than those of C3 plants, thus cooling their surfaces
   c. special leaf cells possess a means of capturing CO_2 even in stress conditions
   d. they carry on normal photosynthesis at very high oxygen levels
   e. they can carry on carbon fixation at night

10. The process of securing carbon from the environment by incorporating it into a stable organic compound is called _____ [p.102]
    a. fluorescence
    b. photolysis
    c. photorespiration
    d. the C4 cycle
    e. carbon fixation

CHAPTER OBJECTIVES/REVIEW QUESTIONS

1. Recognize the difference between autotrophs and heterotrophs [p.93]

2. Understand how light is captured by autotrophs and the types of pigments involved in the process. [pp.94-95]

3. Recognize the importance of Englemann's experiments with absorption spectrums. [p.96]

4. Understand the basic flow of molecules and compounds in the process of photosynthesis. [p.97]

5. Understand the concept of a photosystem. [p.97]

6. Understand the difference between the light-dependent and light-independent reactions. [p.97]

7. Be able to identify the differences between the cyclic and non-cyclic electron flows in photosynthesis. [pp.98-100]

8. Understand the principles of the Calvin-Benson cycle. [p.101]

9. Recognize the evolutionary adaptations presented by the C3, C4 and CAM plants. [p.102]

10. Recognize the relationship between photosynthesis, carbon concentrations, and global warming. [p.103]

CHAPTER SUMMARY

A great one-way flow of energy through the world of life starts after (1) _____ and other pigments absorb the energy of visible light from the sun's rays. In plants, some bacteria, and many protists, that energy ultimately drives the synthesis of (2) _____ and other carbohydrates.

Photosynthesis proceeds through two stages in the (3) _____ of plants and many types of protists. First, pigments embedded in a membrane inside the chloroplast capture (4) _____ energy, which is then converted to (5) _____ energy. Next, that chemical energy drives the synthesis of (6) _____.

In the first stage of (7) _____, sunlight energy is converted to the chemical bond energy of (8) _____. The coenzyme (9) _____ forms in a pathway that also releases oxygen.

The second stage is the (10) _____ part of photosynthesis. Enzymes speed the assembly of sugars from carbon and oxygen atoms, both obtained from (11) _____. The reactions use ATP and NADPH that form in the first stage of (12) _____. ATP delivers energy, and NADPH delivers (13) _____ and hydrogens to the reaction sites. Details of the reactions vary among organisms.

Photosynthesis by (14) _____ removes carbon dioxide from the atmosphere; (15) _____ by all organisms puts it back in. Human activities have disrupted this balance, and so have contributed to (16) _____.

INTEGRATING AND APPLYING KEY CONCEPTS

1. A scientist once proposed that human cells be injected with chloroplasts extracted from living plant cells. Speculate about that possibility in terms of changes in human anatomy, physiology, and behavior. Could this be successful? Why or why not?

2. Scientists are developing a spinach "battery" to regenerate cell phones. What principles of photosynthesis are these scientists using to generate the flow of electrons? What would the inputs to the device be?

3. What is your idea for "soaking up" some of the excess carbon dioxide in the atmosphere?

7

HOW CELLS RELEASE CHEMICAL ENERGY

INTRODUCTION

In this chapter you will explore the mechanisms by which the energy stored in glucose may be released to form ATP. The chapter examines the differences between aerobic and anaerobic pathways, as well as the use of alternative energy sources by the human body.

FOCAL POINTS

- Figure 7.3 [p.108] provides an overview of the connection between aerobic respiration and photosynthesis.
- Figure 7.4 [p.109] provides an overview of aerobic respiration.
- Figure 7.5 [p.110] provides an overview of glycolysis.
- Figure 7.6 [pp.112-113] summarizes the reactions of acetyl-CoA formation and the Krebs cycle.
- Figure 7.9 [p.115] provides a summary of the three stages or aeriobic respiration in a mitochondrion.
- Figure 7.10 [p.119] illustrates alcoholic fermentation.
- Figure 7.12 illustrates how the body can use non-glucose energy sources for energy.

INTERACTIVE EXERCISES

7.1 WHEN MITOCHONDRIA SPIN THEIR WHEELS [p.107]
7.2 EXTRACTING ENERGY FROM CARBOHYDRATES [pp.108-109]
7.3 GLYCOLYSIS – GLUCOSE BREAKDOWN STARTS [pp.110-111]
7.4 SECOND STAGE OF AEROBIC RESPIRATION [pp.112-113]

Selected Terms
Lufts syndrome [p.107], Friedreich's ataxia [p.107], *anaerobic* [p.108], Krebs cycle [p.112]

Boldfaced Terms
[p.108] photoautotroph _____

_____

[p.108] anaerobic _____

_____

[p.108] aerobic _____

_____

[p.108] aerobic respiration _____

_____

[p.109] fermentation _____

_____

[p.109] glycolysis _____

_____

[p.109] pyruvate _____

_____

[p.110] substrate-level phosphorylation _____

_____

Choice

For each statement, choose the most appropriate class of reactions from the list below. [pp.108-109]

    a.   aerobic respiration

    b.   anaerobic reactions

    c.   both a and b

1. _____ Starts with a conversion of glucose to pyruvate

2. _____ The first form of pathways to evolve on the earth

3. _____ Requires oxygen

4. _____ Fermentation pathways are an example

5. _____ Electron transfer phosphorylation is an example

6. _____ The Krebs cycle is an example.

7. _____ The process ends in the cytoplasm

8. _____ The process ends in the mitochondria.

9. _____ Produces ATP.

Labeling

Label each of the numbered components of the diagram below. [p.109]

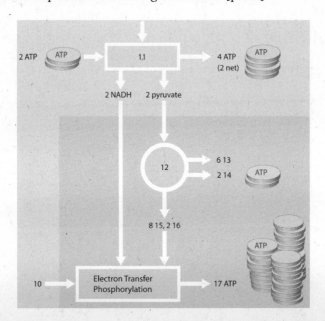

10. _____

11. _____

12. _____

13. _____

14. _____

15. _____

16. _____

17. _____

Concept Map

Complete the following concept map on the process of glycolysis. [pp.110-111]

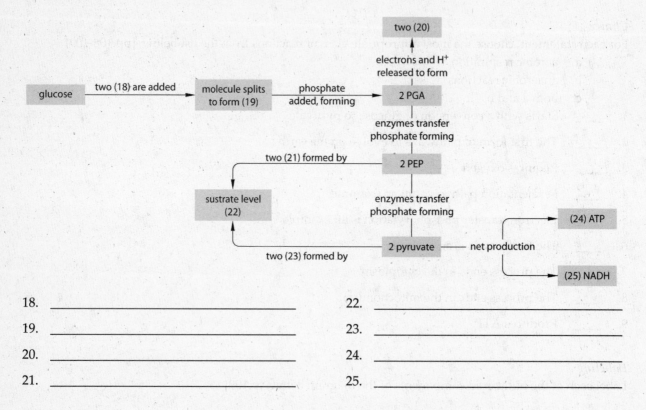

18. _____

19. _____

20. _____

21. _____

22. _____

23. _____

24. _____

25. _____

Labeling

Provide the correct information for each numbered label in the diagram below. Note – some numbers are present at multiple locations in the diagram. [pp.112-113]

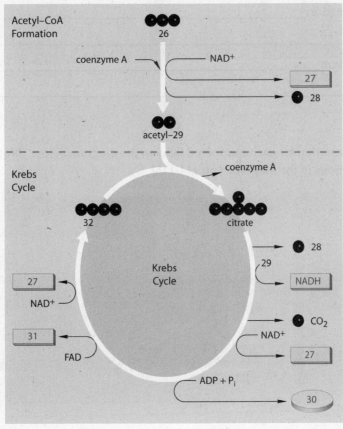

26. _____
27. _____
28. _____
29. _____

30. _____
31. _____
32. _____

Short Answer

33. For each glucose that enters glycolysis, how many of each of the following are produced during acetyl-CoA formation and the Krebs cycle? [p.113]

a. CO_2_____

b. NADH_____

c. $FADH_2$_____

d. ATP _____

7.5 AEROBIC RESPIRATION'S BIG ENERGY PAYOFF [p.114-115]
7.6 FERMENTATION PATHWAYS [pp.116-117]

Selected Terms

Electron transport phosphorylation [p.114], *Saccharomyces cerevisiae* [p.116], *Lactobacillus acidophilus* [p.116]

Boldfaced Terms

[p.116] alcoholic fermentation_____

_____

[p.116] lactate fermentation_____

_____

Labeling

Provide the correct information for each numbered label in the diagram below. Note – some numbers are present at multiple locations in the diagram. [pp.114-115]

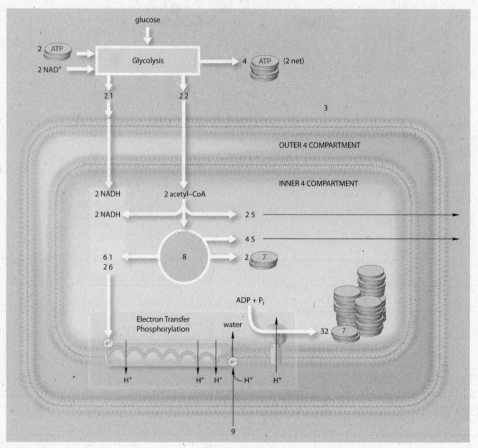

1. _____   6. _____

2. _____   7. _____

3. _____   8. _____

4. _____   9. _____

5. _____

Short Answer

10. For each glucose that enters aerobic respiration, how many of each of the following are produced by the end of the pathway? [p.115]

    a. CO_2 _____

    b. NADH _____

    c. FADH2 _____

    d. net ATP _____

11. What is the purpose of oxygen in the electron transport chain? [pp.114-115] _____

    _____

    _____

    _____

Choice

Choose from the following:

    a. alcoholic fermentation [p.116]
    b. lactate fermentation [p.117]
    c. applies to both types of fermentation

12. _____ Glycolysis is the first stage

13. _____ Yeasts, such as Saccharomyces cerevisea, are famous for their use of this pathway

14. _____ Yields enough energy to sustain many single-celled anaerobic organisms

15. _____ Pyruvate and NADH form, and the net energy yield is 2 ATP

16. _____ Muscle cells use this pathway but not for long; diverting glucose into this pathway would waste too much of its energy for too little ATP

17. _____ The final steps simply regenerate NAD+, the coenzyme that assists the breakdown reactions

18. _____ Each pyruvate molecule that formed in glycolysis is converted to the intermediate acetaldehyde

19. _____ These reactions do not completely degrade glucose to CO_2 and H_2O

20. _____ Lactobaccilus and some other bacteria use only this anaerobic pathway

21. _____ NADH transfers electrons and hydrogen to form ethanol

7.7 ALTERNATIVE ENERGY SOURCES IN THE BODY [pp.118-119]

Selected Terms

insulin [p.118], glucagon [p.118], fatty acids [p.118], triglycerides [p.119]

Choice

For each of the following, choose the appropriate class of molecules to which the statement corresponds. [pp.118-119]

    a.   fats
    b.   proteins
    c.   glucose

1. _____ enzymes covert these to glycerol and fatty acids

2. _____ carbon backbone may become acetyl-CoA, pyruvate or an intermediate of the Krebs cycle

3. _____ the hormones insulin and glucagons are involved

4. _____ ammonia is an end-product of this nutrient's metabolism

5. _____ carbon backbone becomes acetyl-CoA

6. _____ glycerol is converted to PGAL by the liver

7. _____ pancreas and liver are involved in regulating blood levels

8. _____ too much glucose in the bloodstream ends up as this

SELF QUIZ

1. Anaerobic respiration pathways do not use _____. [p.108]
    a.   carbon dioxide
    b.   oxygen

2. The first energy-releasing step in glycolysis splits activated glucose into two molecules of _____. [p.109]
    a.   NAD+
    b.   PGAL
    c.   ATP
    d.   pyruvate
    e.   PEP

3. The *net* energy yield of glycolysis is _____ ATP molecules [pp.108-111]
    a.   three
    b.   four
    c.   thirty-two
    d.   two
    e.   eight

4. Pyruvate is regarded as the end-product of _____. [pp.108-111]
    a.   glycolysis
    b.   acetyl-CoA formation
    c.   fermentation
    d.   electron transfer phosphorylation
    e.   the Krebs cycle

5. During which of the following phases of the energy-releasing pathways is ATP produced directly by substrate-level phosphorylation? [p.111]
    a.   glycolysis
    b.   electron transfer chains
    c.   the Krebs cycle
    d.   formation of acetyl-Co-A
    e.   all of the above

6. Select the process by which NADH and FADH$_2$ transfers electrons along a chain of acceptors to oxygen so as to form water and set up conditions for producing a large number of ATP molecules. [pp.114-115]
   a. glycolysis
   b. the Krebs cycle
   c. acetyl-CoA formation
   d. fermentation pathways
   e. electron transfer phosphorylation

7. The total number of ATP molecules produced by the complete degradation of one glucose molecule is often thirty-six. Which of the following is not correct? [p.115]
   a. glycolysis produces 2 ATP

   b. Krebs cycle produces 2 ATP

   c. electron transfer phosphorylation produces 32 ATP
   d. all of the above are correct

8. Which of the following is incorrect regarding the fermentation reactions? [pp.116-117]
   a. the first step of the reaction is always glycolysis
   b. their purpose is to regenerate NAD+
   c. they are only used by prokaryotic organisms
   d. some produce alcohol as an end-product
   e. all of the above are correct.

9. Of the following, which one is not available as an alternative energy source for the human body? [pp.118-119]
   a. fats
   b. glycogen
   c. proteins
   d. carbohydrates
   e. all of the above are available

10. Complete the following equation: $C_6H_{12}O_6 + O_2 \rightarrow$ _____. [p.108]
   a. 2 pyruvates
   b. $ATP + O_2$
   c. $Acetyl\ CoA + H_2O$
   d. $CO_2 + H_2O$

CHAPTER OBJECTIVES/REVIEW QUESTIONS

1. Explain the similarities between Friedreich's ataxia and Luft's syndrome. [p.107]

2. Understand the basic differences between aerobic and anaerobic pathways. [pp.108-109]

3. Know where the aerobic and anaerobic pathways start and stop in the cell. [p.108]

4. Know the basic formula of aerobic respiration. [p.109]

5. Understand the process of glycolysis including the inputs and outputs of the process. [pp.110-111]

6. Understand the processes of acetyl-CoA formation and the Krebs cycle, including the inputs and outputs of the process. [pp.112-113]

7. Understand the process of electron transfer phosphorylation, including the inputs and outputs of the process. [pp.114-115]

8. Understand the role of oxygen in the Krebs cycle and electron transfer phosphorylation. [pp.112-115]

9. Understand the relationship between glycolysis, acetyl-CoA formation, the Krebs cycle, and electron transfer phosphorylation [p.115]

10. Understand the purpose and significance of lactose and alcoholic fermentation reactions. [pp.116-117]

11. Understand the alternative energy sources that a cell may use to generate ATP. [pp.118-119]

CHAPTER SUMMARY

All organisms produce ATP by various (1) _____ pathways that extract (2) _____ energy from glucose and other organic compounds. (3) _____ respiration yields the most ATP from each glucose molecule. In eukaryotes, it is completed inside the (4) _____.

(5) _____ is the first stage of aerobic respiration and of anaerobic routes, such as fermentation pathways. As enzymes break down glucose to pyruvate, the coenzyme (6) _____ picks up (7) _____ and hydrogen atoms. The net energy yield is (8) _____ ATP.

Aerobic respiration has two more stages. In the (9) _____ and a few reactions before it, pyruvate is broken down to (10) _____, and many coenzymes pick up electrons and hydrogen atoms. In electron transfer (11) _____, coenzymes deliver electrons to transfer chains that set up conditions for ATP formation. (12) _____ accepts electrons at the end of the chains.

Fermentation pathways start with (13) _____. Substances other than oxygen are the final (14) _____ acceptor. Compared with aerobic respiration, the net yield of ATP is (15) _____.

Molecules other than (16) _____ are common energy sources. Different pathways convert (17) _____ and proteins to substances that may enter glycolysis or the Krebs cycle.

INTEGRATING AND APPLYING KEY CONCEPTS

1. What problems may humans and other organisms experience if their mitochondria were defective?
2. Evaluate this statement: Every atom in your body has first passed through the cell(s) of a photosynthetic organism.
3. Some diets propose that you eat 100% protein. Based on the information presented in this chapter, would that be a good idea? Why? Why not?
4. What is the evolutionary benefit of anaerobic respiration in human muscles? Why not only use anaerobic respiration?

8

DNA STRUCTURE AND FUNCTION

INTRODUCTION

The DNA molecule has a history full of scientific intrigue, experimentation and even scandal. While the molecule itself was discovered in the 1800s, it was not until the 1940s that researchers became certain that DNA was the heritable genetic material. Then the race was on to discover the structure and replication scheme of this all important molecule. This chapter describes the experiments that lead to the elucidation of the function of DNA and the scandal which lead to the elucidation of the structure of this molecule. The chapter concludes with a look at DNA technology with respect not to how far we have come in our understanding, but where this understanding may take us.

FOCAL POINTS

- Figure 8.5 [p.126] animates the experiment by Fredrick Griffith in which he discovered "the transforming principle."
- Figure 8.6 [p.127] illustrates the Hershey-Chase experiment.
- Figure 8.7 [p.128] illustrates the four nucleotide bases in the DNA molecule.
- Figure 8.8 [p.129] illustrates the double helix molecule of DNA.
- Figure 8.10 [p.130] illustrates DNA replication.
- Figure 8.11 [p.131] depicts the formation of a newly synthesized strand during replication.

INTERACTIVE EXERCISES

8.1 A HERO DOG'S GOLDEN CLONES [p.123]
8.2 EUKARYOTIC CHROMOSOMES [pp.124-125]

Selected Terms
eukaryotic [p.123], micrometer [p.123], micrograph [p.124], XX [p.124], XY [p.124]

Boldface Terms
[p.123] clone _____

_____

[p.124] chromosome _____

_____

[p.124] sister chromatids _____

_____

[p.124] centromere _____

_____

[p.125] histones _____

_____

[p.125] nucleosome _____

_____

[p.124] chromosome number _____

_____

[p.124] diploid _____

_____

[p.125] karyotype _____

_____

[p.125] autosome _____

_____

[p.125] sex chromosomes _____

_____

Matching [pp.123-125]
Match the following descriptions with the correct terms below.

1. ____ these differ between males and females
2. ____ all except one pair of a diploid cell's chromosomes
3. ____ the finished array of a set of chromosomes
4. ____ a condition of having two sets of chromosomes
5. ____ a strand of organized DNA
6. ____ these are produced after a chromosome is duplicated
7. ____ a genetically identical copy of an organism
8. ____ these appear as "beads on a string"
9. ____ these structures attach two sister chromatids together
10. ____ "spools" of protein

a. karyotype
b. chromosome
c. centromere
d. diploid
e. histones
f. sex chromosomes
g. clone
h. sister chromatids
i. autosomes
j. nucleosomes

8.3 THE DISCOVERY OF DNA'S FUNCTION [pp.126-127]

Selected Terms: DNA [p.126], RNA [p.126], Johann Miescher [p.126], Frederick Griffith [p.126], *Streptococcus pneumoniae* [p.126], Oswald Avery [p.126], Maclyn McCarty [p.126], vaccine [p.126], Hershey and Chase [p.127]

Boldfaced Terms

[p.127] bacteriophages _____

_____

Short Answer

1. Explain how Fredrick Griffith's experiments with Streptococcus pnemnoniae bacteria showed that a cell could be changed forever by materials present in another cell. [p.188]

_____

_____

_____

_____

Complete the Table

2. Complete the table by noting the name of the investigator or the contribution made in the discovery of the structure of DNA.

Investigators	*Contribution*
a. Miescher [p.126]
b. _____ [p.126]	Discovered the transforming principle in *Streptococcus pneumoniae*; live harmless cells were mixed with dead S cells, R cells became S cells
c. _____ [p.126]	Reported that the transforming substance in Griffith's bacteria experiments was probably DNA
d. _____ [p. 127]	Worked with radioactive sulfur (found in protein) and phosphorus (found in DNA) labels; T4 bacteriophage and *E. coli* cells demonstrated that labeled phosphorus was in bacteriophage DNA and contained hereditary instructions for new bacteriophages

8.4   THE DISCOVERY OF DNA'S STRUCTURE [pp.128-129]

Selected Terms: Erwin Chargaff [p.128], Francis Crick [p.128], James Watson [p.128], Rosalind Franklin [p.128], nucleotide [p.128], adenine [p.128], thymine [p.128], guanine [p.128], cytosine [p.128], x-ray crystallography [p.128], A=T & G=C [p.128], DNA double helix [p.129], x-ray diffraction [p.129]

Short Answer

1. List the three parts of every nucleotide. [p.128]

_____

_____

_____

Labeling

For the four nucleotides shown in questions 2-5, label each nitrogen-containing base as *guanine, thymine, cytosine,* or *adenine.* [p. 128]

2. _____   3. _____   4. _____   5. _____

For questions 6-12, label each numbered part of the DNA model as *phosphate, purine, pyrimidine, nucleotide,* or *deoxyribose.* [p.129]

6. _____

7. _____

8. _____

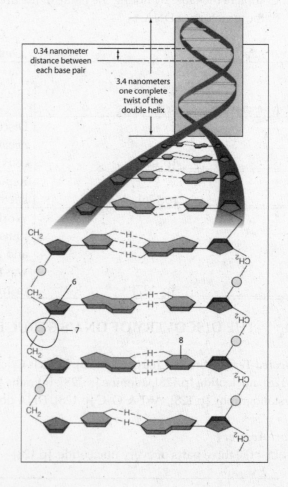

0.34 nanometer distance between each base pair

3.4 nanometers one complete twist of the double helix

True-False

If the statement is true, write a "T" in the answer blank. If the statement is false, correct it by writing the correct word(s) for the underlined word(s) in the answer blank.

9. _____ James <u>Watson</u> shared two crucial insights into the composition of DNA with the scientific community. First, the amount of A relative to G differs from one species to the next. Second, the amounts of T and A in a DNA molecule are exactly the same, and so are the amounts of C and G. [p.128-129]

10. _____ Rosalind Franklin, a colleague of Crick, obtained especially good <u>electron microscope</u> images of DNA fibers. [p.128]

11. _____ Crick and <u>Avery</u> had perceived, DNA consists of two strands of nucleotides, held together at their bases by hydrogen bonds. [pp.128-129]

12. _____ Much of the actual research data that elucidated the chemical structure of DNA came through the work of <u>Rosalind Franklin</u>. [pp.128-129]

8.5 FAME AND GLORY [p.130]
8.6 DNA REPLICATION AND REPAIR [pp.130-131]

Selected Terms: Pauling [p.130], Wilkins [p.130], Nobel Prize [p.130], complementary [p.130], semiconservative [p.131]

Boldfaced Terms
[p.130] DNA replication_____

_____

[p.130] DNA polymerase _____

_____

[p.130] DNA sequence _____

_____

[p.131] DNA ligase _____

_____

[p.131] DNA repair mechanisms _____

_____

[p.131] mutation _____

_____

Choice

1. ___ Before her work at Cambridge University, Rosalind Franklin had refined an x-ray diffraction method while studying the structure of _____. [p.130]
   a. wood
   b. stainless steel
   c. quartz
   d. coal

2. ___ At Cambridge University, Franklin was given the assignment to investigate the structure of DNA in a laboratory down the hall from _____. [p.130]
   a. Francis Crick
   b. Linus Pauling
   c. Maurice Wilkins
   d. James Watson

3. ___ When Watson and Crick finally focused on Franklin's x-ray diffraction image of wet DNA fibers, it gave them which bit of information [p.130]?
   a. DNA had two pairs of chains
   b. two helically twisted chains running in opposing directions
   c. the names of the four nucleotide bases
   d. nucleotide bases are held together by weak hydrogen bonds

Fill in the Blanks

Until Watson and Crick presented their model, no one could explain DNA (4) _____ [p.130], or how the molecule of inheritance is duplicated before a cell divides. (5) _____ [p.130] easily break the hydrogen bonds between the two-nucleotide strands of a DNA molecule. When (2) and other proteins act on DNA, one (6) _____ [p.131] unwinds from the other and exposes stretches of its nucleotide bases. Cells contain stockpiles of free (7) _____ [p.131] that can pair with the exposed bases.

Each parent strand stays intact, and a companion strand is assembled on each one according to this base-pairing rule: A to (8) _____ [p.130], and G to (9) _____ [p.130]. As soon as a stretch of a new, partner strand forms on a stretch of the parent strand, the two twist up together into a (10) _____ [p.131]. Because the parent DNA strand is (11) _____ [p.131] during the replication process, half of every double-stranded DNA molecule is "old" and half is "new." The process is called (12) _____ [p.131] replication. DNA replication uses a team of molecular workers. (13) _____ [p.131] enzymes become active along the length of the DNA molecule. Along with other proteins, some enzymes unwind the strands in both directions and prevent them from (14) _____. [p.131]. (15) _____ [p.131] action jump-starts the unwinding process but isn't necessary to unzip hydrogen bonds between the strands; (16) _____ [p.131] bonds are individually weak.

Enzymes called DNA (17) _____ [p.131] attach short stretches of free nucleotides to unwound parts of the parent template. Free nucleotides themselves drive the strand assembly. Each has three (18) _____ [p.131] groups. DNA polymerase splits off two, and it is this release of (19) _____ [p.130] that drives the attachments.

DNA (20) _____ [p.131] fill in the tiny gaps between the new short stretches to form one continuous strand. Then enzymes wind the template strand and complementary strand together to form a DNA (21) _____ [p.131].

Sometimes a molecule of DNA breaks. (22) _____ [p.131] can fix breaks, and specialized DNA (23) _____ [p.131] can fix mismatched base pairs or replace mutated ones. These repair processes confer a (24) _____ [p.131] advantage on cells. This maintains the integrity of genetic information. When proofreading and repair mechanisms fail, a mistake becomes a _____ [p.131], which is a permanent change in the DNA.

Labeling

25. The term *semiconservative replication* refers to the fact that each new DNA molecule resulting from the replication process is "half-old, half-new." In the chart below, complete the replication required in the middle of the molecule by adding the required letters representing the missing nucleotide bases. Recall that ATP energy and the appropriate enzymes are actually required to complete this process. [p.131]

T - _____   _____ - A
G - _____   _____ - C
A - _____   _____ - T
C - _____   _____ - G
C - _____   _____ - G
C - _____   _____ - G
old new   new old

8.7 USING DNA TO DUPLICATE EXISTING MAMMALS [pp.132-133]
Selected Terms: cloning [p. 132]

Boldface Terms
[p.132] reproductive cloning _____

_____

[p.132] somatic cell nuclear transfer _____

_____

[p.133] therapeutic cloning _____

_____

SELF-QUIZ

1. _____ was the first scientist to isolate
   DNA. [p.126]
   a. Oswald Avery
   b. Linus Pauling
   c. James Watson
   d. Johann Miescher
   e. Frederick Griffith

2. The scientist who demonstrated that
   harmless pneumonia-causing bacteria cells
   had become permanently transformed into
   pathogens through a change in the bacterial
   hereditary material was _____. [p.126]
   a. Oswald Avery
   b. Francis Crick
   c. James Watson
   d. Johann Miescher
   e. Frederick Griffith

3. _____ were the scientists who
   demonstrated that radioactively labeled
   bacteriophages transfer their DNA but not
   their protein coats to their host bacteria.
   [p.127]
   a. Watson and Crick
   b. Hershey and Chase
   c. Hershey and Griffith
   d. Watson and Pauling
   e. Wilkins and Chargaff

4. In 1953, _____ built a model of DNA
   that fit all the pertinent biochemical rules
   and insights they had gleaned from other
   sources. [p.128]
   a. Watson and Crick
   b. Hershey and Chase
   c. Wilkins and Franklin
   d. Watson and Pauling
   e. Wilkins and Chargaff

5. Adenine pairs with _____. [p.128]
   a. thymine
   b. cytosine
   c. guanine

6. _____ discovered that in DNA, the
   amounts of A = T and the amounts of G = C.
   [p.128]
   a. Maurice Wilkins
   b. Rosalind Franklin
   c. Erwin Chargaff
   d. Hershey and Chase
   e. Watson and Crick

7. A single strand of DNA with the base-
   pairing sequence C-G-A-T-T-G is compatible
   only with the sequence _____. [p.129]
   a. C-G-A-T-T-G
   b. G-C-T-A-A-G
   c. T-A-G-C-C-T
   d. G-C-T-A-A-C
   e. G-C-T-A-T-C

8. Rosalind Franklin's x-ray diffraction
   research on DNA established _____.
   [p.128]
   a. phosphate groups project outward
      from the molecule
   b. one pair of chains
   c. one pair of chains oriented in
      opposite directions
   d. a helical structure
   e. all of the above

9. Enzymes known as _____ attach short stretches of free nucleotides to unwound parts of the parent template. [p.130]
   a. semiconservative enzymes
   b. DNA ligases
   c. conservative enzymes
   d. DNA polymerases
   e. replication enzymes

10. Artificial twinning yields genetically identical individuals and is called _____. [p.132]
    a. reproductive cloning
    b. somatic cell nuclear
    c. therapeutic cloning

CHAPTER OBJECTIVES/REVIEW QUESTIONS

1. Summarize the research carried out by Miescher, Griffith, Avery and colleagues, Pauling, and Hershey and Chase; state the specific advances made by each in the understanding of DNA structure. [pp.126-127]

2. The two scientists who assembled the clues to DNA structure and produced the first model were _____ and _____. [pp.128-129]

3. Name the four nucleotides in DNA. [p.128]

4. Sketch a DNA molecule, using letters to represent the nitrogenous bases, phosphates, sugars, and hydrogen bonds. [p.129]

5. Describe the relationship between Rosalind Franklin and Maurice Wilkins. [p.130]

6. List the pieces of information about DNA structure that Rosalind Franklin discovered through her x-ray diffraction research. [p.128]

6. Assume that the two parent strands of DNA have been separated and that the base sequence on one parent strand is A-T-T-C-G-C; the base sequence that will complement that parent strand is _____. [p.128]

7. Describe how double-stranded DNA replicates from stockpiles of nucleotides. [pp.130-131]

8. Define *semiconservative replication*. [p.133]

9. During DNA replication, enzymes called DNA _____ attach short stretches of free nucleotides to unwound parts of the parent template. [p.130]

10. DNA _____ fill in the tiny gaps between the new short stretches of DNA to form one continuous strand; then enzymes wind the template strand and complementary strand together to form a DNA _____ _____. [p.131]

11. DNA _____ can fix breaks in DNA, and specialized DNA _____ can fix mismatched base pairs or replace mutated ones. [p.131]

12. Define and give examples of embryo-splitting and somatic nuclear transfers as cloning methods. [p.132]

CHAPTER SUMMARY

In all living cells, (1) _____ molecules store information that governs heritable traits.

A DNA molecule consists of two chains of (2) _____, (3_____ -bonded together along their length and coiled into a (4) _____ helix. Four kinds of nucleotides make up the chains: adenine, thymine, guanine, and (5) _____.

The (6) _____ in which one kind of nucleotide base follows the next along a DAN strand encodes heritable information. The order of some regions of DNA is unique for each (7) _____.

Like any race, the one that led to the discovery of DNA's (8) _____ had its winners – and its losers.

Before a cell divides, enzymes and other proteins (9) _____ it DNA. Newly formed DNA strands are monitored for (10) _____, most of which are corrected. Uncorrected errors are (11) _____.

Knowledge about the structure and (12) _____ of DNA is the basis of several methods of (13) _____.

INTEGRATING AND APPLYING KEY CONCEPTS

1. Review the stages of mitosis and meiosis, as well as the process of fertilization. Include what has now been learned about DNA replication and the relationship of DNA to a chromosome. As you cover the stages, be sure each cell receives the proper number of DNA threads.

2. What would happen to the number of chromosomes in a eukaryotic cell if that cell did not duplicate its DNA prior to cell division?

3. AT HOME - Using materials that you can find around your home, build a three dimensional model of DNA. Pay particular attention to base pairing and to the formation of the double helix.

4. If a particular species has 23 % of its nucleotides as adenine, what would be the percentages of the other three?

9

FROM DNA TO PROTEIN

INTRODUCTION

Ultimately, most of the structure of a cell and most of the work done by a cell is due to the proteins that the cell produces. A defective protein can be deadly to a cell and perhaps to an entire organism. But where do these proteins originate? How are they made? This chapter explains protein synthesis by describing the 2 cellular processes that are used to make proteins – translation and transcription. Molecular information flows from DNA to RNA and then to protein. The integrity of the DNA molecule is vital to correct protein formation, and this chapter explains why. Changes in the DNA, mutations, may be detrimental because they may cause a protein to be incorrectly made. These same changes, however, are the mechanisms by which cells, and therefore organisms, change and undergo evolution.

FOCAL POINTS

- Figure 9.3 [p.139] illustrates the difference between DNA and RNA.
- Figure 9.4 [p.140] illustrates DNA replication and transcription – the cellular process of RNA synthesis.
- Figure 9.6 [p.141] illustrates post-transcriptional modifications of RNA.
- Figure 9.7 [p.142] illustrates the genetic code.
- Figure 9.8 [p.143] describes the correspondence between DNA, RNA, and proteins.
- Figure 9.11 [p.144]] shows the stages of translation.
- Figure 9.13 [p.146] depicts the different types of mutations.

INTERACTIVE EXERCISES

9.1 RICIN AND YOUR RIBOSOMES [p.137]
9.2 THE NATURE OF GENETIC INFORMATION [p.138-139]
9.3 TRANSCRIPTION [pp.140-141]

Selected Terms
"genetic Terms" [p.138], uracil [p.138], RNA nucleotide [p.138], amino acids [p.138], polypeptide [p.139], base triplets [p.139], 3' to 5' direction [p.141], guanine "cap" [p.141], poly-A tail [p.141]

Boldface Terms
[p.138] genes _____

_____

[p.138] transcription _____

_____

[p.138] ribosomal RNA or rRNA _____

_____

[p.138] transfer RNA or tRNA _____

_____

[p.138] messenger RNA or mRNA _____

_____

[p.139] translation _____

_____

[p.139] gene expression _____

_____

[p.140] RNA polymerase _____

_____

[p.140] promoter _____

_____

[p.141] introns _____

_____

[p.141] exons _____

_____

[p.141] alternative splicing _____

_____

Choice

1. Ricin is a product of the _____ [p. 137].
    a. fava bean
    b. deadly nightshade
    c. *Clostridium* bacterium
    d. castor oil plant

2. Traces of Ricin have shown up _____
    [p.137].
    a. in a Florida home
    b. State Department building
    c. in a U.S. Senate Mailroom
    d. all of the preceding

3. Ricin itself is a _____ [p. 137].
    a. carbohydrate
    b. protein
    c. nucleic acid
    d. fat

4. A dose of Ricin the size of _____ can
    kill you [p.137].
    a. a teaspoon
    b. a tablespoon
    c. a grain of salt
    d. 500 grams

5. Ricin has two polypeptide chains, one helps
    ricin insert itself into cells, the other is an
    enzyme that _____ [p. 137].
    a. destroys mitochondria
    b. wrecks part of the ribosome
    c. inactivates the nucleus
    d. breaks down the cytoskeleton

Complete the Table

6. Three types of RNA are transcribed from DNA in the nucleus. Provide information about these molecules. [p.138]

RNA Molecules	Abbreviation	Description/Function
a. Messenger RNA		
b. Ribosomal RNA		
c. Transfer RNA		

Short Answer

7. List three ways in which a molecule of RNA is structurally different than a molecule of DNA. [p.138]

_____

_____

_____

_____

Fill in the Blanks

Transcription differs from DNA replication in three respects. Part of a (8) _____ [p.140] strand, not the whole molecule is the template. The enzyme (9) _____ [p.140], not DNA (10) _____ [p.141], adds RNA nucleotides to the end of a growing (11) _____ [p.141] strand, in the 5′ to 3′ direction. Also, transcription produces (12) _____ [p.141] free strand of RNA, not a hydrogen-bonded double helix.

A (13) _____ [p.140] region is a particular base sequence coded in the DNA that serves as the "start" signal for (14) _____ [p.140]. RNA synthesis begins when several proteins, including (15) _____ [p.140], attach to a promoter. The (16) _____ [p.141] then swiftly moves down the DNA strand, joining one (17) _____ [p141] after another, using the DNA as a template to make a new (18) _____ [p.141} as it goes. Eventually, it arrives at another particular sequence in the DNA that signals "the end," and the new RNA is released.

Labeling

Select the letters on the sketch that best fits the numbered descriptions. [Refer to Figure 9.6, p.141]

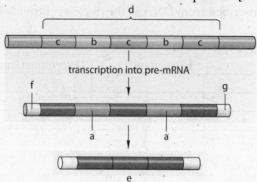

transcription into pre-mRNA

19. ____ Unit of transcription in a DNA strand

20. ____ Enzymes attach a modified guanine "cap" that will bind the mRNA to a ribosome

21. ____ Different enzymes attach a "poly-A tail" of about 100-300 adenine ribonucleotides

22. ____ Introns, or gene sequences that are removed before an mRNA is used for protein synthesis

23. ____ Exons, coding parts of a gene sequence; exons alternate with introns

24. ____ Introns are snipped out before the mRNA leaves the nucleus in mature form; introns remain in the nucleus, where they are recycled

25. ____ Mature mRNA transcript

9.4 RNA AND THE GENETIC CODE [pp.142-143]
9.5 TRANSLATING THE CODE: RNA TO PROTEIN [pp. 144-145]

Selected Terms: stop [p.142], conserved start codon AUG [p.142], stop codons [p.142], [p.143], anticodon [p.143], ribosome [p.143], peptide bond [p.143], translation [p.144], termination [p.144], polysome [p.144]

Boldfaced Terms
[p.142] codons _____

_____

[p.142] genetic code _____

_____

Matching

Match each term with its description.

1. ____ codon [p.142]
2. ____ three bases at a time" [p.142]
3. ____ sixty-one of the base triplets [p.142]
4. ____ the genetic code [p.142]
5. ____ ribosome subunits [p.143]
6. ____ anticodon [p.143]
7. ____ the "stop" codons [p.142]
8. ____ sixty-four (4^3) [p.142]
9. ____ Ribosomal RNA [p.143]
10. ____ peptide bond [p.144]

a. Large and small are assembled in the nucleus from rRNA and proteins; shipped separately to the cytoplasm; intact and functional when translation is to occur
b. Reading frame of the nucleotide bases in mRNA
c. A type of bond between two amino acids
d. UAA, UAG, UGA
e. Deduced the correlation between genes and proteins
f. Refers to the total number of codons
g. Probably evolved from ingested foreign bacteria that managed to survive inside host cells
h. This molecule has enzymatic properties
i. The number of mRNA base triplets that actually specify amino acids
j. The set of sixty-four different codons

Completion

11. Given the following DNA sequence, deduce the composition of the mRNA transcript: [p.140]

TAC   AAG   ATA   ACA   TTA   TTT   CCT   ACC   GTC   ATC

___   ___   ___   ___   ___   ___   ___   ___   ___   ___
(mRNA transcript)

12. Deduce the composition of the tRNA anticodons that would pair with the above specific mRNA codons as these tRNAs deliver the amino acids (to be identified below). [p.143]

___   ___   ___   ___   ___   ___   ___   ___   ___   ___
(tRNA anticodons)

14. From the mRNA transcript in question 12, use Figure 9.7 in the text (the genetic code) of the text to deduce the composition of the amino acids of the polypeptide sequence. [p.142]

___   ___   ___   ___   ___   ___   ___   ___   ___   ___
(amino acids)

15. Write the RNA sequence for a start codon _____. Now write the DNA sequence for this start codon _____ . What amino acid is coded for by this sequence [p.142]? _____

16. Write one of the codons that would signal the end of translation [p.142]. _____

Complete the Table

17.   Complete the following table, which distinguishes the stages of translation.

Translation Stage   *Description*

Translation Stage	Description
[pp.144-145] a.	The ribosome encounters the mRNA's STOP codon; proteins called release factors find to the ribosome and trigger enzyme activity that detaches the mRNA and the polypeptide chain from the ribosome
[pp.144-145] b.	A polypeptide chain is assembled as the mRNA passes between the two ribosomal subunits; molecules of tRNA bring amino acids to the ribosome, then bind to the mRNA in the order specified by its codon; peptide bond formation between the amino acids is catalyzed by part of an rRNA molecule functioning as an enzyme at the center of the large ribosomal subunit
[pp.144-145] c.	Initiator tRNA binds with a small ribosomal subunit; then mRNA's START codon, AUG, joins up with the initiator tRNA's anticodon; a complex is formed by the ribosome, mRNA, and initiator tRNA

9.6 MUTATED GENES AND THEIR PROTEIN PRODUCTS [pp.144-145]

Selected Word:
Mutation [p.146], hemoglobin [p.146], anemia [p.146], beta thalassemia [p.146], frameshift mutations [p.146], sickle hemoglobin [p.147], ionizing radiation & nonionizing radiation [p.147], thymine dimer [p.147]

Boldfaced Terms
[p.146] deletions _____

_____

[p.146] insertions _____

_____

[p.147] base-pair substitution _____

_____

[p.147] transposable elements _____

_____

Choice

1. ___ In genetic mutations called _____, one base is copied incorrectly during DNA replication. [p.147]
    a. transposons
    b. base substitutions
    c. insertions
    d. deletions

2. ___ Insertions and deletions result in _____. [p.146]
    a. transposon mutations
    b. base substitutions
    c. frameshift mutations
    d. a break in RNA molecules

3. ___ When _____ land in a gene, they often block or alter the timing or extent of its activity. [p.147]
    a. base substitutions
    b. frameshift mutations
    c. insertions
    d. transposonable elements

4. ___ _____ breaks chromosomes into pieces. [p.147]
    a. A base substitution
    b. Ionizing radiation
    c. Nonionizing radiation
    d. A frameshift mutation

5. ___ _____ boosts electrons to a higher energy level; when DNA absorbs UV light, cytosine and thymine are susceptible to changing their base-pairing properties (T – T, not A – T). [p.147]
    a. A base substitution
    b. Ionizing radiation
    c. Nonionizing radiation
    d. A frameshift mutation

SELF-QUIZ

1. rRNA _____. [p.138]
    a. serves as the "start" signal for transcription
    b. is a ribosomal component
    c. carries protein-building instructions
    d. delivers amino acids, one at a time to a ribosome
    e. receives RNA polymerase

2. _____ delivers amino acids to ribosomes, where amino acids are linked into the primary structure of a polypeptide. [p.138]
    a. mRNA
    b. tRNA
    c. Introns
    d. rRNA
    e. RNA polymerase

3. Of the following, which *one* is not a part of a mature mRNA transcript. [p.141]
    a. exons
    b. a poly-A tail
    c. a modified guanine "cap"
    d. introns

4. Transfer RNA differs from other types of RNA because it _____. [p.138]
    a. transfers genetic instructions from cell nucleus to cytoplasm
    b. specifies the amino acid sequence of a particular protein
    c. carries an amino acid at one end
    d. contains codons
    e. is a promoter

5. The codons UAA, UAG, and UGA
   _____. [p.142]
   a. all code for glutamine
   b. all code for methionine
   c. are all start signals
   d. all are stop signals
   e. serve as universal codons all amino
      acids

6. The genetic code is a set of _____
   codons, or sets of three ribonucleotide bases.
   [p.142]
   a. thirty-two
   b. sixteen
   c. forty-six
   d. sixty-one
   e. sixty-four

7. In _____, a polypeptide chain is
   assembled as the mRNA passes through the
   two ribosomal subunits. [p.141].
   a. elongation
   b. transcription
   c. initiation
   d. mRNA transcript processing
   e. termination

8. Errors that remain uncorrected in DNA are
   called _____. [p.147]
   a. thymine dimers
   b. mutations
   c. polysomes
   d. exons
   e. codons

9. Frameshift mutations involve _____.
   [p.146]
   a. base substitutions
   b. insertions
   c. transposon
   d. deletions
   e. both b and d

10. Thymine dimers in DNA are induced by
    _____. [p.147]
    a. ionizing radiation
    b. transposons
    c. frameshift mutations
    d. nonionizing radiation
    e. base substitutions

CHAPTER OBJECTIVES/REVIEW QUESTIONS

1. Only plutonium and botulism toxin are more deadly than _____. [p.137]

2. _____ RNA is a component of ribosomes; _____ RNA delivers protein-building instructions; _____ RNA delivers amino acids one at a time to a ribosome. [p.138]

3. State how RNA differs from DNA in structure and function, and indicates what features RNA has in common with DNA. [p.139]

4. Generally describe the process of transcription, and indicate three ways in which it differs from replication. [pp.141-142]

5. What is the function of RNA polymerase? A promoter region? [p.140]

6. Distinguish introns from exons and describe the fate of each. [p.141]

7. What RNA code would be formed from the following DNA code: TAC-CTC-GTT-CCC-GAA? [pp.142-145]

8. Each base triplet in mRNA is called a(n) _____. [p.142]

9. The set of sixty-four different codons is the _____ _____. [p.142]

10. Explain how the DNA message TAC-CTC-GTT-CCC-GAA would be used to code for a segment of protein, and state what its amino acid sequence would be. [pp.142-145]

11. Describe events occurring in the following stages of translation: initiation, elongation, and termination. [pp.144-145]

14. What is the fate of the new polypeptides produced by protein synthesis? [pp.144-145]

15. Briefly describe the spontaneous DNA mutations known as base substitutions, frameshift mutation, and transposable elements. [pp.146-147]

16. Distinguish between mutations caused by ionizing radiation and nonionizing radiation. [p.147]

17. Using a diagram, summarize the steps involved in the transformation of genetic messages into proteins. [Figure 9.8, p.143]

CHAPTER SUMMARY

Life depends on (1) _____ and other proteins. All proteins consist of (2) _____ chains. The chains are sequences of amino acids that correspond to sequences of (3) _____ bases in DNA called (4) _____. The path leading from genes to proteins has two steps: (5) _____ and (6) _____.

During transcription, the two strands of the (7) _____ double helix are unwound in a gene region. Exposed based of one strand become the (8) _____ for assembling a single strand of (9) _____ (a transcript). (10) _____ RNA is the only type of RNA that carries DNA's protein-building instructions.

The nucleotide sequence of RNA is read (11) _____ bases at a time. Sixty-four base triplets that correspond to specific amino acids represent the genetic (12) _____, which ahs been highly (13) _____ over time.

During (14) _____, amino acids become bonded together into a polypeptide chain in a sequence specified by base triplets in (15) _____ RNA. (16) _____ RNAs deliver amino acids one at a time to ribosomes. (17) _____ RNA catalyzes the formation of (18) _____ bonds between the amino acids.

(19) _____ in genes may result in changes in protein structure, protein function, or both. The changes may lead to (20) _____ in traits among individuals.

INTEGRATING AND APPLYING KEY CONCEPTS

1. Genes code for specific polypeptide sequences. Not every substance in living cells is a polypeptide. Explain how genes might be involved in the production of a storage starch (such as glycogen) that is constructed from simple sugars.

2. Explain why a frameshift mutation would be considered much more drastic to the resulting protein than a simple point mutation. Give an example with a hypothetical DNA sequence containing at least four codons, beginning with the correct "Start codon."

10

CONTROLS OVER GENES

INTRODUCTION

This chapter introduces the various mechanisms involved in the regulation of gene expression, or more simply put when certain proteins are made. It begins to answer the questions that if all cells have the same DNA, then why are some cells different from others. A liver cell, for example, certainly looks and functions differently than a skin cell, yet they both have the same DNA. The regulation of gene expression allows cells to be specialized in both structure and function. Examples of gene regulation in both prokaryotic and eukaryotic cells are illustrated in this chapter. Finally, the importance of gene regulation in the processes involved in cancer is explained and emphasized.

FOCAL POINTS

- Figure 10.2 [p. 152] illustrates various levels at which gene expression can be controlled in the eukaryote.
- Figure 10.5 [p. 154] shows different expression of eyes with respect to their gene expression.
- Figure 10.6 [p. 155] illustrates how gene expression controls the development of a fruit fly.
- Figure 10.8 [p.156] shows the development of reproductive organs in human embryos.
- Figure 10.9 [p.157] shows floral development.
- Figure 10.10 [p.158] depicts the control of the lactose operon in bacteria, a model for gene regulation in prokaryotes.

INTERACTIVE EXERCISES

10.1 BETWEEN YOU AND ETERNITY [p.151]
10.2 GENE EXPRESSION IN EUKARYOTIC CELLS [p.152-153]
10.3 THERE'S A FLY IN MY RESEARCH [pp.154-155]
10.4 A FEW OUTCOMES OF GENE CONTROLS [pp.156-157]
10.5 GENE CONTROL IN BACTERIA [pp.158-159]

Selected Terms
radial mastectomy [p.151], regulatory proteins [p.152], methyl group [p.152], methylate [p.152], polytene chromosomes [p.153], *Drosophila melanogastor* [p.154], *Tinman* [p.154], homeodomain [p.154], SRY [p.156], "Barr bodies" [p.156], XX & XY [p.156], "mosaic" [p.156], *Arabidopsis thaliana* [p.157], phenotypic [p.157], ABC master genes [p.157], lac operon [p.158], lactose [p.158], *E. coli* [p.158], allolactose [p.159]

Boldfaced Terms
[p.151] cancer _____

_____

[p.152] differentiation _____

_____

[p.152] activator _____

_____

[p.152] enhancers _____

_____

[p.152] repressor _____

_____

[p.152] transcription factors _____

_____

[p.154] homeotic genes _____

_____

[p.154] master genes _____

_____

[p.154] knockout _____

_____

[p.155] pattern formation _____

_____

[p.156] X-chromosome inactivation _____

_____

[p.156] dosage compensation _____

_____

[p.158] operators _____

_____

[p.158] operon _____

_____

Labeling

For each of the following statements, choose the correct letter from the diagram below. [p.152]

1. ____ pre-mRNA processing.

2. ____ Transport proteins deliver mRNA to the correct region of the cytoplasm.

3. ____ Chemical modifications to the DNA restricts access to genes.

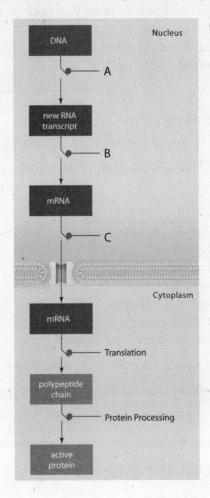

Fill-in-the-Blanks

All cells of your body started out life with the same (4) _____ [p.152], because every one arose by mitotic cell divisions from the same fertilized egg. And they all transcribe many of the same genes, because they are alike in most aspects of (5) _____ [p.152] and basic housekeeping activities.

In other ways, however, nearly all of your body cells become (6) _____ [p.152] in composition, structure and (7) _____ [p.152]. This process of cell (8) _____ [p.152] occurs during the development of multi-celled organisms. Differences arise among cells that use different subset of (9) _____ [p.152]. Specialized tissues and organs are the result.

Choice

For each of the following, chose the stage of control over gene expression that is indicated by the statement. Answers may be used more than once. [p.152]

a. takes place inside the nucleus
b. takes place in the cytoplasm

10. _____ Translation of the mRNA.

11. _____ The new protein can become activated or disabled.

12. _____ The beginning of mRNA transcription,

13. _____ The binding of certain proteins to the mRNA.

14. _____ The binding of transport proteins to the mRNA.

Matching

For each of the following, choose one of the terms from the list below. Some terms may be used more than once. [pp.154-155]

a. X chromosome inactivation
b. homeotic genes
c. dosage compensation
d. all of the above

15. _____ This occurs in female mammals.

16. _____ These code for specific body parts during development of embryos.

17. _____ The master genes of most eukaryotic organisms.

18. _____ This forms a Barr body in the cell.

19. _____ Assures that equal amounts of gene products are present in each sex.

20. _____ A form of selective gene expression.

SELF-QUIZ

1. An arrangement where a promoter and a set of operators control more than one gene is called a _____ [p.152]
   a. translational unit
   b. gene
   c. operon
   d. enhancer
   e. none of the above

2. Operons occur in _____. [p.158]
   a. bacteria
   b. archaea
   c. eukaryotes
   d. all of the above

3. _____ bind to enhancer regulatory regions and speed up transcription. [p.152]
   a. activators
   b. repressors
   c. polymerases
   d. operators

4. A repressor slows or stops _____? [p.152]
   a. translation
   b. transcription
   c. differentiation
   d. X-chromosome inactivation
   e. dosage compensation

5. Barr bodies are produced as a result of
   _____. [p.156]
   a. cell differentiation
   b. translational control
   c. X chromosome inactivation
   d. selective gene expression

6. In multi-celled organisms, cell
   differentiation occurs as a result of _____.
   [p.152]
   a. growth
   b. fertilization
   c. the lac operon
   d. selective gene expression

7. Blocking an RNA polymerase's access to a
   gene is an example of which of the following
   controls? [p.152]
   a. control after translation
   b. control at translation
   c. control at transcript processing
   d. control before transcription

8. An experimental procedure in which genes
   are deleted from a wild-type organism to
   observe the effects on the organism is called
   a _____ experiment. [p.154]
   a. knockout
   b. knockdown
   c. homeotic
   d. X-inactivation
   e. none of the above

9. The emergence of embryonic tissues in
   predictable locations is called _____. [p.155]
   a. X-chromosome inactivation
   b. dosage compensation
   c. pattern formation
   d. selective gene expression

10. The master control genes of development in
    eukaryotic cells are called _____. [p.154]
    a. repressor genes
    b. homeotic genes
    c. proto-oncogenes
    d. activator genes

CHAPTER OBJECTIVES/ REVIEW QUESTIONS

1. Define the role of operators, enhancers and promoters in gene control. [pp.152-153]

2. Define cell differentiation. [p.152]

3. Recognize the various levels of control that are involved in gene expression. [pp.152-153]

4. Recognize the relationship between selective gene expression and cell differentiation. [p.152-153]

5. Define a homeotic gene and the role of the homeodomain. [p.154]

6. How are knockout experiments used to understand the process of gene control? [p.154-155]

7. What is pattern formation? [p.155]

8. Explain the process of X chromosome inactivation and its relationship to dosage compensation.
   [pp.156-157]

9. Define an operon. [p.158]

10. Identify the mechanisms of negative and positive control of the *lac* operon in *E. coli*. [pp.158-159]

CHAPTER SUMMARY

(1) _____ mechanisms govern when, how, and to what extent a cell's genes are (2) _____. They alter gene expression in response to changing (3) _____ both inside and (4) _____ the cell.

Diverse controls govern every step between gene (5) _____ and delivery of the final gene (6) _____ to a targeted location.

In multi-celled species, (7) _____ genes guide the stage-by-stage development of new individuals. (8) _____ gene expression results in cell (9) _____. By this process, different cell lineages become specialized in composition, structure, and (10) _____.

The orderly, (11) _____ expression of certain genes in the basis of the body plan of complex (12) _____ organisms. In female mammals, most of the genes on one of the two X chromosomes are (13) _____ in every cell.

Drosophila research revealed how a (14) _____ body plan emerges. All cells in a developing embryo inherited the same genes, but they selectively (15) _____ or (16) _____ different fractions of those genes.

Prokaryotic gene controls govern responses to short-term changes in (17) _____ availability and other aspects of the (18) _____. The main gene controls bring about fast adjustments in rates of (19) _____.

INTEGRATING AND APPLYING KEY CONCEPTS

1. In the past prescription and over-the counter drugs used to primarily interact solely with receptors located on the surface of the cells. Many newer forms of drugs now actively target gene regulation at the transcriptional or translational level. Why are these drugs typically more effective than their earlier counterparts? What area of gene regulation do you think could be targeted by the next generation of drugs to make them safer and more effective?

2. The lac operon in *E. coli* is said to be an inducible operon, that is it is induced by its substrate lactose. Explain why this type of control mechanism is very efficient for the bacterial cell.

11

HOW CELLS REPRODUCE

INTRODUCTION

For single celled organisms to reproduce, or for the tissues of a multicellular organism to repair themselves, mitosis and cell division must occur. This chapter discusses the mechanisms involved in mitosis and cell division. Appropriate control of cellular division is critical to the health of the organism and the final portion of this chapter is concerned with the regulation of cell growth and the association of improperly regulated growth with the development of cancer in humans.

FOCAL POINTS

- Figure 11.2 [p.164] illustrates and defines the parts of the eukaryotic cell cycle.
- Figure 11.3 [p.165] shows how mitosis maintains the chromosome number.
- Figure 11.5 [pp.166-167] defines the stages of mitosis and shows two chromosomes in a eukaryotic cells undergoing this process.
- Figures 11.6 & 11.7 [p.168] compares the process of cytokinesis in plant and animal cells.
- Figure 11.10 [p.170] compares benign and malignant tumors.

INTERACTIVE EXERCISES

11.1 HENRIETTA'S IMMORTAL CELLS [p.163]
11.2 MULTIPLICATION BY DIVISION [pp.164-165]
11.3 MITOSIS [pp.166-167]

Selected Terms: HeLa cells [p.163], chromosome [p.164], meiosis [p.164], DNA replication [p.164], G1, S, and G2 [p.164], diploid [p.164], XY [p.164], binary fission [p.165], cytoplasmic division [p.165], *mitos* [p.166], spindle [p.166], spindle poles [p.166], centromere [p.166], microtubules [p.166], tubulin subunits [p.166], sister chromatids [p.166]

Boldfaced Terms
[p.164] cell cycle _____

_____

[p.164] interphase _____

_____

[p.164] mitosis _____

_____

[p.164] homologous chromosomes _____

_____

[p.165] asexual reproduction _____

_____

[p.166] prophase _____

_____

[p.166] spindle _____

_____

[p.166] metaphase _____

_____

[p.166] anaphase _____

_____

[p.166] telophase _____

_____

Short Answer

1. Describe the origin and the significance of HeLa cells [p.163]

_____

_____

_____

Matching

Match each term with its description.

2. ____ centromere [p.166]

3. ____ chromosome [p.164]

4. ____ S in the cell cycle [p.164]

5. ____ sister chromatids [p.166]

6. ____ mitosis and meiosis [p.164]

7. ____ G1 & G2 [p.164]

8. ____ homologous chromosomes[p.164]

9. ____ mitosis [p.164]

10. ____ asexual reproduction[p.165]

11. ____ spindle[p.166]

12. ____ binary fission [p.165]

a. These have the same length, shape, and genes
b. Composed of a chromosome and its copy attached to each other until late in the division process
c. A method by which mitosis and cytoplasmic division are used to produce new cells in some organisms such as fungi and protists
d. Growth periods in the cell cycle
e. Formed by microtubules
f. In bacterial cells only, the basis of asexual reproduction
g. DNA synthesis occurs
h. A pronounced constricted region on a chromosome; docking site for certain microtubules that take part in nuclear division
i. Cell divisions by which an organism grows, replaces worn-out cells, and repairs tissues; used by single-celled organisms for asexual reproduction
j. Composed of a molecule of DNA and its proteins
k. The nuclear division mechanisms

Identification

Identify the stage in the cell cycle indicated by the numbers in each diagram. [p.164]

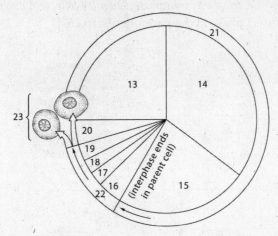

13. _____	17. _____	21. _____
14. _____	18. _____	22. _____
15. _____	19. _____	23. _____
16. _____	20. _____	

Matching

Link each time span identified below with the most appropriate number in the preceding diagram. [p.164]

24. ____ Interval following DNA replication; cell prepares to divide

25. ____ The complete period of nuclear division, followed by cytoplasmic division (a separate event)

26. ____ Interval of cell growth, when DNA replication is completed (chromosomes duplicated)

27. ____ Interval of cell growth, before DNA duplication (chromosomes unduplicated)

28. ____ Usually the longest part of a cell cycle

29. ____ Interval of cytoplasmic division

30. ____ Period that includes G1, S, G2

Identification

Identify each mitotic stage shown. Select from *late prophase, transition to metaphase (prometaphase), cell at interphase, metaphase, early prophase, telophase, interphase-daughter cells,* and *anaphase.* Complete the exercise by entering the letter of the correct phase in the parentheses. [p.167]

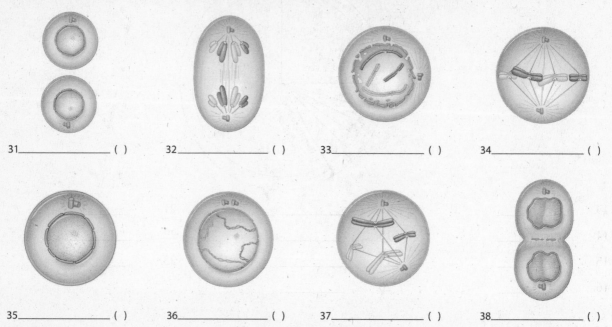

31_____ ( )   32_____ ( )   33_____ ( )   34_____ ( )

35_____ ( )   36_____ ( )   37_____ ( )   38_____ ( )

a. Attachments between two sister chromatids of each chromosome break; the two are now separate chromosomes, which microtubules move to opposite spindle poles

b. Microtubules penetrate the nuclear region and collectively form the bipolar spindle apparatus; microtubules become attached to the two sister chromatids of each chromosome

c. The DNA and its associated proteins have started to condense

d. All the chromosomes are now fully condensed and lined up at the equator of the fully formed microtubule spindle; chromosomes are now in their most tightly condensed form

e. The cell duplicates its DNA and prepares for nuclear division

f. Two daughter cells have formed, each is diploid with two of each type of chromosome, just like the parent cell's nucleus

g. Chromosomes continue to condense; new microtubules become assembled; They move one of two centriole pairs toward the opposite end of the cell; the nuclear envelope begins to break up

h. There are two clusters of chromosomes, which decondense; patches of new membrane fuse to form a new nuclear envelope; mitosis is complete

11.4 CYTOKINESIS: DIVISION OF THE CYTOPLASM [p.168]

Selected Terms: cell cortex [p.168], cytoskeletal elements [p.168], vesicle [p.168], actin & myosin {p.168], contractile ring [p.168]

Boldfaced Terms

[p.168] cytokinesis _____

_____

[p168] cleavage furrow _____

_____

[p.168] cell plate _____

_____

Choice
Choose from the following:
   a.   plant cells
   b.   animal cells

1.   ____   Formation of a cell plate [p.168]

2.   ____   As the microfilament ring shrinks in diameter, it pulls the cell surface inward [p.168]

3.   ____   Cellulose is deposited inside the sandwich; in time, these deposits will form two cell walls [p.168]

4.   ____   The primary wall of the growing cell is still thin; new material is deposited on it [p.168]

5.   ____   A cleavage furrow [p.168]

6.   ____   A ring of actin and myosin filaments attached to the plasma membrane contracts [p.168]

7.   ____   Contractions continue; the cell is pinched in two [p.168]

8.   ____   As mitosis ends, vesicles cluster at the spindle equator [p.168]

9.   ____   Cellulose deposits build up at the plate; in time, they are thick enough to form a cross-wall through the cell [p.168]

11.5 CONTROL OVER CELL DIVISION [p.169]
11.6 CANCER: WHEN CONTROL IS LOST [p.170}

Selected Terms: free radicals [p.169], apoptosis [p.169], "checkpoint" genes [p.169], epidermal growth factor (EGF) [p.169], tumor suppressors [p.170], human papillomavirus (HPV) [p.170], benign [p.170], *malignant* neoplasm [p.170], chemotherapy [p.170]

Boldfaced Terms
[p.169] neoplasms _____

_____

[p.169] growth factors _____

_____

[p.170] tumor _____

_____

[p.170] oncogene _____

_____

[p.170] proto-oncogenes _____

_____

[p.170] cancer_____

_____

[p.171] metastasis _____

_____

Short Answer

1. List the four characteristics displayed by *all* cancer cells. [p.170]

_____

_____

_____

_____

Matching

Match the description with its term below.

2. ____ the process in which malignant cells break loose from a tumor

3. ____ a malignant neoplasm that gets progressively worst and is dangerous to health

4. ____ any gene that transforms a normal cell into a tumor

5. ____ checkpoint genes encoding proteins that promote mitosis

6. ____ a neoplasm that forms a lump in the body

7. ____ an accumulation of cells that lost control over how they grow and divide

8. ____ programmed cell death

9. ____ molecules that stimulate cells to divide and differentiate

a. proto-ongogenes
b. growth factor
c. apoptosis
d. metastasis
e. tumor
f. cancer
g. neoplasm
h. oncogene

SELF-QUIZ

1. The replication of DNA occurs _____.
   [p.164]
   a between the growth phases of
     interphase
   b. immediately before prophase of
      mitosis
   c. during prophase of mitosis
   d. during anaphase of mitosis
   e. in daughter cells during telophase

2. The time that a cell forms until it divides is
   called the _____. [p.164]
   a. G₂ section
   b. S section
   c. G₁ section
   d. cell cycle

3. In the cell life cycle of a particular cell,
   _____. [p.164]
   a. mitosis occurs immediately after G1
   b. G2 comes after S
   c. G1 comes after S
   d. mitosis and S comes before G1
   e. S occurs immediately prior to
      mitosis

4. The correct order of the stages of mitosis is
   _____. [p.167]
   a. prophase, metaphase, telophase,
      anaphase
   b. telophase, anaphase, metaphase,
      prophase
   c. telophase, prophase, metaphase,
      anaphase
   d. anaphase, prophase, telophase,
      metaphase
   e. prophase, metaphase, anaphase,
      telophase

5. During _____, sister chromatids of
   each chromosome are separated from each
   other, and those former partners, now
   chromosomes, are moved toward opposite
   spindle poles. [p.167]
   a. prophase
   b. metaphase
   c. anaphase
   d. telophase
   e. transition to metaphase

6. In the process of cytokinesis, cleavage
   furrows are associated with _____ cell
   division, and cell plate formation is
   associated with _____ cell division.
   [p.168]
   a. animal; animal
   b. plant; animal
   c. plant; plant
   d. animal; plant

7. Nuclear division is called _____. [p.166]
   a. cytoplasmic division
   b. mitosis

8. Diploid refers to _____. [p.164]
   a. having two chromosomes of each
      type in somatic cells
   b. twice the parental chromosome
      number
   c. half the parental chromosome
      number
   d. having one chromosome of each
      type in somatic cells
   e. The chromosome conditions in a sex
      cell

9. The stage of mitosis in which the
   chromosomes are aligned midway between
   the spidle poles is _____.
   a. prophase
   b. metaphase
   c. anaphase
   d. teolphase

10. When plant cell divide a _____ forms between the two new cells. [p.168]
    a. cleavage furrow
    b. cell plate

11. A _____ is an accumulation of cells that lost control over how they divide and grow. [p.169]
    a. neoplasm
    b. tumor

12. During _____, sister chromatids of each chromosome are separated from each other, and those former partners, now chromosomes, are moved toward opposite poles. [p.167]
    a. prophase
    b. metaphase
    c. anaphase
    d. telophase
    e. prometaphase

CHAPTER OBJECTIVES/REVIEW QUESTIONS

1. Define *HeLa cells*; explain their origin. [p.163]

2. Mitosis and meiosis refer to the division of the cell's _____. [p.164]

3. List and describe, in order, the various activities occurring in the eukaryotic cell life cycle. [p.164]

4. Interphase of the cell cycle consists of G1, _____, and G2. [p.164]

5. S is the time in the cell cycle when _____ replication occurs. [p.164]

6. Define *binary fission* [p.165]

7. The _____ number of chromosomes in your body cells is 46. [p.164]

8. Describe, in detail, the cellular events occurring in the prophase, metaphase, anaphase, and telophase of mitosis. [pp.166-167]

9. Compare and contrast cytokinesis as it occurs in plant and animal cell division; use the following concepts: *cleavage furrow, microfilaments at the cell's midsection,* and *cell plate formation.* [p.168]

10. Be able to define *growth factors, oncogenes, neoplasms, malignant neoplasm, metastasis.* [pp.169-171]

11. List the four characteristics displayed by *all* cancer cells. [p.170]

CHAPTER SUMMARY

Individuals of a (1) _____ have a characteristic number of (2) _____ in each of their cells. The chromosomes differ in length and (3) _____, and they carry different portions of the cell's (4) _____ information. Division mechanisms parcel out the information to each (5) _____ cell, along with enough (6) _____ for that cell to start up its own operation.

A cell (7) _____ starts when a daughter cell forms and ends when that cell completes its own division. A typical cell cycle goes through interphase, (8) _____, and cytoplasmic division. In interphase, a cell increases its (9) _____ and number of components, and copies its (10) _____.

Mitosis divides the (11) _____, not the cytoplasm. It has four sequential stages: prophase, metaphase, (12) _____, and telophase. A (13) _____ spindle forms. It moves the cell's duplicated chromosomes into two parcels, which end up in two genetically identical (14) _____.

After (15) _____ division, the (16) _____ divides and typically puts a nucleus in each daughter cell. The cytoplasm of an (17) _____ cell is simply pinched into two. In (18) _____ cells, a cross-wall forms in the cytoplasm and divides it.

Built-in mechanisms monitor and control the (19) _____ and (20) _____ of cell division. On rare occasions, the surveillance mechanisms fail, and cell division becomes uncontrollable. (21) _____ formation and (22) _____ are the outcome.

INTEGRATING AND APPLYING KEY CONCEPTS

1. Runaway cell division is characteristic of cancer. Imagine the various points of the mitotic process that might be sabotaged in cancerous cells in order to halt their multiplication. Then try to imagine how one might discriminate between cancerous and normal cells to guide those methods of sabotage most effective in combating cancer.

2. The side effects of cancer chemotherapy in humans are very significant and most often result in nausea and hair loss. Think about cell division, and more importantly, the regulation of cell division, and hypothesize why these two particular side effects are so common.

12

MEIOSIS AND SEXUAL REPRODUCTION

INTRODUCTION

Survival and success of a species is certainly possible without sexual reproduction. The bacteria as a group are probably the most successful on the planet and reproduce strictly by asexual means. However, many species, such as humans, are dependent upon sexual reproduction for the continuation of the species. This chapter details the cellular mechanisms that allow for sexual reproduction and the subsequent variation that accompanies this process. It is this variation that accounts for the success of sexually reproducing species, such as humans.

FOCAL POINTS

- Figure 12.3 [p.176] illustrates the reproductive organs of a human female, a human male, and a flower.
- Figure 12.4 [p.177] depicts how meiosis halves the chromosome number.
- Figure 12.5 [pp.178-179] shows meiosis.
- Figure 12.6 [pp.180-181] illustrates crossing over.
- Figure 12.8 [p.182] diagrams that compare the general life cycle of both plants and animals.
- Figures 12.9 & 12.10 [p.183] illustrates the general mechanisms for sperm and egg formation in animals.
- Figure 12.11 [p.184] compares the processes of mitosis and meiosis.

INTERACTIVE EXERCISES

12.1 WHY SEX? [p.175]
12.2 SEXUAL REPRODUCTION AND MEIOSIS [pp.176-177]
12.3 THE PROCESS OF MEIOSIS [pp.178-179]

Selected Terms

Asexual reproduction [p.176], clone [p.176], nuclear division [p.176], gene [p.176], mutations [p.176], beta globin [p.176], homologous partner [p.177], sperm [p.177], egg [p.177], 23 pair [p.177], homologous chromosomes [p.177], haploid nuclei [p.177], meiosis I & II [p.177], sister chromatids [p.177], diploid [p.177], prophase I, metaphase I, anaphase I, telophase I, [p.178], spindle [p.178], nuclear envelope [p.178], prophase II, metaphase II, anaphase II, telophase II, [p.179], centriole [p.179], division of the cytoplasm [p.179]

Boldfaced Terms

[p.175] sexual reproduction _____

_____

[p.176] somatic _____

_____

[p.176] allele _____

_____

[p.176] meiosis _____

_____

[p.177] germ cells _____

_____

[p.177] gametes _____

_____

[p.177] haploid _____

_____

[p.177]fertilization _____

_____

[p.177] zygote _____

_____

Choice

Choose from the following: [pp.175-177]

a. asexual reproduction
b. sexual reproduction

1. _____ Offspring inherit different combinations of alleles

2. _____ One parent alone produces offspring

3. _____ Each offspring inherits the same number and kinds of genes as its parent

4. _____ Commonly involves two parents

5. _____ The production of "clones"

6. _____ Involves meiosis, formation of gametes, and fertilization

7. _____ Offspring are genetically identical copies of the parent

8. _____ Produces the variation in traits that is the foundation of evolutionary change

9. _____ Change can only occur by rare mutations

10. _____ The first cell of a new individual has pairs of genes on pairs of chromosomes; usually, one of each pair is maternal and the other paternal in origin

Dichotomous Choice

Circle one of two possible answers given in parentheses in each statement. [pp.176-177]

11. (Meiosis/Mitosis) divides chromosomes into separate parcels not once but twice prior to cell division.

12. Sperms and eggs are also known as (germ/gamete) cells.

13. (Haploid/Diploid) germ cells produce haploid gametes.

14. (Haploid/Diploid) cells possess pairs of homologous chromosomes.

15. (Meiosis/Mitosis) produces haploid cells.

16. Identical alleles are found on (homologous chromosomes/sister chromatids).

17. Two attached DNA molecules are known as (sister chromatids/homologous chromosomes).

18. Two attached sister chromatids represent (one/two) chromosome(s).

19. One pair of duplicated chromosomes would be composed of (two/four) chromatids.

20. With meiosis, chromosomes proceed through (one/two) divisions to yield four haploid nuclei.

21. Germ cell DNA is replicated during (prophase I of meiosis I/interphase preceding meiosis I).

22. During meiosis I, each duplicated (chromosome/chromatid) lines up with its partner, homologue to homologue, and then the partners are moved apart from one another.

23. Cytoplasmic division following meiosis I results in two (diploid/haploid) daughter cells.

24. During (meiosis I/meiosis II), the two sister chromatids of each chromosome are separated from each other, and each sister chromatid is now a separate chromosome.

25. If human body cell nuclei contain twenty-three pairs of homologous chromosomes, each resulting gamete will contain (twenty-three/forty-six) chromosomes.

Identification

Identify each of the meiotic stages shown below by entering the correct stage of either meiosis I or meiosis II in the blank. Choose from *prophase I, metaphase I, anaphase I, telophase I, prophase II, metaphase II, anaphase II,* and *telophase II*. Complete the exercise by matching and entering the letter of the correct stage description in the parentheses. [pp.178-179]

a. Motor proteins projecting from the microtubules move the chromosomes and spindle poles apart; chromosomes are tugged into position midway between the spindle poles; the spindle becomes fully formed

b. Microtubules have moved one member of the centriole pair to the opposite spindle pole in each of two daughter cells; microtubules attach to the chromosomes, which motor proteins slide toward the spindle's equator

c. Four daughter nuclei form; when cytoplasmic division is over, each daughter is haploid (n); all chromosomes are in the unduplicated state

d. Some microtubules extend from the spindle poles and overlap at its equator; these lengthen and push the poles apart; other microtubules extending from the poles shorten and pull each chromosome away from its homologous partner; these motions move the homologous partners to opposite poles

e. At the end of interphase, chromosomes are duplicated and in threadlike form; now they start to condense; each pairs with its homologue, and the two usually cross over and swap segments; newly forming spindle microtubules become attached to each chromosome

f. In each of two cells, the microtubules, motor proteins, and duplicated chromosomes interact, which positions all of the duplicated chromosomes midway between the two spindle poles

g. The cytoplasm of the cell divides at some point; there are now two haploid (n) cells with one of each type of chromosome that was present in the parent (2n) cell; all chromosomes are still in the duplicated state

h. Attachment between the sister chromatids of each chromosome breaks, and the two are moved to opposite spindle poles; each former "sister" is now a chromosome on its own

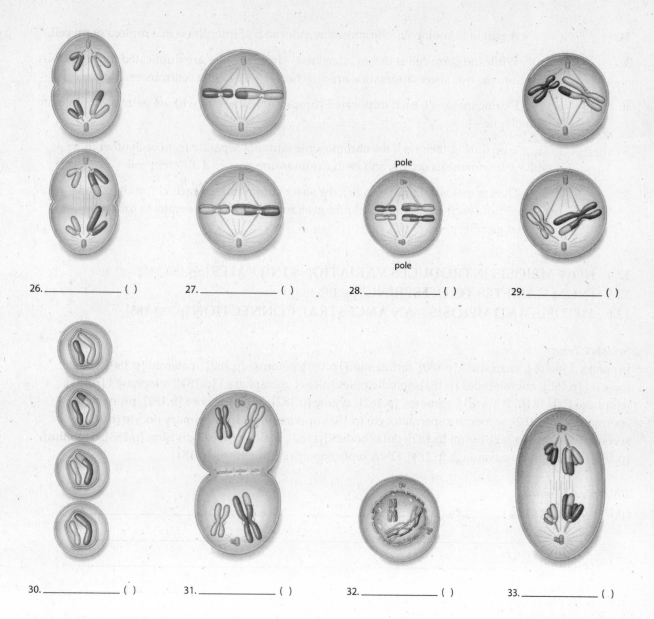

26. _____ ( )   27. _____ ( )   28. _____ ( )   29. _____ ( )

pole

pole

30. _____ ( )   31. _____ ( )   32. _____ ( )   33. _____ ( )

Matching

Following careful study of the major stages of meiosis shown in Figure 12.5 of the main text [pp.178-179], apply what you have learned by matching the following written descriptions with the appropriate sketch.. Assume that the cell in this model initially has one pair of homologous chromosomes (one from a paternal source and one from a maternal source) and crossing over does not occur. Complete the exercise by indicating the diploid ($2n = 2$) or haploid ($n = 1$) chromosome number of the cell in the parentheses.

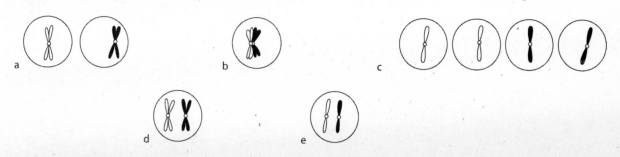

34. ____ ( ) A pair of homologous chromosomes prior to S of interphase in a diploid germ cell.

35. ____ ( ) While the germ cell is in S of interphase, chromosomes are duplicated through DNA replication; the two sister chromatids are attached at   the centromere.

36. ____ ( ) During meiosis I, each duplicated chromosome lines up with its partner, homologue to homologue.

37. ____ ( ) Also during meiosis I, the chromosome partners separate from each other in anaphase I; cytokinesis occurs, and each chromosome goes to a different cell.

38. ____ ( ) During meiosis II (in two cells), the sister chromatids of each chromosome are separated from each other; four haploid nuclei form; cytokinesis results in four cells (potential gametes).

12.4 HOW MEIOSIS INTRODUCES VARIATIONS IN TRAITS [pp.180-181]
12.5 FROM GAMETES TO OFFSPRING [pp.182-183]
12.6 MITOSIS AND MEIOSIS – AN ANCESTRAL CONNECTION [p.184-185]

Selected Terms
prophase I [p.180], anaphase I [p.180], fertilization [p.180], paternal [p.180], maternal [p.180], genetic mosaics [p.180], microtubules [p.181], spindle poles [p.181], metaphase I [p.181], telophase I [p.181], telophase II [p.181], 2^{23} [p.181], gametes, [p.182], zygote [p.182], haploid spore [p.182], primary spermatocyte [p.182], secondary spermatocyte [p.182], spermatid [p.182], primary oocyte [p.182], secondary oocyte [p.182], ovum [p.182], polar bodies [p.182], male germ cell division [p.183], oogonium [p.183], evolutionary advantage [p.184], DNA replication prior to mitosis [p.184]

Boldfaced Terms
[p.180] crossing over _____

_____

[p.182] sporophyte _____

_____

[p.182] gametophyte _____

_____

[p.182] sperm _____

_____

[p.182] egg _____

_____

Matching

Match each term with its description. [pp.180-184]

1. _____ 2^{23}
2. _____ paternal chromosomes
3. _____ prophase I
4. _____ non-sister chromatids
5. _____ crossing over
6. _____ function of meiosis
7. _____ maternal chromosomes
8. _____ metaphase I

a. Random positioning of maternal and paternal chromosomes at the spindle equator
b. A time of much gene shuffling
c. Reduction of the a parental chromosomal number by half
d. Break at the same places along their length and then exchange corresponding segments—including genes
e. Twenty-three chromosomes inherited from your mother
f. Combinations of maternal and paternal chromosomes possible in gametes from one germ cell
g. Twenty-three chromosomes inherited from your father
h. Breaks up old combinations of alleles and puts new ones together in pairs of homologous chromosomes

Choice

Choose from the following: [pp.182-183]

    a. animals
    b. plants
    c. animals and plants

9. _____ Meiosis results in the production of haploid spores

10. Germ cells give rise to gametes

11. Meiosis results in the production of haploid gametes

12. A diploid germ cell becomes an oocyte, or immature egg

13. A spore develops into a haploid gametophyte

14. A haploid gametophyte produces haploid gametes by mitosis

15. _____ Three polar bodies are formed that do not function as gametes

16. A haploid spore divides by mitosis to produce a gametophyte

17. A gamete-producing body and a spore-producing body develop during the life cycle

18. _____ The egg receives most of the cytoplasm

Sequence

Arrange the following entities in correct order of development, entering a 1 by the stage that appears first and a 4 by the stage that completes the process of spermatogenesis. Complete the exercise by indicating in parentheses whether each cell is *n* or *2n* in the parentheses. [pp.182-183]

19. _____ (   ) primary spermatocyte

20. _____ (   ) sperm

21. _____ (   ) spermatid

22. _____ (   ) secondary spermatocyte

Matching

Choose the most appropriate answer to match with each oogenesis concept.[pp.182-183]

23. _____ primary oocyte

24. _____ oogonium

25. _____ secondary oocyte

26. _____ ovum and three polar bodies

27. _____ first polar body

a. The cell in which synapsis, crossing over, and recombination occur

b. A cell that is equivalent to a diploid germ cell

c. A haploid cell formed after division of the primary oocyte that does not form an ovum at second division

d. Haploid cells, but only one of which functions as an egg

e. A haploid cell formed after division of the primary oocyte, the division of which forms a functional ovum

Short Answer

29. List the various mechanisms that contribute to the huge number of new gene combinations that may result from fertilization. [pp.180-181]

_____

_____

_____

_____

_____

SELF QUIZ

1. Which of the following is not a characteristic of sexual reproduction? [pp.175-177]
   a. sperm and eggs
   b. zygotes
   c. meiosis
   d. clones
   e. germ cells

2. Which of the following does *not* occur in prophase I of meiosis? [p.178]
   a. chromosome duplication
   b. a cluster of four chromatids
   c. homologues pairing tightly
   d. crossing over
   e. chromosomes condensing

3. "The two identical DNA molecules stay attached at the centromere. This is a description of _____ [pp.178-179]
   a. homologous chromosomes
   b. a diploid chromosome
   c. sister chromatids
   d. two homologous chromosomes
   e. chromosome duplication

4. Two sister chromatids of each chromosome are separated from each other during _____. [pp.178-179]
   a. anaphase I
   b. metaphase II
   c. metaphase I
   d. telophase II
   e. anaphase II

5. The first haploid cells to appear during meiosis are seen at _____. [pp.178-179]
   a. Telophase II
   b. Telophase I
   c. Anaphase II
   d. Anaphase I
   e. Prophase II

6. Crossing over is one of the most important events in meiosis because _____. [p.180]
   a. it leads to genetic recombination
   b. homologous chromosomes must be separated into different daughter cells
   c. the number of chromosomes allotted to each daughter cell must be halved
   d. homologous chromatids must be separated into different daughter cells
   e. nonsister chromatids do not break during the process

7. Crossing over _____. [p.180]
   a. generally results in pairing of homologues and binary fission
   b. is accompanied by gene-copying events
   c. involves breakages and exchanges between sister chromatids
   d. is a molecular interaction between two of the nonsister chromatids of a pair of homologous chromosomes
   e. results in the recombination of genes located on sister chromatids

8. In plants, gametes are produced by _____. [p.182]
   a. meiosis in gametophyte plants
   b. mitosis in sporophyte plants
   c. germ cells
   d. mitosis in gametophyte plants
   e. meiosis in sporophyte plants

9. In female animals, a diploid germ cell develops into a(n) _____. [p.182]
   a. primary spermatocyte
   b. primary oocyte
   c. egg
   d. secondary oocyte
   e. first polar body

10. The sporophyte of a plant is typically? [p.182]
   a. haploid
   b. diploid
   c. not functional
   d. unicellular

CHAPTER OBJECTIVES/REVIEW QUESTIONS

1. List advantages of sexual reproduction in a species. [p.175]

2. List the characteristics of asexual reproduction in a species. [p.176]

3. Define *germ cells, meiosis, gametes, zygote, homologous chromosomes,* and *alleles.* [p.178]

4. If a cell has a _____ chromosome number, it has a pair of each type of chromosome, often from two parents. [pp.176-177]

5. Gametes have a _____ chromosome number, or half the parental number. [p.177]

6. At what point in the cell cycle does a germ cell duplicate its DNA? [pp.176-177]

7. The two nuclear divisions of meiosis are called meiosis _____ and meiosis _____ [pp.178-179]

8. Be able to tell the story of the stages of meiosis; include proper terminology and chromosome movement descriptions. [pp.178-179]

9. Describe the ways that meiosis adds variation in traits. [pp.180-18]

10. List the events that provide millions of chromosome combinations in gametes. [pp.180-181]

11. List and define the terms that describe gamete formation in plants and animals. [pp.182-183]

CHAPTER SUMMARY

By (1) _____ reproduction, one parent alone transmits its (2) _____ information to its offspring. By (3) _____ reproduction, offspring typically inherit information from two parents that differ in their (4) _____. Alleles are different forms of the same (5) _____; they specify different versions of a (6) _____.

(7) _____ cells have a pair of each type of chromosome, one maternal and one paternal. (8) _____, a nuclear division mechanism, reduces the chromosome number. It occurs only in cells set aside for (9) _____ reproduction. Meiosis sorts our a reproductive cell's chromosomes into four (10) _____ nuclei, which are distributed to daughter cells by way of (11) _____ division.

During meiosis, each pair of maternal and paternal (12) _____ swaps segments and exchanges (13) _____. The pairs get randomly (14) _____, so forthcoming gametes end up with different mixes of maternal and paternal chromosomes. (15) _____ also governs which gametes combine during fertilization. All three events contribute to (16) _____ in traits among offspring.

In animals (17) _____ form by different mechanisms in males and females. In most plants, (18) _____ formation and other events intervene between meiosis and gamete formation.

Recent (19) _____ evidence suggests that meiosis originated through mechanisms that already existed for (20) _____ and, before that, for repairing damaged (21) _____.

INTEGRATING AND APPLYING KEY CONCEPTS

During the year 2001, several scientists from several countries, including the United States, claimed to have the knowledge and technology to clone human beings. Other scientists suspect that human cloning may have already been achieved in secret. Those who criticize human cloning experiments suggest that each successful human clone would be accompanied by repeated failures in the form of severely deformed and dead human embryos. If indeed human cloning is currently possible, speculate about the public acceptance of such experimentation and the effects of reproduction without sex on human populations. What are your thoughts on cloning?

13

OBSERVING PATTERNS IN INHERITED TRAITS

INTRODUCTION

Among species that undergo sexual reproduction, it has long been observed that certain traits may be predictably passed from parent to the offspring. The mechanism responsible for this, however, was not understood until the 1860s, when Gregor Mendel conducted his famous and enlightening experiments on peas. This chapter outlines the influential work of Mendel, and the impact this work had on modern genetics and our understanding of inheritance. The chapter then goes on to examine the effect that nature can have on human inherited traits.

FOCAL POINTS

- Figure 13.3 [p.191] illustrates the loci of a few human genes and their descriptions.
- Figure13.4 [p.191] is an animated look at a pair of homologous chromosomes.
- Figure 13.5 [p.192] illustrates the segregation of alleles.
- Figure 13.6 [p.193] shows the step by step construction of a Punnett square.
- Table 13.1 [p.193] shows Mendel's seven pea plant traits.
- Figure 13.7 [p.194] illustrates the process of independent assortment.
- Figure 13.8 [p.195] animates a dihybrid cross.
- Figure 13.9 [p.196] illustrates the combinations of alleles that are the basis of human blood type.
- Figure 13.10 [p.196] illustrates incomplete dominance.
- Figure 13.11 [p.197] shows epistasis in dogs.
- Figure 13.13 [p.198] demonstration of continuous variation of a trait.
- Figure 13.14 [p.198] shows an example of environmental effects on the phenotype.

INTERACTIVE EXERCISES

13.1 MENACING MUCUS [p.189]
13.2 MENDEL, PEA PLANTS, AND INHERITANCE PATTERNS [pp.190-191]
13.3 MENDEL'S LAW OF SEGREGATION [pp.192-193]
13.4 MENDEL'S THEORY OF INDEPENDENT ASSORTMENT [pp.194-195]

Selected Terms
cystic fibrosis [p.189], chloride ions [p.189], hereditary [p.190], Charles Darwin [p.190], blending inheritance [p.190], natural selection [p.190], traits [p.190], Gregor Mendel [p.190], garden pea [p.190], *Pisum sativum* [p.190], gene [p.191], DNA [p.191], chromosome [p.191], allele [p.192], F_1 & F_2 [p.193], chromatids [p.194], linked genes [p.195]

Boldfaced Terms
[p.191] locus_____

_____

[p.191] homozygous _____

_____

[p.191] genotype _____

_____

[p.191] phenotype _____

_____

[p.191] hybrids _____

_____

[p.191] heterozygous _____

_____

[p.191] dominant _____

_____

[p.191] recessive _____

_____

[p.192] Punnett square _____

_____

[p.192] testcrosses _____

_____

[p.192] monohybrid experiment _____

_____

[p.193] law of segregation _____

_____

[p.194] dihybrid experiment _____

_____

[p.194] law independent assortment _____

_____

[p.195] linkage group _____

_____

Matching [p.191-193]

Select from the following choices below by placing the letter in the bland next to its descriptor.

1. _____ condition of having the same alleles for a given gene

2. _____ an individual's observable traits

3. _____ a particular set of alleles that an individual carries

4. _____ condition of having different alleles for a given gene

a. heterozygous
b. geneotype
c. homologous
d. phenotype

Genetics Problems [pp.192-195]

5. In garden pea plants, tall (*T*) is dominant over dwarf (*t*). Use the Punnett square method [Figure 13.5] to determine the genotype and phenotype probabilities of offspring from the above cross, *Tt* X *tt*.

6. Using the gene symbols (tall – *T* and dwarf pea plants - *t*) apply the six Mendelian ratios listed above to complete the following table of single-factor crosses by inspection. State results as phenotype and genotype ratios. [pp.192-195]

Cross	Phenotype Ratio	Genotype Ratio
a. *Tt* X *tt*		
b. *TT* X *Tt*		
c. *tt* X *tt*		
d. *Tt* X *Tt*		
e. *tt* X *Tt*		
f. *TT* X *tt*		
g. *TT* X *TT*		
h. *Tt* X *TT*		

7. [*Hint*] When working genetic problems that deal with two gene pairs, use a fork-line device (see the diagram below) to visualize the independent assortment of gene pairs located on nonhomologous chromosomes as gametes.] Assume that in humans, pigmented eyes (*B*) (an eye color other than blue) are dominant over blue (*b*), and right-handedness (*R*) is dominant over left-handedness (*r*). To solve this problem, you will cross the parents: *BbRr* X *BbRr*. A sixteen-block Punnett square is required with gametes from each parent arrayed on two sides of the Punnett square [refer to textbook Figure 13.8, p.195.

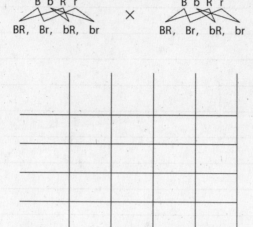

BbRr   ×   BbRr

BR, Br, bR, br   BR, Br, bR, br

8. Array the gametes at the right on two sides of the Punnett square; combine these haploid gametes to form diploid zygotes in the squares. In the blank spaces below the Punnett square, enter the probability ratios derived in the Punnett square for the phenotypes listed: [pp.194-195]

_____ a. pigmented eyes, right-handed

_____ b. pigmented eyes, left-handed

_____ c. blue-eyed, right-handed

_____ d. blue-eyed, left-handed

9. In horses, black coat color is influenced by the dominant allele (*B*), and chestnut coat color is influenced by the recessive allele (*b*). Trotting gait is due to a dominant gene (*T*), pacing gait to the recessive allele (*t*). A homozygous black trotter is crossed to a chestnut pacer. [pp.194-195]

a. What will be the appearance of the F_1 and F_2 generations? _____

_____

b. Which phenotype will be most common? _____

c. Which genotype will be most common? _____

d. Which of the potential offspring will be certain to breed true? _____

13.5 BEYOND SIMPLE DOMINANCE [pp.196-197]
13.6 COMPLEX VARIATION IN TRAITS [pp.198-199]

Selected Terms
multiple traits [p.196], ABO blood [p.196], transfusion [p.196], fibrillin [p.197], Marfan Syndrome [p.197], melanin [p.197], environmental factors [p.198], *Daphnia pulex* [p.199], yarrow [p.199], serotonin [p.199]

Boldfaced Terms
[p.196] codominance _____

_____

[p.196] multiple allele system _____

_____

[p.196] incomplete dominance _____

_____

[p.196] epistasis _____

_____

[p.196] pleiotropic _____

_____

[p.196] continuous variation _____

_____

[p.198] bell curve _____

_____

Genetics Problems

Study the example of ABO blood typing in the textbook, Figure 13.9. Note that it is a multiple allele system with codominance and then solve the following problems. [p.196]

1. AO x AB = _____

2. BO x AO = _____

3. AB x AB^B = _____

4. Study the examples of snapdragons given in the text, which demonstrate incomplete dominance; then determine the phenotypes and genotypes of the offspring of the following crosses. R = red flower color, R' = white flower color, and RR' is pink. Use Figure 13.10 from the textbook for more information.[p.196]

Cross	Phenotypes	Genotypes
a. RR x $R'R'$ =		
b. RR' x RR' =		

5. In poultry, an interaction occurs in which two genes produce a phenotype that neither gene can produce alone. The two interacting genes (R and P) produce comb shape in chickens. Here are the possible genotypes and phenotypes [p.197]:

Genotypes	Phenotypes
$R\_\_P\_\_$	walnut comb
$R\_\_pp$	rose comb
$rrP\_\_$	pea comb
$rrpp$	single comb

What is this process called?
[Hint: where a blank appears in the genotypes above, either the dominant or the recessive symbol in that blank yields the same phenotype.]

What are the genotype and phenotype ratios of the offspring of a heterozygous walnut-combed male and a single-combed female? [p.197]_____

_____

_____

6. In the inheritance of the coat (fur) color of Labrador retrievers, allele *B* specifies black that is dominant to brown (chocolate), *b*. Allele *E* permits full deposition of color pigment but the presence of two recessive alleles, *ee*, reduces deposition, and a yellow coat results. Additional information can be found on Figure 13.11 in the textbook. [p.197]

Predict the phenotypes of the coat color and their proportions resulting from the following cross:
BbEe x *Bbee* = _____

_____

Complete the Table
7. Complete the following table by supplying the type of inheritance illustrated by each example. Choose from these gene interactions: *pleiotropy, multiple allele system, incomplete dominance, codominance,* and *epistasis.* [pp.196-197]

Type of Inheritance	Example
[p.160] a.	Pink-flowered snapdragons produced from red-and white-flowered parents
[p.160] b.	AB type blood from a gene system of three alleles, *A, B,* and *O*
[p.160] c.	A gene with three or more alleles such as the *ABO* blood typing alleles
[p.161] d.	Black, brown, or yellow fur of Labrador retrievers and comb shape in poultry
[p.161] e.	The multiple phenotypic effects of the gene causing human sickle-cell anemia

Choice
Choose from the following:
    a.   environment as a primary effect
    b.   a number of genes affect a trait

8. ____ Height of human beings [p.198]

9. ____ Continuous variation in a trait [p.198]

10. ____ Pointed or non-pointed heads in Daphnia pulex [p.199]

11. ____ The range of eye colors in the human population [p.198]

12. ____ Heat-sensitive version of one of the enzymes required for melanin production in Himalayan rabbits [p.198]

13. ____ Three yarrow cuttings planted at three elevations [p.199]

14. ____ Coat Color in Siamese cats [p.198]

Short Answer

14. What is the take-home lesson when dealing with questions of heritable and environmental factors that give rise to variations in traits? [pp.198-199]

_____

_____

_____

SELF QUIZ

1. The best statement of Mendel's principle of independent assortment is that _____. [pp.194-195]
   a. one allele is always independently dominant to another
   b. independent hereditary units from the male and female parents are blended in the offspring
   c. the two hereditary units that influence certain independent traits to separate during gamete formation
   d. genes on pairs of homologous chromosomes are distributed into one gamete or another independently of genes on pairs of other chromosomes
   e. The two genes of each pair are separated from each other during meiosis, so they end up in different gametes

2. All the different molecular forms of the same gene are called _____. [pp.190-191]
   a. chiasmata
   b. alleles
   c. autosomes
   d. genes
   e. mutations

3. In the F_2 generation of a monohybrid cross involving complete dominance, the expected phenotypic ratio is _____. [pp.192-193]
   a. 3:1
   b. 1:1:1:1
   c. 1:2:1
   d. 1:1
   e. 1:3:1

4. In the F_2 generation of a cross between a red-flowered snapdragon (homozygous) and a white-flowered snapdragon, the expected phenotypic ratio of the offspring is _____. [p.196]
   a. 3/4 red, 1/4 white
   b. 100 percent red
   c. 1/4 red, 1/2 pink, 1/4 white
   d. 100 percent pink
   e. 1/2 red, 1/2 white

5. In a testcross, F_1 hybrids are crossed to an individual known to be _____ for the trait. [pp.192-193]
   a. heterozygous
   b. homozygous dominant
   c. homozygous
   d. homozygous recessive
   e. heterozygous dominant

6. The tendency for dogs to bark while trailing is determined by a dominant gene, *S*, while silent trailing is due to the recessive gene, *s*. In addition, erect ears, *D*, is dominant over drooping ears, *d*. What combination of offspring would be expected from a cross between two erect-eared barkers who are heterozygous for both genes? [pp.192-193]
   a. ¼ erect barkers, ¼ drooping barkers, ¼ erect silent, ¼ drooping silent
   b. 9/16 erect barkers, 3/16 drooping barkers, 3/16 erect silent, 1/16 drooping silent
   c. 1/2 erect barkers, 1/2 drooping barkers
   d. 9/16 drooping barkers, 3/16 erect barkers, 3/16 drooping silent, 1/16 erect silent
   e. 3/4 erect barkers, 1/4 drooping barkers

7. A man with type A blood could be the father of _____. [p.196]
   a. a child with type A blood
   b. a child with type B blood
   c. a child with type O blood
   d. a child with type AB blood
   e. all of the above

8. Alleles at one chromosome locus may affect two or more traits in good or bad ways. This outcome of the activity of one gene's product is _____. [p.197]
   a. pleiotropy
   b. epistasis
   c. a mosaic effect
   d. continuous variation
   e. gene interaction

9. Suppose two individuals, each heterozygous for the same characteristic, are crossed. The characteristic involves complete dominance. The expected genotypic ratio of their progeny is _____. [pp.192-193]
   a. 1:2:1
   b. 1:1
   c. 100 percent of one genotype
   d. 3:1
   e. 1:1:1:1

10. If the two homozygous classes in the F_1 generation of the cross in exercise 9 are allowed to mate, the observed genotypic ratio of the offspring will be _____. [pp.192-193]
   a. 1:1
   b. 1:2:1
   c. 100 percent of one genotype
   d. 3:1
   e. 1:1:1:1

11. Applying the types of inheritance learned in this chapter in the text, the skin color trait in humans exhibits _____. [pp.198-199]
   a. pleiotropy
   b. epistasis
   c. environmental effects
   d. continuous variation
   e. gene interactions

CHAPTER OBJECTIVES/REVIEW QUESTIONS

1. What was the prevailing method of explaining the inheritance of traits before Mendel's work with pea plants? [p.190]

2. Garden pea plants are naturally _____-fertilizing, but Mendel took steps to _____-fertilize them for his experiments. [p.190]

3. _____ are units of information about specific traits; they are passed from parents to offspring. [p.190]

4. What is the general term applied to the location of a gene on a chromosome? [p.191]

5. Define *allele*; how many types of alleles are present in the genotypes *Tt? tt? TT?* [p.191]

6. When two alleles of a pair are identical, it is a _____ condition; if the two alleles are different, this is a _____ condition. [p.191]

7. Distinguish a dominant allele from a recessive allele. [p.191]

8. _____ refers to the genes present in an individual; _____ refers to an individual's observable traits. [p.191]

9. Offspring of _____ crosses are heterozygous for the one trait being studied. [p.192]

10. Be able to use the Punnett-square method of solving genetics problems. [p.192]

11. Define the *testcross* and cite an example. [p.192]

12. Mendel's theory of _____ states that during meiosis, the two genes of each pair separate from each other and end up in different gametes. [pp.192-193]

13. Be able to solve dihybrid genetic crosses. [pp.194-195]

14. Mendel's theory of _____ _____ states that gene pairs on homologous chromosomes tend to be sorted into one gamete or another independently of how gene pairs on other chromosomes are sorted out. [pp.194-195]

15. Define *multiple allele system* and cite an example. [p.196]

16. Distinguish between incomplete dominance, epistasis, pleiotrophy and codominance. [p.196-197]

17. Explain why Marfan Syndrome is a good example of pleiotropy. [p.197]

18. List possible explanations for less predictable trait variations that are observed. [pp.198-199]

19. List two human traits that are explained by continuous variation. [p.198]

20. Himalayan rabbits, yarrow plants, and garden hydrangeas are good examples of environmental effects on _____ expression. [p.198]

CHAPTER SUMMARY

Gregor Mendel gathered the first indirect (1) _____ evidence of the genetic basis of (2) _____. His meticulous work tracking traits in many generations of pea plants gave him clues that (3) _____ traits are specified in (4) _____. The units, which are distributed into gametes in predictable patterns, were later identified as (5) _____.

Some experimenters yielded evidence of gene (6) _____. When one chromosome separates from its (7) _____ partner during meiosis, the pairs of alleles on those chromosomes also separate and end up in different (8) _____.

Other experiments yielded evidence of (9) _____ assortment. During meiosis, the members of a pair of homologous chromosomes are (10) _____ into gametes independently of how all other pairs are distributed.

Not all traits have clearly dominant or (11) _____ forms. One allele of a pair may be fully or partially dominant over its non-identical partner, or (12) _____ with it. Two or more gene pairs often influence the same trait, and some single genes influence many (13) _____. The (14) _____ also influences variation in gene expression.

INTEGRATING AND APPLYING KEY CONCEPTS

Solve the following genetics problem:

1. In garden peas, one pair of alleles controls the height of the plant and a second pair of alleles controls flower color. The allele for tall (D) is dominant to the allele for dwarf (d), and the allele for purple (P) is dominant to the allele for white (p). A tall plant with purple flowers crossed with a tall plant with white flowers produces 3/8 tall purple, 3/8 tall white, 1/8 dwarf purple, and 1/8 dwarf white. What are the genotypes of the parents?

2. A child is born with blood type B. The mother also has type B blood. Two men claim to be the father: James has type A blood and Mark is Type AB. Could either of these men be the father of this child? If so, which one could possibly be the father? Give all possible phenotypes of the individuals involved.

3. Explain why people with one copy of mutated CFTR gene are healthy, while people with two copies of that gene are ill with cystic fibrosis.

14

HUMAN INHERITANCE

INTRODUCTION

Human inheritance, health and human disease have been topics of interest and discussion since the advent of humans. These issues still pervade society today, for example, certain genetic diseases are currently the subject of very popular television shows. Research has shown us that many diseases are inherited on chromosomes. This chapter discusses these genetic diseases with respect to how they are passed on and how these certain genetic characteristics actually cause disease. The final part of the chapter describes the technology involved in detecting genetic abnormalities and the ethical consequences of this knowledge.

FOCAL POINTS

- Figure 14.2 [p.204] shows a typical pedigree used to study disease patterns in families.
- Table 14.1 [p.205] summarizes some of the patterns of inheritance for some genetic abnormalities and disorders.
- Figure 14.3 [p.206] illustrates autosomal dominance inheritance.
- Figure 14.5 [p.207] illustrates autosomal recessive inheritance.
- Figure 14.6 [p.208] shows the inheritance pattern of a sex linked condition.
- Figure 14.7 [p.208] illustrates what a person who has red-green color blindness sees.
- Figure 14.8 [p.209] shows a classic case of X-linked recessive inheritance.
- Figure 14.9 [p.210] shows a large-scale change in chromosome structure.
- Figure 14.12 [p.212] shows an example of nondisjunction.
- Figure 14.13 [p.213] depicts a human karyotype, or chromosome map of Down syndrome.

INTERACTIVE EXERCISES

14.1 SHADES OF SKIN [p.203]
14.2 HUMAN GENETIC ANALYSIS [pp.204-205]
14.3 AUTOSOMAL INHERITANCE PATTERNS [p.206-207]

Selected Words
melanosomes [p.203], melanin [p.203], variations [p.203], folate [p.203], SLC24A5 [p.203], polydactyly [p.204], polygenic [p.205], autosomal dominant [p.206], achondroplasia [p.206], Huntington's disease [p.206], Hutchinson-Gilford progeria [p.207], albinism [p.207], tyrosinase [p.207], carrier [p.207], lamin [p.207], Tay-Sachs disease [p.207], autosomal recessive [p.207]

Boldfaced Terms
[p.204] pedigrees _____

_____

Short Answer

1. Compare autosomal dominant with autosomal recessive traits. [pp.206-207]

    _____

    _____ _____

2. Huntington disorder is a rare form of autosomal dominant inheritance, *H*; the normal gene is *h*. The disease causes progressive degeneration of the nervous system with onset exhibited near middle age. An apparently normal man in his early twenties learns that his father has recently been diagnosed as having Huntington disorder. What are the chances that the son will develop this disorder? [pp.206-207] _____

3. The autosomal allele that causes albinism (n) is recessive to the allele for normal pigmentation, (N). A normal pigmented woman whose father has albinism marries a man who is an albino whose parents are of normal pigmented skin. They have three children, two with normal skin pigmentation and one who is an albino.

    List the genotypes for each person involved. [p.207]

    _____

    _____

14.4 X-LINKED INHERITANCE PATTERNS [pp.208-209]
14.5 HERITABLE CHANGES IN CHROMOSOME STRUCTURE [p.210-211]
14.6 HERITABLE CHANGES IN CHROMOSOME NUMBER [p.212-213]

Selected Words

X-linked recessive allele [p.208], hemophilia A [p.209], red-green color blindness [p.208], receptors [p.209], Duchenne muscular dystrophy [p.209], cri-du-chat [p.210], chromosomae abnormalities [p.210], prophase I of meiosis [p.210], crossing over [p.210], reciprocal translocation [p.210], balanced translocation [p.210], karyotyping [p.211], DNA sequence comparisons [p.211], globin chain gene [p.211], telomere [p.211], n +1 gamete [p.212], trisomic condition [p.212], 2n + 1 [p.212], n-1 [p.212], 2n -1 [p.212], monosomic [p.212], Down syndrome [p.212], Turner syndrome [p.212], XO [p.212], *XXX* condition [p.213], *XXY* [p.213], Klinefelter syndrome [p.213], XYY condition [p.213]

Boldfaced Terms

[p.210] duplications _____

_____

[p.210] deletion _____

_____

[p.210] inversion _____

_____

[p.210] translocation _____

_____

[p.211] nondisjunction _____

_____

[p.211] aneuploidy _____

_____

[p.211] polyploid _____

_____

Dichotomous Choice

Circle the word in parentheses that make each statement correct. Refer to the Punnett-square table shown in text Figure 14.6. [p.208]

1. Male humans transmit their Y chromosome only to their (sons/daughters). [p.208]

2. Male humans receive their X chromosome only from their (mothers/fathers). [p.208]

3. Human mothers and fathers each provide an X chromosome for their (sons/daughters). [p.208]

Labeling

On rare occasions, chromosome structure becomes abnormally rearranged. Such changes can have profound effects on the phenotype of an organism. Label these diagrams of abnormal chromosome structure as a *deletion*, a *duplication*, an *inversion*, or a *translocation*. Please consult your textbook pp.210-211 for more information.

4. _____ ( )   5. _____ ( )   6. _____ ( )   7. _____ ( )

    [p.176]   [p.176]   [p.176]   [p.176]

Short Answer

8. Which of the abnormal chromosomal rearrangements in the preceding diagrams results in the cri-du-chat syndrome? [p.210] _____

9. Provide an example showing that changes in chromosome structure have become adaptive? [p.211]

_____

_____

Genetics Problems

10. A color-blind man and a woman with normal vision whose father was color-blind have a son. Color blindness, in this case, is caused by an X-linked recessive gene. If only the male offspring are considered, what is the probability that their son is color-blind? [p.208]

_____

11. Hemophilia A is caused by an X-linked recessive gene. A woman who is seemingly normal but whose father was a hemophiliac marries a normal man. What proportion of their sons will have hemophilia? What proportion of their daughters will have hemophilia? What proportion of their daughters will be carriers? [p.209] _____

12. The pedigree below shows the pattern of inheritance of color blindness in a family (persons with the trait are indicated by black circles). What is the chance that the third-generation female indicated by the arrow (below) will have a color-blind son if she marries a normal male? A color-blind male? [p.208]

_____

Complete the Table

13. Complete the table by describing the mechanisms of chromosome number change in organisms for each category. [p.212-213]

Category of Change	Description
a. Aneuploidy	
b. Polyploidy	
c. Nondisjunction	

Short Answer

14. If a nondisjunction occurs at anaphase I of the first meiotic division, what will be the proportion of abnormal gametes (for the chromosomes involved in the nondisjunction)? [pp.212-213]

_____

15. If a nondisjunction occurs at anaphase II of the second meiotic division, what will be the proportion of abnormal gametes (for the chromosomes involved in the nondisjunction)? [p.212-213]

_____

16. List generally known polyploid organisms. [p.212]

_____

_____

17. Define the following terms: *trisomic*, and *monosomic*. [p.212]

_____

_____

_____

_____

Choice

Choose from the following: [pp.212-213]

      a. Down syndrome
      b. Turner syndrome
      c. Klinefelter syndrome
      d. XXY condition
      e. XYY condition
      f. XXX syndrome

18. ____ About one of every 500 to 1,000 males inherits one Y and two or more X chromosomes, mainly by nondisjunction

19. ____ Most don't have functional ovaries and can't produce enough sex hormones

20. ____ The testes and the prostate gland usually are smaller than average; hair is sparse, the voice is pitched high, and the breasts are a bit enlarged

21. ____ Tend to be taller than average, with mild mental impairment, but most are phenotypically normal

22. ____ Most affected individuals show moderate to severe mental impairment and heart defects; as a group, they tend to be cheerful and sociable people

23. ____ Occurs at a frequency of about 1 in 1,000 live births; adults are fertile; no physical or medical problems

Choice

24. ____ Which diagram represents a chromosome that has undergone crossover and recombination? (Assume that the organism involved is heterozygous with the genotype) [p.176]

     a. | A B | b. | a b | c. | B A | d. | a B |

14.7 GENETIC SCREENING [pp.214-215]

Selected Words

genetic abnormality [p.214], genetic disorder [p.214], genetic disease [p.214], prenatal diagnosis [p.214], embryo [p.214], fetus [p.214], tyrosine [p.214], phenylalanine [p.214], *obstetric sonography* [p.214], ultrasound [p.214], *fetoscopy* [p.214], *phenylketonuria* or PKU [p.214], amniocentesis [p.215], *chorionic villi sampling* (CVS) [p.215], preimplantation diagnosis [p.215]

Matching

The following standardized symbols are used in the construction of pedigree charts. Choose the appropriate description for each. You may consult your textbook on Figure 14.2 for more information. [p.204]

1. _____ ■ ●

2. _____ ●

3. _____ ◆

4. _____ ■

5. _____ I, II, III, IV...

6. : _____ ■●

7. _____ ●■■●

    a. individual showing the trait being studied
    b. male
    c. generation
    d. offspring in order of birth, from left to right
    e. female
    f. sex not specified
    g. marriage/mating

Short Answer

8. Distinguish between these terms: *genetic abnormality*, *genetic disorder*, *syndrome*, and *disease*. [pp.214-215]

_____

_____

_____

_____

_____

_____

Complete the Table [p.205]

9. Complete the table below by indicating whether the genetic disorder listed is due to inheritance that is autosomal recessive, autosomal dominant, X-linked recessive, or due to changes in chromosome number or changes in chromosome structure.

Genetic Disorder/Abnormality	Inheritance Pattern or Cause
a. Galactosemia	
b. Achondroplasia	
c. Hemophilia	
d Huntington disorder	
e. Turner and Down syndrome	
f. Cri-du-chat syndrome	
g. XYY condition	
h. Color blindness	
i. XXX syndrome	
j. Fragile X syndrome	
k. Progeria	
l. Klinefelter syndrome	
m. Phenylketonuria	

SELF QUIZ

1. A _____ is the loss of some portion of a chromosome. [p.210]
    a. inversion
    b. duplication
    c. translation
    d. deletion

2. Chromosomes other than those involved in sex determination are known as _____. [p.206]
    a. nucleosomes
    b. heterosomes
    c. alleles
    d. autosomes
    e. heterozygous chromosomes

3. _____ is when part of a sequence of a chromosome is oriented in the opposite direction. _____. [p.210]
    a. deletion
    b. inversion
    c. translocation
    d. duplication

4. Repetitions of a DNA sequence on a chromosome is called a _____. [p.210]
    a. inversion
    b. translocation
    c. duplication
    d. deletion

5. The Cri-du-chat disorder is known to be caused by _____. [p.205]
    a. sex-linked inheritance
    b. autosomal recessive inheritance
    c. an inversion
    d. a deletion
    e. a change in chromosome number

6. Red-green color blindness is a sex-linked recessive trait in humans. A color-blind woman and a man with normal vision have a son. What are the chances that the son is color-blind? If the parents ever have a daughter, what is the chance for each birth that the daughter will be color-blind? (Consider only the female offspring.) [pp.208-209]
    a. 100 percent, 0 percent
    b. 50 percent, 0 percent
    c. 100 percent, 100 percent
    d. 50 percent, 100 percent
    e. none of the above

7. Suppose that a hemophilic male (X-linked recessive allele) and a female carrier for the hemophilic trait have a nonhemophilic daughter with Turner syndrome. Nondisjunction could have occurred in _____. [pp. 209;212]
    a. both parents
    b. neither parent
    c. the father only
    d. the mother only
    e. the nonhemophilic daughter

8. Nondisjunction involving the X chromosome occurs during oogenesis and produces two kinds of eggs, XX and O (no X chromosome). If normal Y sperm fertilize the two types, which genotypes are possible? [p.212]
    a. XX and XY
    b. XXY and YO
    c. XYY and XO
    d. XYY and YO
    e. YY and XO

9. Of all phenotypically normal males in prisons, the type once thought to be genetically predisposed to becoming criminals was the group with _____. [p.213]
    a. XXY disorder
    b. XYY disorder
    c. Turner syndrome
    d. Down syndrome
    e. Klinefelter syndrome

10. Amniocentesis is _____. [p.215]
    a. a surgical means of repairing deformities
    b. a form of chemotherapy that modifies or inhibits gene expression or the function of gene products
    c. used in prenatal diagnosis; a tiny sample of amniotic fluid is drawn; cell samples are analyzed for many severe genetic disorders
    d. a form of gene-replacement therapy
    e. a diagnostic procedure; cells for analysis are withdrawn from the chorion

CHAPTER OBJECTIVES/REVIEW QUESTIONS

1. A _____ is a chart showing genetic connections among individuals; be familiar with the standardized symbols used in such charts. [p.204]

2. Distinguish between human sex chromosomes and autosomes. [pp.206-209]

3. Explain how sex determination occurs in humans. [p.206]

4. Carefully characterize patterns of autosomal recessive inheritance, autosomal dominant inheritance, and X-linked recessive inheritance. [pp.206-209]

5. Describe the characteristics of Hutchinson-Gilford progeria syndrome. [p.206]

6. A(n) _____ is a loss of a chromosome segment; a(n) _____ is a gene sequence separated from a chromosome but then was inserted at the same place, but in reverse; a(n) _____ is a repeat of several gene sequences on the same chromosome; a(n) _____ is the transfer of part of one chromosome to a nonhomologous chromosome. [pp.210]

7. _____, one of the genes in a human Y chromosome, is the master gene for male sex determination. [p.211]

8. Cite evidence that tends to support the idea that chromosome structure does evolve. [pp.210-211]

9. _____ is the failure of the chromosomes to separate in either meiosis or mitosis. [p.212]

10. When gametes or cells of an affected individual end up with one extra or one less than the parental number of chromosomes, it is known as _____; relate this concept to monosomy and trisomy. [p.212]

11. Trisomy 21 is known as _____ syndrome; Turner syndrome has the chromosome constitution _____; XXY chromosome constitution is _____ syndrome; taller than average males with sometimes mild mental impairment have the _____ condition. [pp.205;212-213]

12. Explain the procedures used in three types of prenatal diagnosis: amniocentesis, chorionic villi analysis, and fetoscopy; compare the risks. [pp.214-215]

13. A procedure known as preimplantation diagnosis relies on _____ _____ fertilization. [p.182]

CHAPTER SUMMARY

All animals have a pair of (1) _____ chromosomes that are identical in length, shape, and which (2) _____ they carry. Sexually-reproducing species also have a pair of (3) _____ chromosomes. The members of this pair differ between females and males. A (4) _____ on one of the human sex chromosomes dictates the male sex. (5) _____, a diagnostic tool, reveals changes in the structure or number of an individual's (6) _____. Many genes on autosomes are expressed in (7) _____ patterns of simple (8) _____. Some (9) _____ are affected by genes on the X chromosome. Inheritance patterns of such traits differ in (10) _____ and (11) _____. On rare occasions, a chromosome may undergo permanent change in its (12) _____, as when a segment is deleted, (13) _____, inverted, or (14) _____. On rare occasions, the (15) _____ of autosomes or sex chromosomes changes. In humans, the change usually results in a (16) _____ disorder. Various analytical and (17) _____ procedures often reveal genetic disorders.

INTEGRATING AND APPLYING KEY CONCEPTS

1. The parents of a young boy bring him to their doctor. They explain that the boy does not seem to be going through the same vocal developmental stages as his older brother. The doctor orders a common cytogenetics test to be done, and it reveals that the young boy's cells contain two X chromosomes and one Y chromosome. Describe the test that the doctor ordered and explain how and when such a genetic result, XXY, most logically occurred.

2. Solve the following genetics problem. Show rationale, genotypes, and phenotypes. A husband sues his wife for divorce, arguing that she has been unfaithful. His wife gave birth to a girl with a fissure in the iris of her eye, an X-linked recessive trait. Both parents have normal eye structure. Can the genetic facts be used to argue for the husband's suit? Explain your answer.

15

BIOTECHNOLOGY

INTRODUCTION

The information that you have learned so far allows an almost complete understanding of the structure of DNA, the ways that DNA is used to produce proteins and the processes involved with the cellular treatment of DNA. As scientists began to decipher this knowledge, along with it came the development of techniques by which we could intentionally manipulate DNA and create organisms that contained desired DNA, often from an entirely different species. The development of this technology holds much promise for the medical field and the agricultural industry, in particular. However, the intentional manipulation of DNA is not without risk and controversy. This chapter first discusses some of the various technologies involved in this field. It then goes on to cover the potential benefits of genetic engineering, and the potential concerns.

FOCAL POINTS

- Figure 15.2 [p. 220] illustrates the use of enzymes in the creation of a recombinant DNA molecule.
- Figure 15.3 [p. 220] shows a representative plasmid which is often used as a DNA vector.
- Figure 15.4 [p. 221] animates the formation of a recombinant DNA molecule.
- Figure 15.5 [p.222] illustrates nucleic acid hybridization.
- Figure 15.6 [p.223] animates the polymerase chain reaction which is used to amplify DNA.
- Figure 15.7 [p.224] shows DNA sequencing.

INTERACTIVE EXERCISES

15.1 PERSONAL DNA TESTING [p.219]
15.2 CLONING DNA [pp.220-221]
15.3 FROM HAYSTACKS TO NEEDLES [pp222-223]
15.4 DNA SEQUENCING [pp.224-225]
15.5 GENOMICS [pp.226-227]

Selected Terms
nucleotide [p.219], SNP [p.219], APOE [p.219], lipoprotein [p.219], Alzheimer's disease [p.219], personal genetic testing [p.219], [p. 221], "sticky" end [p.220], complementary [p.220], intron [p.221], DNA ligase [p.221], reverse of RNA transcription [p.221], DNA polymerase [p.221], genomic library [p. 222], cDNA [p.222], radioactive phosphate group [p.222], hybridize [p.223], Taq polymerase [p. 223], *Thermus aquaticus* [p. 223], hydroxyl group [p.224], 3'carbon [p.224], A, G, C, T [p.224], dideoxynucleotides [p.224], James Watson [p.225], Francis Collins [p.225], Craig Venter [p.225], human genome project [p.225], cryptic data [p.226], APOA5 [p.226], lipoprotein [p.226], knock out [p.226], TTTTC [p.227], DNA fingerprinting [p.227]

Boldfaced Terms
[p.220] restriction enzyme _____

_____

[p.220] recombinant DNA _____

_____

[p.220] DNA cloning _____

_____

[p.220] plasmid _____

_____

[p.221] cloning vectors _____

_____

[p.221] reverse transcriptase _____

_____

[p.222] genome _____

_____

[p. 222] DNA libraries _____

_____

[p. 222] probe _____

_____

[p.222] nucleic acid hybridization _____

_____

[p. 222] polymerase chain reaction (PCR) _____

_____

[p.223] primers _____

_____

[p.224] DNA sequencing _____

_____

[p.224] electrophoresis _____

_____

[p.226] genomics _____

_____

[p.227] DNA profiling _____

_____

[p.227] short tandem repeats _____

_____

Matching

For each of the following, match the word or phrase with its correct definition. [pp.220-221]

1. _____ reverse transcriptase
2. _____ restriction enzyme
3. _____ DNA ligase
4. _____ cDNA
5. _____ plasmid

a. Seals the nicks in a recombinant DNA molecule.
b. A small circular piece of DNA located outside the chromosome in bacteria.
c. A viral enzyme that uses mRNA to form cDNA.
d. Complementary DNA that is made from mRNA.
e. This cuts double-stranded DNA at a specific base sequence.

Matching [pp.222-223]

Match the term with the definition below.

6. _____ genome
7. _____ cDNA
8. _____ probe
9. _____ nucleic hybridization
10. _____ PCR
11. _____ primers

a. base pairing between DNA (or from DNA and RNA) from more than one source
b. the entire set of genetic materials.
c. short single strands of DNAthat base pair with certain DNA sequences.
d. technique used to mass produce copies of particular sections of DNA without having to clone it in a living cell.
e. a fragment of DNA or RNA labeled with a tracer.
f. a set of cells that host various cloned DNA fragments.

Complete the Table

For each of the following, provide the role of the procedure in the study of molecular biology.

Procedure	Role
electrophoresis [p.224]	12.
PCR [p.222]	13.
nucleic acid hybridization [p.222]	14.
DNA sequencing [p.224]	15.

Matching

Match each of the following terms to its correct definition.

16. _____ probe [p.222]
17. _____ A, T, C, and C [p.224]
18. _____ cDNA library [p.221]
19. _____ Thermus aquaticus [p.223]
20. _____ primers [p.223]
21. _____ Taq polymerase

a. short single strands of DNA that base pair with certain DNA sequences
b. a collection of cells that house cloned fragments of DNA.
c. the heat tolerant organism that provides the polymerase for PCR reactions.
d. a radioisotope labeled fragment of DNA uses in nucleic acid hybridization
e. enzyme obtained from Thermus aquaticus
f. the DNA nucleotides

Short Answer

22. The following figure is a DNA fingerprint from a crime scene. Please describe how DNA fingerprinting is used to convict suspects, and tell which of the following suspects is probably guilty of this crime. Consult your text book, p. 227 for additional information.

_____

_____

_____

_____

_____

_____

_____

_____

_____

_____

_____

_____

15.6 GENETIC ENGINEERING [p.228]
15.7 DESIGNER PLANTS [p.228-229]
15.8 BIOTECH BARNYARDS [p.230-231]
15.9 SAFETY ISSUES [p. 231]
15.10 GENETICALLY MODIFIED HUMANS? [p.232-233]

Selected Terms

Agrobacterium tumefaciens [p.228], *Bacillus thuringensis* [p. 228], transgenic crops [p.228], Ti [p.228], "Frankenfood" [p.229], Golden rice [p.229], APHIS [p.229], transgenic goat [p.230], lysozyme [p.230], human interleukin-2 [p.230], multiple sclerosis, cystic fibrosis, diabetes, cancer, Huntington's disease [p.231], Francis Crick [p.231], Paul Berg [p.231], Parkinson's disease [p.232], SCID-X1 [p.232]

Boldfaced Terms

[p.228] genetic engineering _____

_____

[p. 228] transgenic _____

_____

[p.228] genetically modified organisms _____

_____

[p.231], xenotransplantation _____

_____

[p.232] gene therapy_____

_____

[p.232] eugenics _____

_____

True-False

If the statement is true, place a "T" in the space provided. If false, correct the underlined word so that the statement becomes true. [pp.228-233]

1. _____ Transferring a gene from one species to another is an example of xenotransplantation.

2. _____ *Agrobacterium tumefaciens* is commonly used in the process of human gene therapy.

3. _____ Frankenfood is the term that has been used by groups in media hype regarding genetically modified food.

4. _____ *Bacillus thuringensis* is commonly used as an insecticide.

Choice

Choose which area of research is best described by each statement. Some answers may be used more than once. [pp.228-233]

      a. genetic engineering
      b. human gene therapy
      c. xenotransplantation
      d. eugenics
      e. transgenic

5. ____ The replacement of a defective gene with a healthy one to fight a specific disease.

6. ____ A source of drugs and protein for medical studies.

7. ____ An idea of selecting the most desirable human traits.

8. ____ The development of transgenic plants and animals.

9. ____ The movement of organs from one species to another.

10. ____ Moving a gene from one species in to another species creates an organism that is called?

SELF QUIZ

1. The molecule _____ cuts double-stranded DNA at specific locations. [p.220]
   a. plasmid
   b. DNA ligase
   c. restriction enzymes
   d. *Taq* polymerase
   e. none of the above

2. The movement of a gene from one species to another is an example of _____. [p.228]
   a. xenotransplantation
   b. genetic engineering
   c. human gene therapy
   d. eugenic engineering
   e. none of the above

3. A small circular piece of bacterial DNA that contains only a few genes is called a _____. [p.220]
   a. ligase
   b. restriction enzyme
   c. tandem repeat
   d. plasmid
   e. cDNA

4. An idea of selecting the most desirable human traits is called eugenics. [p.232]
   a. True
   b. False

5. Which of the following is used to manufacture a cDNA library? [p.221]
   a. reverse transcriptase
   b. restriction enzymes
   c. DNA ligase
   d. *Taq* polymerase
   e. PCR

6. The _____ procedure can make billions of copies of a DNA sequence in a short period of time. [p.222]
   a. nucleic acid hybridization
   b. DNA fingerprinting
   c. automated DNA sequencing
   d. polymerase chain reaction
   e. genetic engineering

7. A radioisotope labeled probe is used in _____. [p222]
   a. nucleic acid hybridization
   b. polymerase chain reaction
   c. automated DNA sequencing
   d. a cDNA library
   e. the use of restriction enzymes

8. Identifying an individual by their DNA is called _____. [p.227]
   a. structural genomics
   b. functional genomics
   c. DNA profiling
   d. eugenic engineering
   e. DNA hybridization

9. The separation of DNA fragments based on their size is called _____. [p.224]
   a. DNA fingerprinting
   b. polymerase chain reaction
   c. nucleic acid hybridization
   d. genetic engineering
   e. electrophoresis

10. Tandem repeats are used primarily for which of the following? [p.227]
    a. DNA fingerprinting
    b. nucleic acid hybridization
    c. polymerase chain reaction
    d. automated DNA sequencing
    e. all of the above

CHAPTER OBJECTIVES/ REVIEW QUESTIONS

1. Define *recombinant DNA*. [p.220]

2. Understand the difference in function between a DNA ligase and a restriction enzyme. [p.220]

3. Recognize how plasmids can be used as cloning vectors. [p.220]

4. Understand the relationship between cDNA and mRNA. [p.221]

5. Understand the concept of a gene library and how it can be screened using nucleic acid hybridization. [p. 222]

6. Understand the process of the polymerase chain reaction. [pp.222-223]

7. How does electrophoresis separate DNA fragments? [p.224]

8. Why are tandem repeats important in the process of DNA fingerprinting? [p.226]

9. Understand the goals of the Human Genome Project. [p.225]

10. How can DNA fingerprinting be used to distinguish between individuals? [pp.226-227]

11. Describe DNA profiling and short tandem repeats. [p.227]

12. What is genetic engineering? [228]

13. Understand what a genetically modified organism (GMO) is and how they are produced. [228]

14. Be able to discuss the up and down sides of a GMO. [p.229]

15. What is xenotransplantation? [p.231]

16. Describe the safety issues surrounding GMOs? [p.231]

17. How can gene therapy be used to fight human disease? [pp.232-233]

CHAPTER SUMMARY

Researchers routinely make (1) _____ DNA by isolating, cutting, and joining DNA from different (2) _____. Plasmids and other (3) _____ can carry foreign DNA into host cells.

Researchers isolate and make copies of DNA in order to study it. (4) _____ copies particular (5) _____ of DNA in sufficient quantity for experiments and other practical purposes.

(6) _____ reveals the linear order of nucleotides in a fragment of DNA. A DNA (7) _____ is an individual's unique array of DNA sequences.

Genomics is the study of (8) _____. Comparisons of the genomes of different (9) _____ offer practical benefits.

Genetic engineering – the directed modification of an organism's (10) _____ - is now used routinely in research and medical applications. It continues to raise many (11) _____ questions.

INTEGRATING AND APPLYING KEY CONCEPTS

1. You have been assigned as part of a research team to produce a new transgenic plant. You have determined that a specific gene of interest needs to be copied from one plant, cloned, and then inserted into another species of plant. Using the procedures provided in this chapter, outline how that procedure could be performed.

2. A group of concerned citizens comes to you with questions about the safety of your genetically engineered plant. They say to you "genetic pollution is forever". What do you say to these citizens about the safety of your plant and why it should be produced? What are the possible benefits of genetic engineering?

3. Suppose your child had a disease that gene therapy could help live a better and healthier life. Would you consider gene therapy as a possible treatment? Why? Why not?

16

EVIDENCE OF EVOLUTION

INTRODUCTION

This chapter presents a brief history of evolutionary thought from the time of the ancient Greeks up to the discoveries of Charles Darwin and Alfred Wallace. The chapter then presents supporting evidence of how natural selection has influenced life on Earth. As you proceed through the chapter, note how each class of evidence supports the others and strengthens support for natural selection being the mechanism of evolutionary change.

FOCAL POINTS

- Figure 16.1 [p.237] illustrates evidence of a large asteroid's collision with Earth and a mass extinction.
- Figures 16.2 & 16.3 [p.238] shows similar looking birds and plants that live on different continents.
- Figure 16.4 [p.239] illustrates vestigial body parts of a snake and a human.
- Section 16.3 [pp.240-241] reviews early beliefs and Charles Darwin's early life.
- Section 16.4 [p.242-243] reviews natural selection with respect to Charles Darwin and Alfred Wallace.
- Table 16.1 [p.243] lists the principles of natural selection.
- Figure 16.10 [p.245] reviews radiometric dating.
- Figure 16.11 [p.246] overlays major biological events with the geologic time scale and references the Grand Canyon.
- Figure 16.13 [p.249] illustrates the drifting continents.
- Figure 16.16 [p.250] demonstrates morphological divergences among vertebrate forelimbs.
- Figure 16.15 [p.251] illustrates morphological convergence of flight.
- Figure 16.16 [p.252] illustrates comparisons of vertebrate embryos.

INTERACTIVE EXERCISES

16.1 REFLECTIONS OF A DISTANT PAST [p.237]
16.2 EARLY BELIEFS, CONFOUNDING DISCOVERIES [pp.238-239]
16.3 A FLURRY OF NEW THEORIES [pp.240-241]
16.4 DARWIN, WALLACE, AND NATURAL SELECTION [pp.242-243]

Selected Terms
asteroid [p.237], Barringer Crater, Arizona [p.237], K-T boundary [p.237], 65.5 million years ago [p.237], continuum of organization [p.238], "great chain of being" [p.238], Wallace [p.238], flightless birds [p.239], comparative morphology [p.239], geology [p.239], extinct species [p.239], Cuvier [p.240], Lamarch [p.240], "fluida" [p.240], Darwin [p.240], Henslow [p.240], *Beagle* [p.240], Lyell [p.240], Principles of Geology [p.240], "Transmutation of Species [p.241], 6,000 years[p.241] glyptodons [p.242], Malthus [p.242], limited resources [p.242], Galapagos archipelago [p.242], finches [p.242], Amazon basin [p.243], Malay Archipelago [p.243], *On the Origin of Species* [p.243]

Boldfaced Terms

[p.237] mass extinction _____

_____

[p.238] naturalist _____

_____

[p.238] biogeography _____

_____

[p.239] comparative morphology _____

_____

[p.239] fossils _____

_____

[p.240] catastrophism _____

_____

[p.240] evolution _____

_____

[p.240] lineage _____

_____

[p.241] theory of uniformity _____

_____

[p.242] artificial selection _____

_____

[p.242] fitness_____

_____

[p.242] adaptation _____

_____

[p. 242] adaptive trait _____

_____

[p.243] natural selection_____

_____

Matching

Choose the most appropriate answer for each term. [pp.238-239]

1. ____ comparative morphology
2. ____ biogeography
3. ____ fossils
4. ____ chain of being
5. ____ species
6. ____ naturalists

a. a theory that all species were formed at the same time and were linked together
b. each kind of being in the Chain of Beings
c. the study of patterns of distribution of species
d. people who observe life from a scientific perspective
e. the rock-like remains of past organisms
f. the study of body plans and structure among groups of organisms

Choice

For each of the following, choose one of the individuals from the list below. [pp.238-241]

    a. George Cuvier
    b. Jean Lamarck
    c. Charles Lyell
    d. Aristotle

7. ____ theory of uniformity

8. ____ inheritance of acquired characteristics

9. ____ catastrophism

10. ____ Chain of Beings

11. ____ natural processes have more of an effect on Earth than did catastrophes

12. ____ force for change was an intense drive for perfection

13. ____ first to suggest that the environment is a factor in lines of descent

14. ____ one of the first naturalists

Choice

For each of the following, choose a term from the list below that best completes the sentence. A term may be used more than once. [pp.242-243]

    a. individual(s)
    b. population(s)
    c. alleles
    d. natural selection

15. ____ Environmental agents of selection act on the range of variations, and the ____ may evolve as a result.

16. ____ Natural ____ have inherent reproductive capacity to increase numbers through successive generations.

17. ____ _____ share a pool of heritable information about traits, encoded in genes.

18. ____ _____ is the outcome of differences in reproduction among individuals of a population that vary in shared characteristics.

19. ____ No _____ can indefinitely grow in size.

20. ____ Variations in traits start with _____, slightly different molecular forms of genes that arise through mutations.

21. ____ Sooner or later, _____ will end up competing for dwindling resources.

16.5 FOSSILS – EVIDENCE OF ANCIENT LIFE [pp.244-245]
16.6 PUTTING TIME IN PERSPECTIVE [pp.246-247]

Selected Terms
fossils [p.244], trace fossils [p.244], sedimentary rock [p.244], glaciers [p.244], radioisotope [p.245], radioactive decay [p.245], carbon 13 & carbon 14 [p.245], zircon crystal [p.245], rock layers [p.246], fossil record [p.246], Grand Canyon [p.247]

Boldfaced Terms
[p.245] half-life_____

_____

[p.245] radiometric dating _____

_____

[p.248] geologic time scale _____

_____

Matching
Match each term with the correct statement. [pp.244-247]

1. ____ fossilization
2. ____ fossil record
3. ____ stratification (rock layering)
4. ____ trace fossils
5. ____ half-life
6. ____ radiometric dating
7. ____ geologic time scale

a. the formation of sedimentary rocks
b. indirect evidence of past life, such as tracks and burrows
c. a vertical series of fossils that serves as a record of past life
d. the chronology of Earth's history
e. the use of parent to daughter elements to determine sample age
f. the amount of time required to reduce the amount of radioisotope in half
g. the process by which minerals replace bones and hardened tissues

Short Answer
8. Carbon-14 has a half-life of 5,370 years. Assume that a sample of organic material contains 4.0 grams of carbon-14. How many grams would be left after three half-lives? How many years would have elapsed?_____

_____

_____

Choice

Refer to Figure 16.11 in the text [p.246]. For each of the following events, choose the most appropriate time period from the list provided. Some answers may be used more than once.

a. Cenozoic
b Proteozoic
c. Archean
d. Paleozoic
e. Mesozoic

9. ____ origin of the vascular plants

10. ____ gymnosperms dominant the land plants

11. ____ origin of the angiosperms

12. ____ origin of the reptiles

13. ____ origin of the amphibians

14. ____ origin of eukaryotic cells

15. ____ origin of modern humans

16. ____ origins of life

17. ____ origin of the mammals

16.7. DRIFTING CONTINENTS, CHANGING SEAS [pp.248-249]
16.8 SIMILARITIES IN BODY FORM AND FUNCTION [pp.250-251]
16.9 SIMILARITIES IN PATTERNS OF DEVELOPMENT [pp.252-253]

Selected Terms

supercontinent [p.248], continental drift [p.248], *Glossopteris* [p.249], Permian-Triassic [p.249], *Lystrosaurus* [p.249], pterosaurs [p.250], biochemistry [p.250], *Apetalal* [p.252], *Brassica oleracea* [p.252], *Arabidopsis* [p.252], antennapedia [p.252], *Hox* [p.252], *Dlx* [p.253]

Boldfaced Terms

[p.248] Pangea _____

_____

[p.249] plate tectonics _____

_____

[p.249] Gondwana _____

_____

[p.250] homologous structures _____

_____

[p.250] morphological divergence _____

_____

[p.251] analogous structures _____

_____

[p.251] morphological convergence _____

_____

Choice

For each of the following, use your knowledge to identify the category of evidence to which the statement best applies. [pp.248-251]

    a. plate tectonics
    b. morphological divergence and convergence
    c. patterns of development
    d. DNA and other genetic information

1. ____ molecular clocks that record the accumulation of neutral mutations

2. ____ homologous structures

3. ____ fossils in Africa and South America

4. ____ embryonic studies of early embryo stages in vertebrates

5. ____ studies of mitochondrial patterns of inheritance

6. ____ comparison of cytochrome c proteins sequences

7. ____ analogous structures

8. ____ the movement of continents over long periods of time

9. ____ examinations of conserved genes

10. ____ studies of *Hox* and *Dlx*

11. ____ studies of Gondwana and Pangea

12. ____ comparison of iron compasses in North America and Europe

---

SELF QUIZ

1. Which of the following individuals proposed the inheritance of acquired characteristics? [p.240]
   a. Cuvier
   b. Aristotle
   c. Darwin
   d. Lamarck
   e. Lyell

2. Stratification is an important component in the study of _____. [p.239]
   a. comparative morphology
   b. evolutionary trees
   c. biogeography
   d. fossil record

3. A mechanism for how evolution that was formulated by Darwin and Wallace is called _____. [p.243]
   a. natural selection
   b. catastrophism
   c. stratification
   d. artificial selection

4. A half-life is _____. [p.245]
   a. the time between two mass extinction events
   b. the time required for half of a radioactive isotope to decay
   c. half the lifespan of a species
   d. a unit of the geological time scale

5. The study of Pangea and Gondwana are associated with _____. [pp.248-249]
   a. biogeography
   b. DNA, RNA and proteins
   c. development
   d. plate tectonics
   e. fossil record

6. Evolution of similar body parts in different lineages is an example of _____.[p.250]
   a. catastrophism
   b. morphological divergence
   c. morphological convergence
   d. speciation

7. Similar body parts that evolved from a common ancestor are called _____ structures. [p.250]
   a. analogous
   b. morphological convergent
   c. homologous

8. Most fossils are found in _____ rocks. [p.244]
   a. metamorphic
   b. igneous
   c. sedimentary

9. *On the Origin of Species* was written by _____. [p.243]
   a. Charles Darwin
   b. Alfred Wallace
   c. Charles Lyell
   d. Thomas Malthus

10. Which of the following was Charles Lyell's contribution to the development of evolutionary thought? [p.243]
    a. catastrophism
    b. Chain of Beings
    c. stratification
    d. theory of uniformity
    e. inheritance of acquired characteristics

CHAPTER OBJECTIVES/REVIEW QUESTIONS

1. Explain how biogeography, comparative morphology and fossils contributed to the study of evolution. [pp.237-239]

2. Give the basic premises of Cuvier's and Lamarck's theories. [p.240]

3. Explain the theory of uniformity and why it played an important role in the study of evolution. [p.241]

4. Describe the evidence that Darwin and Wallace used to develop their theories of natural selection. [pp.240-243]

5. Explain the process of fossilization and the importance of stratification. [pp.244-245]

6. Explain radiometric dating. [p.245]

7. Understand the geologic time scale. [pp.246-247]

8. Understand the importance of plate tectonics in studying evolution. [pp.248-249]

9. Distinguish between morphological convergence and divergence and give an example of each. [pp.250-251]

10. Recognize the importance of developmental studies in evolutionary research. [pp.252]

11. Understand the significance of the *Hox* gene. [p.252]

CHAPTER SUMMARY

Long ago, Western scientists started to catalog previously unknown species and think about their (1) _____ distribution. They discovered similarities and differences among major groups, including those represented as (2) _____ in layers of sedimentary rock.

Evidence of (3) _____, or changes in lines of descent, gradually accumulated. (4) _____ and Alfred Wallace independently developed a theory of (5) _____ to explain how (6) _____ trans that define each species evolve.

The fossil record offers (7) _____ evidence of past changes in lines of descent.

Correlating evolutionary theories with (8) _____ history helps explain the distribution of (9) _____, past and present.

Species of different (10) _____ often have similar body parts that may be evidence of descent from a shared (11) _____.

(12) _____ expressions help us discover and confirm relationships among species and lineages.

INTEGRATING AND APPLYING KEY CONCEPTS

1. In previous chapters you defined a species as a group of interbreeding organisms that are reproductively isolated from other groups. What problem does this definition hold for scientists who study macro-evolutionary processes? How would you propose that the definition be changed?

2. As a researcher you have discovered two groups of plants on two isolated islands in the Pacific Ocean. These plants have some similar characteristics, but also some different ones. You are not sure whether they represent morphological convergence or divergence evolution. How would you test this?

3. How do think two individuals (Darwin and Wallace) could come up with the same idea of a mechanism for natural selection independent of each other?

17

PROCESSES OF EVOLUTION

INTRODUCTION

This chapter examines what biologists know about how the process of evolution occurs. It presents a series of models that biologists use to examine evolutionary change over time. The chapter also introduced the concepts of variation, speciation, adaptation, and extinction – all of which are potential outcomes of evolutionary change.

FOCAL POINTS

- Table 17.1 [p.258] presents sources of variation in traits among individuals of a species.
- Figure 17.3 [p.260] shows the frequency of wing color alleles among all of the individuals in a hypothetical population of butterflies.
- Figure 17.4 [p.261] presents the three modes of natural selection.
- Figure 17.5 [p.262] shows a graph of directional selection.
- Figure 17.8 [p.264] shows stabilizing selection.
- Figure 17.10 [p.265] shows disruptive selection.
- Figure 17.13 [p.267] shows distributions of malarial cases with respect to Africa and in relationship to sickle cell anemia.
- Figure 17.17 [p.270] illustrates the differences between prezygotic and postzygotic isolation mechanisms.
- Figure 17.20 [p.272] demonstrates allopatric speciation.
- Figure 17.21 [p.273] shows allopatric speciation of Hawaiian honeycreepers.
- Figure 17.22 [p.274] illustrates polyploidy in the history of modern wheat.
- Table 17.2 [p.275] shows a comparison of speciation models.
- Figure 17.25 [p.276] illustrates adaptive radiation.
- Figure 17.27 [p.278] shows a cladogram.

INTERACTIVE EXERCISES

17.1 RISE OF THE SUPER RATS [p.257]

17.2 INDIVIDUALS DON'T EVOLVE, POPULATIONS DO [pp.258-259]

Selected Terms
Rattus [p.257], mammalian pest [p.257], pathogen [p.257], warfarin [p.257], vitamin K [p.257], variation [p.258], interbreeding [p.258], alleles [p.258], morphological traits [p.258], *morpho-* [p.258], crossing over [p.258], independent assortment [p.258], fertilization [p.258], transposition [p.258], duplication [p.258], natural selection [p.259], mutation [p.259]

Boldfaced Terms

[p.258] population_____

_____

[p.258] lethal mutation_____

_____

[p.258] neutral mutation_____

_____

[p.259] gene pool_____

_____

[p.259] allele frequencies_____

_____

[p.259] microevolution_____

_____

[p.259] genetic equilibrium_____

_____

Matching

Match each of the following terms with the correct statement. [pp.258-259]

1. ____ population
2. ____ physiological traits
3. ____ gene pool
4. ____ neutral mutation
5. ____ allele frequencies
6. ____ lethal mutation
7. ____ morphological traits
8. ____ genetic equilibrium
9. ____ microevolution

a. one group of individuals of the same species in a specified area
b. a mutation that results in death for the organism
c. a mutation that has no influence on survival or reproduction
d. the sum of the genes in a population
e. when a population is not evolving with respect to that gene
f. traits that are based on metabolic activities
g. the relative abundance of alleles of a given gene among individuals of a population
h. traits that are based on the form of the species
i. small changes in the allele frequencies of a population over time

17.3 A CLOSER LOOK AT GENETIC EQUILIBRIUM [pp.260-261]
17.4 NATURAL SELECTION REVISITED [p.261]

Selected Terms

Hardy-Weinberg equilibrium [p.260], stabilizing selection [p.261], disruptive selection [p.261], peppered moth [p.262], rock pocket mice [p.262], antibiotic resistance [p.263]

Boldfaced Terms

[p.261] natural selection_____

_____

[p.262] directional selection _____

_____

Genetics Problem

In a population of the fruit fly *Drosophila melanogaster* the frequency of individuals with a certain autosomal recessive wing disorder is 16%. Using this information, calculate the following: [pp.260-261]

1. The value of q^2 _____

2. The value of q_____

3. The value of p _____

4. The frequency of homozygous dominant individuals_____

5. The frequency of heterozygous individuals _____

6. Assuming that the population is in Hardy-Weinberg equilibrium, what will the frequency of the recessive mutation be in the gene pool in the next generation? (HINT: This is represented by the value q) _____

Short Answer

7. List the five conditions of a stable population. [p.260]_____

_____

_____

_____

_____

_____

Identification

Identify the type of selection in each of the graphs below. [p.261]

8. _____   9. _____   10. _____

17.5 DIRECTIONAL SELECTION [pp.262-263]
17.6 STABILIZING AND DISRUPTIVE SELECTION [pp.264-265]

Selected Terms
directional selection [p.262], sociable weavers [p.264], *Philetairus socius* [p.264], black-bellied seedcracker [p.273], *Pyrenestes ostrinus* [p.265], dimorphism [p.265]

Boldfaced Terms
[p.264] stabilizing selection _____

_____

[p.265] disruptive selection _____

_____

Choice
For each of the following statements, choose the appropriate form of selection from the list below.
   a. stabilizing selection [p.264]
   b. disruptive selection [p.265]
   c. directional selection [pp.262-263]

1. ____ birth weight in human babies

2. ____ the effects of predation on peppered moths

3. ____ antibiotic resistance in bacteria

4. ____ bill size in the black-bellied seedcracker

5. ____ body weight in sociable weavers

6. ____ favors the most common phenotype

7. ____ allele frequencies shift in one direction

8. ____ phenotypes at both ends of the range are favored

9. ____ coat color in rock pocket mice

17.7 FOSTERING DIVERSITY [p.266-267]
17.8 GENETIC DRIFT [pp.268-269]
17.9 GENE FLOW [p.269]

Selected Terms
malaria [p.267], HbA & HbS alleles [p.267], *Plasmodium* [p.267], homozygous [p.267], heterozygotes [p.267], hemoglobin [p.267], allele frequencies [p.268], *Ellis-van Creveld syndrome* [p.269]

Boldfaced Terms
[p.266] sexual selection_____

_____

[p.267] balanced polymorphism _____

_____

[p.268] genetic drift_____

_____

[p.268] fixed _____

_____

[p.268] bottleneck _____

_____

[p.268] founder effect _____

_____

[p.269] inbreeding_____

_____

[p.269] gene flow_____

_____

Matching
Match each of the following definitions to the correct term. [pp.266-269]

1. ____ a species that has distinct male and female phenotypes

2. ____ environmental conditions favor heterozygous individuals

3. ____ a drastic reduction in population size

4. ____ the movement of alleles between populations

5. ____ those better at securing mates are more successful genetically

6. ____ when a very small population forms a new population

7. ____ random change in allele frequencies over time

8. ____ when two or more alleles persist at high frequencies in a population

a. sexual dimorphism
b. genetic drift
c. sexual selection
d. founder effect
e. balancing selection
f. balanced polymorphism
g. bottleneck
h. gene flow

True-False

If the statement is true, place a 'T' in the blank. If the statement is false, correct the underlined term so that the statement is correct.

9. _____ The random changes of allele frequencies associated with genetic drift frequently lead to a <u>gain</u> in genetic diversity. [p.276]

10. _____ <u>Sexual selection</u> helps maintain the Hb^A/Hb^S heterozygous condition in humans. [p.275]

11. _____ Bottleneck and founder effects are examples of <u>genetic drift</u>. [p.276]

12. _____ <u>Gene flow</u> counteracts the effect of mutation, natural selection, and genetic drift. [p.277]

13. _____ The high rate of Ellis-van Creveld syndrome in Amish populations is due to gene flow. [p.277]

14. _____ Inbreeding <u>increases</u> the genetic diversity of a population. [p.277]

17.10 REPRODUCTIVE ISOLATION [pp.270-271]
17.11 ALLOPATRIC SPECIATION [pp.272-273]
17.12 SYMPATRIC AND PARAPATRIC SPECIATION [pp.274-275]

Selected Terms

species [p.270], prezygotic [p.270], postzygotic [p.270], gamete incompatibility [p.270], hybrid embryos [p.270], hybrid sterility [p.270], temporal isolation [p.270], mechanical isolation [p.270], ecological mechanism [p.270], behavioral isolation [p.270], gamete incompatibility [p.271], hybrid inviability [p.271], fertile [p.271], fitness [p.271], anther [p.271], stigma [p.271], archipelagos [p.272], geographic barrier [p.272], polyploidy [p.274], cichlid [p.274], Lake Vitoria [p.274], greenish warblers [p.274], *Phylloscopus trochiloides* [p.274], walking worms [p.275], *Tasmanipatus barrette* [p.275], *T. anophthalmus* [p.275]

Boldfaced Terms

[p.270] speciation _____

_____

[p.270] reproductive isolating _____

_____

[p.272] allopatric speciation_____

_____

[p.274] sympatric speciation _____

_____

[p.275] parapatric speciation _____

_____

Labeling

Provide the indicated label for each of the numbered items in the diagram below. [p.270]

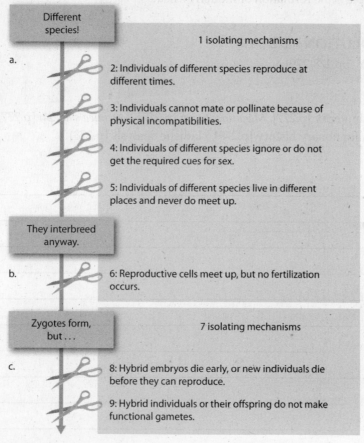

Different species!

1 isolating mechanisms

a.

2: Individuals of different species reproduce at different times.

3: Individuals cannot mate or pollinate because of physical incompatibilities.

4: Individuals of different species ignore or do not get the required cues for sex.

5: Individuals of different species live in different places and never do meet up.

They interbreed anyway.

b.

6: Reproductive cells meet up, but no fertilization occurs.

Zygotes form, but...

7 isolating mechanisms

c.

8: Hybrid embryos die early, or new individuals die before they can reproduce.

9: Hybrid individuals or their offspring do not make functional gametes.

No offspring, sterile offspring, or weak offspring that die before reproducing

1. _____

2. _____

3. _____

4. _____

5. _____

6. _____

7. _____

8. _____

9. _____

Choice

For each of the following, choose the appropriate model of speciation from the list below.

    a.   allopatric speciation [pp.272-273]

    b.   sympatric speciation [pp.274-275]

    c.   parapatric speciation [p.274-275]

10. ____ The formation of hybrids is an example

11. ____ New species form within the home range of an existing species

12. ____ Physical separation between populations

13. ____ Polyploidy is a leading cause

14. ____ An archipelago is an example

15. ____ Occurs when a population extends across a broad region with diverse habitats

16. ____ An example is the formation of modern wheat.

17.13 MACROEVOLUTION [pp.276-277]
17.14 PHYLOGENY [pp.278-279]

Selected Terms
feathers [p.276], blue butterfly [p.277], *Maculinea arion* [p.277], *Myrmica sabuleti* [p.277], Linnaeus [p.278], taxonomy [p.278], evolutionary history [p.278], cladistic analysis [p.278]

Boldfaced Terms
[p.276] exaptation _____

_____

[p.276] stasis _____

_____

[p.276] extinct _____

_____

[p.276] adaptive radiation _____

_____

[p.276] key innovation _____

_____

[p.277] coevolution _____

_____

[p.278] phylogeny _____

_____

[p.278] cladistics _____

_____

[p.278] character _____

_____

[p.278] cladogram _____

_____

[p.278] clade _____

_____

[p.258] monophyletic group _____

_____

[p.278] evolutionary trees _____

_____

[p.278] sister groups _____

_____

Matching
Match each term to the correct definition.

1. ____ sister groups [p.278]

2. ____ clade [p.278]

3. ____ evolutionary trees [p.278]

4. ____ cladogram [p.278]

5. ____ cladistics [p.278]

6. ____ phylogeny [p.278]

7. ____ stasis [p.276]

8. ____ key innovation [p.276]

9. ____ adaptive radiation [p.276]

a. the two lineages that arise at the same node on a cladogram
b. a burst of divergences from a single lineage
c. a method of grouping species on the basis of their shared characteristics
d. the evolutionary history of a species or group of them
e. meaning little or no change in a lineage over long periods of time
f. a new trait that allows its bearer to exploit a habitat more efficiently or in a novel way
g. a summary of our best data supported hypothesis about how a group of species evolved
h. a diagram that shows a network of evolutionary relationships
i. a group of species that share a set of characters

SELF QUIZ

1. The sum of all of the genes of a population is called _____. [p.259]
   a. the gene pool
   b. microevolution
   c. a bottleneck
   d. macroevolution
   e. none of the above

2. If the alleles frequencies for a specific gene are not changing, the population is said to be _____ for that gene. [p.259]
   a. evolving
   b. going extinct
   c. in genetic drift
   d. in genetic equilibrium

3. Which of the following is not correct regarding a stable population? [p.260]
   a. there is no mutation
   b. the population size is small
   c. there is equal reproductive ability
   d. mating is random
   e. there is no gene flow

4. Antibiotic and insecticide resistance are examples of _____ selection. [p.263]
   a. sexual
   b. directional
   c. stabilizing
   d. disruptive

5. The form of selection that tends to preserve intermediate phenotypes is known as _____ selection. [pp.264-265]
   a. balancing
   b. directional
   c. stabilizing
   d. disruptive

6. The preservation of a heterozygous condition in a population is called _____. [p.267]
   a. genetic drift
   b. gene flow
   c. stabilizing selection
   d. a balanced polymorphism
   e. sexual dimorphism

7. The random changes in allele frequencies over time are called _____. [p.268]
   a. gene flow
   b. equilibrium
   c. genetic drift
   d. balancing selection

8. Which of the following is not a prezygotic isolation mechanism? [p.270]
   a. hybrid sterility
   b. temporal isolation
   c. behavioral isolation
   d. ecological isolation
   e. all of the above are prezygotic

9. Which of the following forms of speciation is the result of geographic separation of two species? [p.272]
   a. parapatric
   b. allopatric
   c. sympatric
   d. dispatric

10. When populations inhabiting the same geographic region speciate in the absence of a physical barrier between them, it is called _____ [p.274]
    a. allopatric speciation
    b. sympatric speciation
    c. parapatric speciation

CHAPTER OBJECTIVES/REVIEW QUESTIONS

1. Be able to define gene pool and population [p.258]

2. Distinguish between a lethal and neutral mutation [p.258]

3. Understand the concept of genetic equilibrium [p.259]

4. Be able to use the Hardy-Weinberg equation to determine allele and genotype frequencies in a population. [pp.268-269]

5. Understand the three models of selection and be able to give an example of each. [pp.260-261]

6. Understand the significance of sexual selection. [p.266]

7. Define a balanced polymorphism and give an example. [p.267]

8. Understand how genetic drift, gene flow, and inbreeding influence allele frequencies in a population. [pp.268-269]

9. Know the major forms of prezygotic and postzygotic isolation. [pp.270-271]

10. Distinguish between allopatric, sympatric, and parapatric speciation and give an example of each. [pp.272-275]

11. Define exaptation, stasis, extinction, adaptive radiation, coevolution, and key innovation. [p.276]

12. Define phylogeny, cladistics, cladogram, clade, evolutionary trees, and sister groups. [p.278]

CHAPTER SUMMARY

(1) _____ evolve. Individuals of a population differ in which (2) _____ they inherit, and thus in (3) _____. Over generations, any allele may increase or decrease in (4) _____ in a population. Such shifts occur by the micro-evolutionary processes of (5) _____, natural selection, (6) _____ drift, and gene (7) _____.

Sexually reproducing species consist of one or more populations of individuals that (8) _____ successfully under (9) _____ conditions, produce (10) _____ offspring, and are reproductively (11) _____ from other species. The origin of new species varies in details and duration. Typically, it starts after gene (12) _____ ends between parts of a population. Micro-evolutionary events occur (13) _____ and lead to genetic (14) _____ of the subpopulations. Such divergences are reinforced as (15) _____ isolation mechanisms evolve.

Genetic change above the population level is called (16) _____. Recurring patterns of macroevolution include preadaptation, (17) _____ radiations, coevolution, and (18) _____.

An evolutionary (19) _____ is a heritable aspect of form, function, (20) _____, or development that increases an individual's capacity to (21) _____ and reproduce in a particular (22) _____.

INTEGRATING AND APPLYING KEY CONCEPTS

1. Do you think that it would be possible for a viral infection to completely eliminate our species from the planet? What type of evolutionary force would this represent? How would the species genetically respond to this?

2. Why is it important for scientists to assign mathematical models, such as the Hardy-Weinberg equation, to natural phenomena? What are the benefits and limitations to doing this?

3. Develop a cladogram using your classmate's shoes. Think about how you could do this: shoe strings or not, high tops – low tops, ribbed bottom – smooth bottom, leather – not leather

18

LIFE'S ORIGIN AND EARLY EVOLUTION

INTRODUCTION

This chapter provides a quick tour through the history of life, starting with the conditions on a very young Earth over 4 billion years ago. As you progress through this chapter, take note of the major evolutionary events that occurred as life changed to adapt to changes in the environment of the planet.

FOCAL POINTS

- Figure 18.3 [p.284] provides an artist's depiction of early Earth.
- Figure 18.4 [p.285] is a diagram of an apparatus designed by Miller and Urey for testing organic chemical reactions with respect o the beginning of life on Earth.
- Figure 18.6 [p.286] is a proposed sequence for the evolution of cells.
- Figures 18.9 & 18.10 [pp.288-289] look at fossils of early life.
- Figure 18.14 [pp.292-293] provides an overview of the major milestones in the history of life.

INTERACTIVE EXERCISES

18.1 LOOKING FOR LIFE [p.283]
18.2 EARTH'S ORIGIN AND EARLY CONDITIONS [pp.284]
18.3 THE SOURCE OF LIFE'S BUILDING BLOCKS [p.285]
18.4 FROM POLYMERS TO CELLS [pp.286-287]

Selected Terms
Chile's Atacama Desert [p.283], *Bacillus infernus* [p.283], universe [p.284], element [p.284], Milky Way [p.284, asteroid [p.284], volcanoes [p.284], meteorite [p.284], iron oxidation (rusting) [p.284], Miller & Urey [p.285], methane, ammonia, & hydrogen [p.285], amino acids [p.285], geothermal energy [p.285], Huber & Wachter [p.285], electrode [p.285]

Boldfaced Terms
[p.283] astrobiology _____

_____

[p.284] big bang theory _____

_____

[p.285] hydrothermal vent _____

_____

[p.286] protocell _____

_____

[p.287] RNA world _____

_____

[p287] ribozyme _____

_____

Choice

For each of the following, choose the most appropriate stage of physical/chemical evolution of life.

    a.   conditions on the early Earth [p.284]
    b.   origin of the molecules of life [p.285]
    c.   origin of agents of metabolism [p.285]
    d.   origin of the plasma membrane [pp.286-287]
    e.   origin of self-replicating genetic systems. [p.287]

1. ____ Proto-cells that were little more than membrane sacs protecting information-storing templates and various metabolic agents of the environment.

2. ____ Simple systems of RNA, enzymes, and coenzymes have been created in the lab

3. ____ Four billion years ago the Earth was a thin-crusted inferno

4. ____ One hypothesis has simple organic molecules forming in outer space.

5. ____ Life may have started as an RNA world.

6. ____ Atmosphere was a mix of hydrogen, nitrogen, carbon monoxide and carbon dioxide.

7. ____ Simple metabolic pathways evolved at hydrothermal vents

8. ____ The Stanley Miller experiments demonstrate that some amino acids could form when atmospheric gases were stimulated by lightening.

9. ____ Metabolism first formed on clay-rich tidal flats

10. ____ Ribozymes in an RNA world

Labeling

Provide the missing term for each of the numbered items in the diagram below. [pp.285-287]

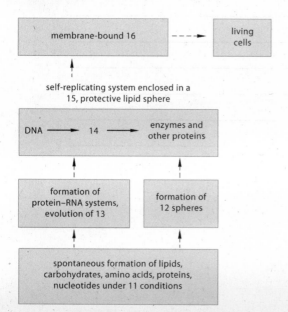

11. _____

12. _____

13. _____

14. _____

15. _____

16. _____

Short Answer

17. Briefly list the three possible ideas about the origin of organic molecules on ancient Earth. [p.285]

_____

_____

_____

_____

18. Define astrobiology and state a possible reason why it is important to scientists.

_____

_____

_____

_____

18.5 LIFE'S EARLY EVOLUTION [pp.288-289]
18.6 EVOLUTION OF ORGANELLES [pp.290-291]
18.7 TIME LINE FOR LIFE'S ORIGIN AND EVOLUTION [pp.292-293]

Selected Terms

fossil [p.288], microfossil [p.288], organic compounds [p.288], sediments [p.288], ozone [p.289], UV radiation [p.289], vertebrate animals [p.289], *Nitrosococcus ocean* [p.290], *Gemmata obscuriglobus* [p.290], rickettsias [p.290], *Amoeba proteus* [p.291], Kwang Jeon [p.291], glaucophytes [p.291], peptidoglycan [p.291], complex carbohydrates [p.292], lipids [p.292], molecular evolution [p.292], proteins [p.292], nucleic acids [p.292], atmospheric oxygen [p.292], archaeans [p.292], eukaryotic [p.292], photosynthesis [p.292], cyclic and non-cyclic pathways [p.292], cyanobacteria [p.292], aerobic respiration [p.292], endomembrane system [p.292], domains of life [p.292], anaerobic [p.293], symbiotic relationship [p.293], mitochondria [p.293], heterotrophic [p.293], chloroplast [p.293]

Boldfaced Terms

[p.288] stromatolites _____

_____

[p.289] biomarker _____

_____

[p.290] endosymbiosis _____

_____

Matching

Match each of the following with the correct statement.

1. ____ endosymbiosis [pp.290-291]

2. ____ biomarker [pp.288-289]

3. ____ eukaryotic cells [pp.290-291]

4. ____ stromatolites [pp.288-289]

5. ____ prokaryotic cells [pp.288-289]

a. an interaction of two organisms to the benefit of both

b. these are fossilized remains of early autotrophs

c. the earliest form of cells

d. a compound made only by a particular type of cell

e. these originated in response to increased oxygen levels

Short Answer

6. List three outcomes of the formation of an oxygen-rich atmosphere. [pp.288-289]

_____

_____

_____

_____

Choice

For each of the following, choose the most appropriate category from the list below. [pp.290-291]

    a. origin of the nucleus and ER

    b. origin of the chloroplast

    c. origin of the mitochondria

7. ____ Allowed for the protection of the genetic material.

8. ____ Bacterial origins of these organelles are supported by studies of Amoeba proteus.

9. ____ Cyanobacteria origins of these are supported by studies of the glaucophytes

10. ____ These made the production of ATP by aerobic pathways possible

11. ____ Evolved in response to an oxygen-rich atmosphere.

12. ____ Formed areas in which specific substances could be concentrated

Labeling

Match each number in the diagram below with the correct event in the history of life. [pp.292-293]

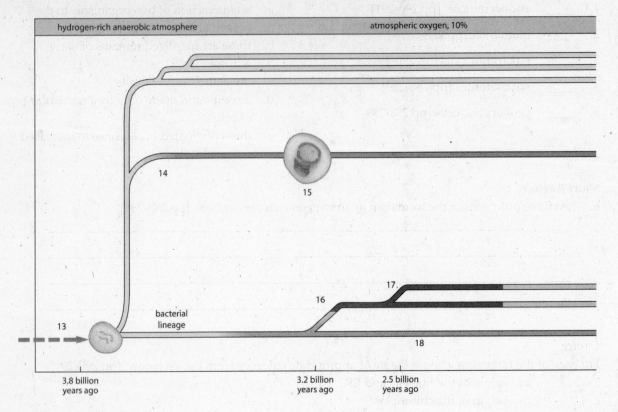

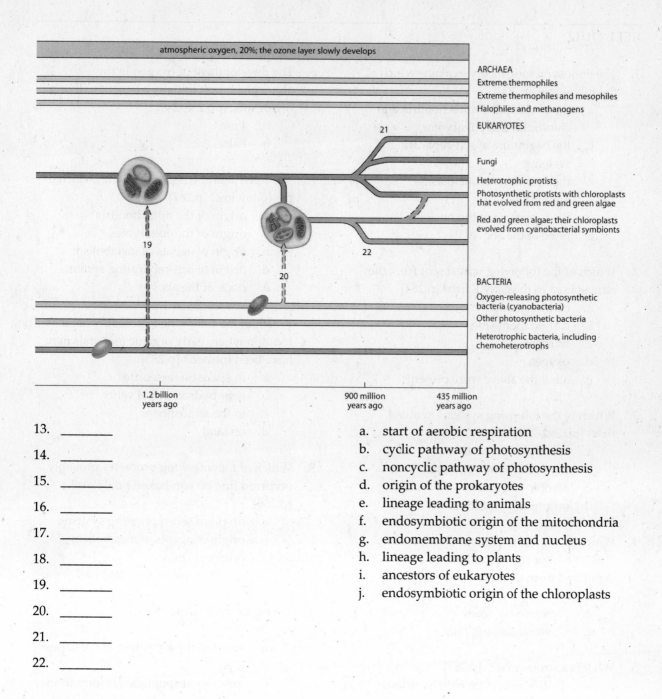

atmospheric oxygen, 20%; the ozone layer slowly develops

ARCHAEA
Extreme thermophiles
Extreme thermophiles and mesophiles
Halophiles and methanogens

EUKARYOTES

21

Fungi

Heterotrophic protists

Photosynthetic protists with chloroplasts
that evolved from red and green algae

Red and green algae; their chloroplasts
evolved from cyanobacterial symbionts

19

20

22

BACTERIA
Oxygen-releasing photosynthetic
bacteria (cyanobacteria)
Other photosynthetic bacteria

Heterotrophic bacteria, including
chemoheterotrophs

1.2 billion
years ago

900 million
years ago

435 million
years ago

13. _____

14. _____

15. _____

16. _____

17. _____

18. _____

19. _____

20. _____

21. _____

22. _____

a. start of aerobic respiration
b. cyclic pathway of photosynthesis
c. noncyclic pathway of photosynthesis
d. origin of the prokaryotes
e. lineage leading to animals
f. endosymbiotic origin of the mitochondria
g. endomembrane system and nucleus
h. lineage leading to plants
i. ancestors of eukaryotes
j. endosymbiotic origin of the chloroplasts

Life's Origin and Early Evolution **177**

SELF QUIZ

1. The endosymbiotic theory explains which of the following? [p.290]
    a. the evolution of mitochondria and chloroplasts in eukaryotes
    b. the beginning of self-replicating systems
    c. the formation of the plasma membrane
    d. the formation of the nucleus
    e. none of the above

2. Which of the following was absent from the atmosphere of the early Earth? [p.284]
    a. carbon dioxide
    b. carbon monoxide
    c. nitrogen
    d. oxygen
    e. all of the above were present

3. Which of the following systems evolved first? [pp.292-293]
    a. cyclic photosynthesis
    b. noncyclic photosynthesis
    c. aerobic respiration
    d. origins of the chloroplast

4. Which of the following was discovered in 1966 to harbor rod shaped bacteria and benefitted from that relationship? [p.291]
    a. *Ameoba proteus*
    b. *Nitrosococcus ocean*
    c. *Gemmata obscuriglobus*

5. What is a stromatolite? [p.288]
    a. the first organism with a nucleus
    b. a biomarker
    c. the first sexually-reproducing organism
    d. fossilized remains of ancient autotrophs

6. The development of oxygen in the atmosphere was initially a benefit to living organisms. [pp.288-289]
    a. True
    b. False

7. The RNA world is associated with which of the following? [p.287]
    a. origin of the mitochondria
    b. origin of the eukaryotes
    c. origin of agents of metabolism
    d. origin of self-replicating systems
    e. none of the above

8. Which of the following is probably not a location where early organic materials may have been formed? [p.285]
    a. in space on meteorites
    b. near hydrothermal vents
    c. in the atmosphere
    d. on land

9. Which of the following processes probably occurred first on sun-baked tidal flats? [p.286]
    a. origin of self-replicating systems
    b. origin of agents of metabolism
    c. origin of DNA
    d. origin of the nucleus and ER

10. An oxygen-rich atmosphere _____. [pp.288-289]
    a. allowed the formation of the ozone layer
    b. prevented spontaneous formation of life
    c. allowed the evolution of aerobic respiration
    d. all of the above

CHAPTER OBJECTIVES/REVIEW QUESTIONS

1. What is the significance of organisms such as *Bacillus infernus* [p.283]

2. What is the significance of the big bang? [p.284]

3. List the major components of the Earth's early atmosphere. [pp.284-285]

4. What is the significance of Stanley Miller's experiments? [p.285]

5. What is the role of tidal flats and hydrothermal vents in the evolution of life? [pp.285-286]

6. Explain how metabolic pathways may have initially evolved. [p.286]

7. What were the first plasma membranes made of? [pp.286-287]

8. Explain what is meant by an RNA world? [p.287]

9. What is a stromatolite? [p.288]

10. What effects did the accumulation of oxygen in the atmosphere have on early life? [pp.288-289]

11. Why did the nucleus and ER evolve in eukaryotes? [p.290]

12. What is the significance of *Amoeba proteus*? [p.291]

13. Explain the importance of the endosymbiotic theory. [pp.290-291]

14. Place important events in the evolution of life in chronological order. [pp.292-293]

CHAPTER SUMMARY

When Earth first formed about (1) _____ years ago, conditions were too harsh to support life. Over time, its crust cooled, (2) _____ formed, and (3) _____ compounds of the sort now found in living cells may have formed spontaneously or arrived in (4) _____.

Laboratory (5) _____ and advanced computer simulations support the hypothesis that forerunners of living cells arose through known physical and (6) _____ processes, such as the tendency of lipids to assemble into (7) _____ -like structures when mixed with water.

The first cells probably were (8) _____ prokaryotes. Some gave rise to (9) _____, others to archaeans and to ancestors of (10) _____ cells. (11) _____ bacteria started releasing oxygen into the (12) _____. Oxygen accumulated over time and became a global (13) _____ pressure.

The nucleus, ER, and other (14) _____ organelles are defining features of eukaryotic cells. Some may have evolved from infoldings of the (15) _____. Mitochondria and (16) _____ appear to be descended from bacterial cells that became modified after taking up permanent residence in host cells.

A (17) _____ for milestones in the history of life offers insight into shared connections among all (18) _____.

INTEGRATING AND APPLYING KEY CONCEPTS

1. Some scientists have suggested that it may be possible to transform Mars into an Earth-like planet using microorganisms from Earth. Mars has a thin carbon dioxide atmosphere, but no oxygen. Assuming that water exists in sufficient quantities, what organisms would you first introduce on the planet? Given the history of the Earth, how long would it take to establish an oxygen-rich atmosphere?

2. Jupiter's moon Europa and Saturn's moon Titan are believed to hold important clues about conditions of the early Earth. Why would it be important for us to go to these locations to understand the evolution of life?

19
VIRUSES, BACTERIA, AND ARCHAEANS

INTRODUCTION

"Out of sight, out of mind" is often the mindset of people when it comes to the microorganisms, and sometimes just "mere chemicals" with which we share this planet. This chapter begins with the viruses and the viroids; mere chemicals that have come to have important implications if the realm of human and plant health. We tend to overlook the tiny organisms that out number us by the billions. These tiny organisms that aid us digestion, decomposition, Vitamin K and pharmaceutical production, that help us remain healthy, that are vital in environmental cycling of elements are often thought of only when they make us sick. In reality, it is the small minority of these organism that are disease causing, or pathogenic. The vast majority of microbes are not only beneficial, but necessary, to us. This chapter describes some of the microorganisms in our environment and our interactions with them. Two domains of life, the Bacteria and the Archae, are discussed in this chapter. The metabolic diversity as well as the pathogenic potential of these prokaryotes is explained.

FOCAL POINTS

- Figure 19.2 [p.298] illustrates models of viral structures.
- Figure 19.3 [p.299] shows the bacteriophage replication pathways.
- Figure 19.4 [p.299] shows the replication cycle of HIV.
- Figure 19.6 [p.302] illustrates bacterial cell shapes.
- Table 19.1 [p.302] reviews the characteristics of the prokaryotic cell.
- Figure 19.7 [p.303] binary fission, the process by which prokaryotic cells divide.
- Figure 19.12 [p.306] shows a comparison of a two domain and a three domain tree of life.
- Table 19.3 [p. 315] lists the eight deadliest infectious diseases and their impact on human health.

INTERACTIVE EXERCISES

19.1 EVOLUTION OF A DISEASE [p.297]
19.2 VIRAL STRUCTURE AND FUNCTION [pp.298-299]
19.3 VIRAL EFFECTS ON HUMAN HEALTH [pp.300-301]
19.4 VIROIDS: TINY PLANT PATHOGENS [p.301]

Selected Terms
HIV [p.297], SIV [p.297], SFV [p.297], noncellular agent [p.298], host [p.298], herpesvirus [p.298], adenovirus [p.98], DNA virus [p.298], RNA virus {p.298], viral coat [p.299], viral protein [p.299], nonenveloped virus [p.300], HPV [p.300], enveloped virus [p.300], latent state [p.300], chicken pox [p.300], influenza [p.300], mumps [p.300], measles [p.300], German measles [p.300], West Nile virus [p.300], SARS [p.300], coronavirus [p.300], H1N5 & H1N1 [p.300], Theodor Diener [p.301], plasmodesmata [p.301], phloem [p.301]

Boldfaced Terms
[p.298] virus_____

_____

[p.298] bacteriophages _____

_____

[p.298] lytic pathway _____

_____

[p.298] lysogenic pathway _____

_____

[p.300] pathogen _____

_____

[p.300] emerging diseases _____

_____

[p.300] vector _____

_____

[p.300] endemic disease _____

_____

[p.300] epidemic _____

_____

[p.300] pandemic _____

_____

[p.301] viroids _____

_____

Fill-in-the-Blanks

Today, we define a virus as a noncellular (1) _____ [p.298] particle that consists of DNA

or (2) _____ [p.298] enclosed within a (3) _____ [p.298] coat. The coat protects the genetic

material as the virus journeys to a new (4) _____ [p.298] cell. (5) _____ [p.298] have a

complex coat, (6) a rodlike _____ [p.298], with the DNA encased in a protein (7) _____

[p.298]. Viral replication begins when a virus attaches to (8) _____ on the host's plasma

membrane.

Labeling

In the diagram below, choose the most appropriate label for each number. [p.299]

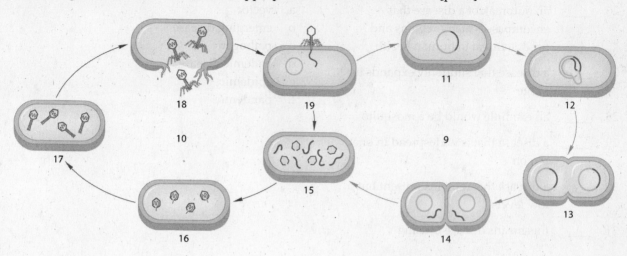

9. ____

10. ____

11. ____

12. ____

13. ____

14. ____

15. ____

16. ____

17. ____

18. ____

19. ____

a. Viral DNA integrates into the host chromosome

b. After cell division, each daughter cell has recombinant DNA

c. The viral DNA directs the host cell to make viral proteins and replicate viral DNA

d. Lysis of the host cell

e. Virus injects genetic material into a suitable host cell.

f. The lysogenic pathway

g. Tail fibers and other parts are added to the coats.

h. Viral proteins are assembled into coats.

i. The viral DNA is excised from the bacterial chromosome.

j. The lytic pathway

l. The bacterial chromosome with the integrated viral DNA is replicated.

Choice

For each of the following statements, choose one of the following answers.

a. viroids [p.301]

b. viruses [pp.298-299]

c. none of the above

20. ____ Circles of RNA with no protein coat

21. ____ Is classified as a living organism

22. ____ Infects citrus, apples, coconuts, avocados, and chrysanthemums

23. ____ Infectious DNA or RNA contained within a protein coat

24. ____ Have a lytic and lysogenic pathway

25. ____ Its RNA does not encode for any protein

Viruses, Bacteria, and Archaeans **183**

Matching

Match the terms below with the following. [pp.300-301]

26. ____ an outbreak of a disease that encompasses many regions and poses a threat to human health

27. ____ a disease that suddenly expands its range

28. ____ an example would be a mosquito

29. ____ a disease that is widespread in one region

30. ____ a disease that remains present but a low levels

31. ____ this means disease causing

a. vector
b. emerging disease
c. pathogen
d. endemic disease
e. epidemic
f. pandemic

19.5 BACTERIAL STRUCTURE AND FUNCTION [p.302]
19.6 BACTERIAL REPRODUCTION AND GENE EXCHANGE [p.303]

Selected Terms

coccus [p.302], bacillus [p.302], spirillum [p.302], cell wall [p.302], capsule [p.302], flagella [p.302], microtubules [p.302], photoautotrophs [p.302], photosynthesis [p.302], chemoautotrophic [p.302], phoyoheterotrophs [p.302], chemoheterotrophs [p.302], asexual reproduction [p.303], antibiotic resistance [p.303]

Boldfaced Terms

[p.302] bacterial chromosome _____

_____

[p.302] nucleoid _____

_____

[p.302] pili _____

_____

[p.303] binary fission _____

_____

[p.303] horizontal gene transfer _____

_____

[p.303] conjugation _____

_____

[p.303] plasmid _____

_____

Labeling

On the below diagram of a prokaryotic cell, label each of the indicated components. Then, match each structure to its correct function. [p.302]

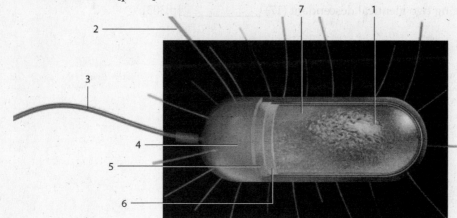

1. _____ (   )

2. _____ (   )

3. _____ (   )

4. _____ (   )

5. _____ (   )

6. _____ (   )

7. _____ (   )

a. this protects the cell from phagocytes
b. the genetic material of a bacterium
c. this is involved in bacterial conjugation.
d. this helps the cell resist rupturing as internal pressure increases
e. this contains the ribosomes, the site of protein synthesis.
f. this rotates like a propeller for locomotion.
g. the membrane of the cell.

Matching

For each of the following, choose the type of metabolic diversity that is best indicated by the statement. Some answers may be used more than once. [p.302]

8. ____ Use another organism as its energy and carbon source.

9. ____ Use the sun's energy, but break down organic compounds for carbon

10. ____ Use photosynthesis and carbon dioxide.

11. ____ Oxidize inorganic compounds such as iron and sulfur.

12. ____ These can either be saprophytic or parasitic

a. photoheterotroph
b. chemoheterotroph
c. photoautotrophs
d. chemoautotrophs

Fill-in-the-Blanks

Most often, a bacterium cell reproduces by a division mechanism called (13) _____ [p.303]. Briefly, a parent cell replicates its (14) _____ [p.303] molecule, and the replica docks at the plasma membrane next to the parent molecule. A new (15) _____ [p.303] cell wall material get added between them. The membrane and (16) _____ [p.303] extend into the cell's midsection and divides producing two identical descendant (17) _____ [p.303].

Draw
Draw the three typical shapes of bacterial cells. [p. 302]
18. coccus

19. bacillus

20. spirillum

19.7 BACTERIAL DIVERSITY [pp.304-305]
19.8 THE ARCHAEANS [pp.306-307]

Selected Terms
thermal pools & hydrothermal vents [p.304], PCR [p.304], myxobacteria [p.304], biotechnology [p.304], spore [p.304], *E.coli* 0157:H7 [p.304], rickettsias [p.304], *Thermus aquaticus* [p.304], *Spirulina* [p. 304], *Rhizobium* [p.304], *Salmonella* [p.304], *Campylobacter* [p.304], *Escherichia coli* [p.304], *Vibrio cholerae* [p.304], *Heliobacter pylori* [p.304], *Thiomargarita namibiensis* [p.304], *Lactobacillus* [p.304], *L. acidophilus* [p. 304], *Clostridium* [p.305], *Bacillus* [p.305], *Bacillus anthracis* [p.305], *Clostridium tetani* [p.305], *C. botulinium* [p.305], pathogenic [p.305], syphilis [p.305], Lyme disease [p.305], *Streptococcus* [p.305], *Mycobacterium tuberculosis* [p.305], *Staphylococcus* [p.305], "flesh-eating" [p.305], Chlamydia [p.305], methicillin [p.305], Botox [p.305], clade [p.306], Carl Woese [p.306], histone [p.306]

Boldfaced Terms
[p.304] cyanobacteria _____

_____

[p.304] nitrogen fixation _____

_____

[p.304] proteobacteria _____

_____

[p.304] gram staining _____

_____

[p.304] normal flora _____

_____

[p.305] endospores _____

_____

[p.305] spirochetes _____

_____

[p.305] chlamydias _____

_____

[p.306] methanogens _____

_____

[p.306] extreme halophiles_____

_____

[p.306] extreme thermophiles _____

_____

Matching

Choose the correct term for each of the following. Some answers may be used more than once. [pp.308-309]

1. ____ The most diverse bacterial lineage, also includes nitrogen fixers.

2. ____ Some of these produce endospores to survive hostile conditions.

3. ____ These are usually parasites living inside eukaryotic cells.

4. ____ These have a spring-like appearance.

5. ____ Photoautotrophs that cycle key nutrients in ecosystems.

6. ____ Some members form spores on a stalk when conditions are unfavorable.

7. ____ The walls of this group can be stained purple using a specific dye.

a. proteobacteria
b. chlamydias
c. spirochetes
d. cyanobacteria
e. gram staining bacteria

Complete the Table

For each of the following species, provide the general classification of bacteria to which the species belongs. Choose from: cyanobacteria, gram staining bacteria, proteobacteria, chlamydia, spirochetes and archaea [p. 304-307]

Species	Classification
Spirulina	8.
E. coli	9.
Rhizobium	10.
Lactobacillus acidophilus	11.
Methanogens	12.
Salmonella	13.
Thiomargarita namibiensis	14.
Campylobacter	15.
Vibrio cholerea	16.
Heliobacter pylori	17.
Syphilis	18.
Bacillus anthracis	18.
Mycobacterium tuberculosis	19.
Clostridium botulinum	20.

Choice

For each of the following, choose which group of Archaea bacteria is best described by the statement. Some questions may use more than answer. [pp.306-307]

    a.   extreme halophiles
    b.   methanogens
    c.   extreme thermophiles

21. _____ these live in or near deep-sea hydrothermal vents.

22. _____ are strict anaerobes.

23. _____ these are the source of DNA polymerases in PCR reactions.

24. _____ these live in hot springs

25. _____ live in salt deposits

SELF QUIZ

1. Which viral pathway has the genetic material from the virus integrating into the bacterial chromosome? [p.298]
    a.   lysogenic pathway
    b.   lytic pathway
    c.   biogenic pathway
    d.   auto-assembling pathway

2. In the _____, viral DNA becomes integrated into the host chromosme and a latent period precedes formation of new viruses. [p.298]
    a.   lysogenic pathway
    b.   lytic pathway

3. In the _____, viral genes enter a host and immediately direct it to make new viral particles. [p.298]
    a.   lysogenic pathway
    b.   lytic pathway

4. Which of the following is classified as a living organism? [pp.298-301]
    a.   viroids
    b.   viruses
    c.   none of the above are considered living organisms.

5. All of the following may be found in a virus, except [pp.298-299]:
   a. protein coat
   b. genetic material
   c. ribosomes
   d. envelope
   e. tail fibers

6. A disease that occurs in several countries of the world at the same time is called a _____. [p.300]
   a. sporadic outbreak
   b. epidemic
   c. endemic outbreak
   d. pandemic

7. Which of the following gets their energy from photosynthesis and their carbon from carbon dioxide? [p.302]
   a. chemoheterotrophs
   b. chemoautotrophs
   c. photoheterotrophs
   d. photoautotrophs

8. Which of the following would not be present in a bacterial cell? [p.302]
   a. ribosomes
   b. polysaccharide coating
   c. membrane-bound organelles
   d. cell wall

9. The most diverse group of bacteria on the planet are the _____. [pp.304-305]
   a. cyanobacteria
   b. chlamydias
   c. proteobacteria
   d. gram staining bacteria
   e. archaea

10. Methanogens, extreme halophiles and extreme thermophiles are all examples of which of the following bacterial groups? [pp.306-307]
   a. proteobacteria
   b. cyanobacteria
   c. spirochetes
   d. gram-positive
   e. archaea

CHAPTER OBJECTIVES/ REVIEW QUESTIONS

1. Know the major components of a typical virus. [pp.298-299]

2. Be able to label the key steps of the lysogenic and lytic pathways and know the steps in animal virus replication. [p.298]

3. Understand the difference between and infection and a disease, and between epidemic and pandemic. [p.300]

4. Define a viroid and recognize its possible evolutionary source. [p.301]

5. Understand the physical structure of a typical bacteria. [p.302]

6. Know the major forms of metabolic diversity for the bacteria, and the energy and carbon source for each. [p.302]

7. Define antibiotic resistance. [p.303]

8. Understand the processes of binary fission and lateral gene transfer as they relate to bacterial genetics. [p.303]

9. Recognize the major groups of bacteria. [pp.304-305]

10. Understand the relationship of the Archae to the bacteria and eukaryotes. [p.306-307]

11. Understand the environments in which the archea are found. [pp.306-307]

CHAPTER SUMMARY

We divide all prokaryotic cells into the domains (1) _____ and (2) _____. These single-celled organisms are structurally simple; they do not have a (3) _____ or the diverse cytoplasmic (4) _____ found in most eukaryotic cells. Collectively, they show great (5) _____ diversity.

Bacteria and archaeans alone reproduce by prokaryotic (6) _____. Exchanges of chromosomal DNA and (7) _____ DNA often occur within and among species.

Bacteria are the most studied (8) _____ species. They are the most abundant and widely distributed organisms. Archaeans, discovered more recently, are less well known. Many are adapted for (9) _____ environments.

Viruses are noncellular particles that cannot (10) _____ themselves without taking over the metabolic machinery of a host cell. A (11) _____ coat encloses their DNA or RNA and a few enzymes. Some viruses become enveloped in membrane when they bud from host cells.

_____ (12) are even simpler than viruses. They are short sequences of infectious (13) _____.

INTEGRATING AND APPLYING KEY CONCEPTS

1. Imagine that you are working on a NASA project in exobiology, which is the study of potential life outside of the Earth. You send a probe to Europa, one of Jupiter's moons, and discover a new species of bacteria. Europa is too far away from the Sun to get any solar input, so what type of metabolism could it have? What group of Earth bacteria do you think it would be most like? Defend your answer.

2. Where might the next emerging pathogen be?

3. Antibiotic resistance is becoming a public health emergency in the United States and abroad. Do your own research to learn why the following activities are beneficial (and of absolute necessity) in the fight against antibiotic resistance.

    a. Do not take antibiotics for the cold virus.
    b. Take the entire antibiotic dose in your prescription.
    c. Do not share antibiotics.
    d. Avoid eating meat that has been fed antibiotics.

20

THE PROTISTS

INTRODUCTION

This chapter begins our exploration of the eukaryotic organisms. The Protists are a group of organisms that, although often unicellular, has a tremendous impact on human life. Many of these simple organisms are photosynthetic, producing oxygen and removing carbon dioxide from the atmosphere, possibly helping to reduce global warming. Members of this group of organisms also are often the base of aquatic food chains. Many industrial, food and cosmetic products are made from these organisms. However, not all protists are so beneficial to us. The more notable organisms in this group are responsible for some of the world's most deadly and prevalent diseases. Malaria, for example, infects over 3 million people worldwide and is caused by protists of the *Plasmodium* genus. This chapter surveys this large and biologically important group of organisms and prepares the student for the study of the more complex eukaryotic species in the following chapters.

FOCAL POINTS

- Figure 20.2 [p.312] illustrates the diversity of the protists.
- Figure 20.4 [p.313] illustrates the protist, *Euglena*, and its organelles.
- Figure 20.7 [p.315b] illustrates the protist, *Paramecium,* and its organelles.
- Figure 20.8 [p.316] depicts the life cycle of the protist that causes malaria, *Plasmodium*.
- Figure 20.11 [p.318] animates the life cycle of the red algae, *Porphyra*. This is a precursor to the alternation of generations you will see in the plant life cycles in the next chapter.
- Figure 20.15 [p.321] illustrates the life cycle of *Dictyostelium discoideum*.

INTERACTIVE EXERCISES

20.1 HARMFUL ALGAL BLOOMS [p.311]
20.2 A COLLECTION OF LINAGES [p.312]
20.3 FLAGELLATED PROTOZOANS [p.313]

Selected Terms
Multicellular autotroph [p.311], "red tide"[p.311], *Karenia brevis* [p.311], dinoflagellate [p.311], brevetoxin [p.311], silica shelled [p.312], monophyletic [p.312], chloroplast [p.312], colonial [p.312], diplomonads & parabasalids [p.313], cysts [p.313], tsetse fly [p.313], Chagas disease [p.313], Desert sandfly [p.313], vector [p.313], leishmaniasis [p.313]eyespot [p.313] *primary endosymbiosis* [p.312], *secondary endosymbiosis* [p.312], *Giradia lamblia* [p.313], *Trichomonas vaginalis* [p.313], *Trypanosoma brucei* [p.313]

Boldfaced Terms
[p.311] protists _____

_____

[p.311] algal bloom _____

_____

[p.311] toxin _____

_____

[p.313] flagellated protozoans _____

[p.313] pellicle _____

_____

[p.313] trypanosomes _____

_____

[p.313] euglenoids _____

_____

[p.313] contractile vacuole _____

_____

Fill in the Blank [p.312]

Protists are a collection of (1) _____, rather than a clade, or monophyletic (2)

_____.

In fact, some protists are more closely related to plants, fungi, or (3) _____ than to other

protists. Most protest lineages include only (4) _____ species. However, some lineages include (5)

_____ cells and true multicellularity. (6) _____ organisms consist of cells that cannot

survive and reproduce on their own. Protists have diverse life styles, but most reproduce both (7)

_____ and sexually.

Diagram

Match the letters on the diagram on the next page of the *Euglena* with the following structures. [p.313]

8. ____ long flagellum

9. ____ chloroplast

10. ____ Golgi body

11. ____ eyespot

12. ____ pellicle

13. ____ endoplasmic reticulum (ER)

14. ____ mitochondrion

15. ____ nucleus

16. ____ Since the *Euglena* has a chloroplast, would it be a heterotroph or an autotroph?

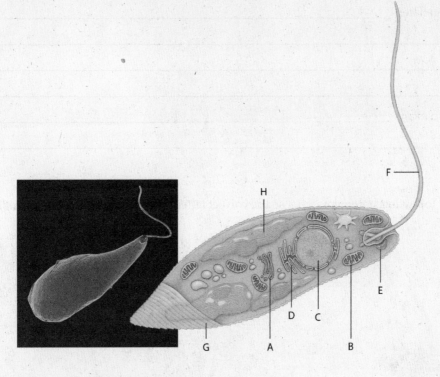

20.4 MINERAL-SHELLED PROTOZOANS [p.314]
20.5 THE ALVEOLATES [p.314]
20.6 MALARIA AND THE NIGHT FEEDING MOSQUITOS [p.316]

Selected Terms

marine protists [p.314], sieve-like shells [p.314], calcium carbonate [p.314], chalk & limestone [p.314], silica shell [p.314], planktonic [p.314], "alveolus" [p.314], "whirling flagellate" [p.314], cellulose [p.314], reef-building coral [p.314], bioluminescent [p.314], *Didinium* [p.315], *Paramecium* [p.315], gullet [p.315], endocytosis [p.315], macronucleus [p.315] micronuclei [p.315], free-living [p.315], diarrhea [p.315], parasitic p.315], apical [p.315], sporozoans [p.315], *Plasmodium* [p.316], merozoites [p.316], gametocytes [p.316], sporozoites [p.316], sickle-cell anemia [p.316], antimalarial [p.316]

Boldfaced Terms

[p.314] foraminiferans _____

_____

[p.314] plankton _____

_____

[p.314] radiolarians _____

_____

[p.314] alveolates _____

_____

[p.314] dinoflagellates _____

_____

[p.315] ciliates _____

_____

[p.315] apicomplexans _____

_____

Labeling
For each of the following statements, choose the correct letter on the diagram of *Plasmodium's* lifecycle.
[p.316]

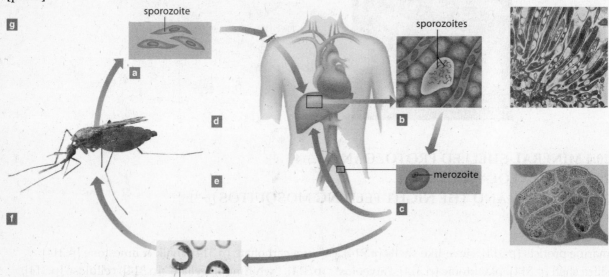

1. ____ *Plasmodium* zygotes develop in the gut of female mosquito

2. ____ Sporozoites asexually reproduce in the liver cells.

3. ____ Some of the merozoites enter the liver and cause more malaria episodes

4. ____ Mosquito bites a human and the bloodstream carries the sporozoites to the liver.

5. ____ Some merozoites develop into male and female gametophytes and enter the bloodstream.

6. ____ Female mosquito sucks blood from an infected human.

7. ____ The merozoites invade red blood cells and reproduce asexually.

Choice

For each of the following, choose the appropriate group of protists that is described by the statement.
[pp.314-316]

    a. ciliated protozoans
    b. foraminiferans
    c. apicomplexans
    d. radiolarians
    e. dinoflagellates

8. _____ Body plan includes cellulose plates beneath the plasma membrane and two flagella.

9. _____ Species that cause malaria belong to this group.

10. _____ This group has silica shells.

11. _____ Some members of this group are bioluminescent.

12. _____ Often contain two different sized nuclei, and can reproduce both sexually and asexually.

13. _____ May cause red tide in coastal marine environment.

14. _____ An example is *Karenia brevis*.

15. _____ Possess a unique microtubular device to penetrate a host cell.

16. _____ Are a component of plankton in the ocean.

17. _____ Some of these are photosynthetic and are associated with coral reefs.

18. _____ Have a shell of calcium carbonate and live on the ocean floor.

19. _____ These are sometimes referred to as sporozoans.

20. _____ Two examples are *Didinium* and *Paramecium*.

20.7 STRAMENOPHILES [p.317]
20.8 RED ALGAE AND GREEN ALGAE [p.318-319]

Selected Terms

"straw-haired" [p.317], filamentous [p.317], *Phytophthora* [p.317], "plant destroyer" [p.317], famine [p.317], fucoxanthin [p.317], giant kelp [p.317], alginic acid [p.317], coralline algae [p.318], chloroplyll a & b [p.318], phycobilins [p.318], agar [p.318], polysaccharide [p.318], gelatin[p.318], *Porphyra [p.318]*, carrageenan [p.318], *Chlamydomonas* [p.318], *Volvox* [p.318], sea lettuce [p.318], Melvin Calvin [p.319], *Chlorella* [p.319], starch [p.319], *Chara* [p.319]

Boldfaced Terms

[p.317] water molds _____

_____

[p.317] diatoms _____

_____

[p.317] brown algae _____

_____

[p.318] red algae _____

_____

[p.318] alternation of generation _____

[p.318] gametophyte _____

_____

[p.318] sporophyte _____

_____

[p.318] green algae _____

_____

Choice

For each of the following, choose the correct group of protists that is being described by the statement. Some statements may have more than one answer. [pp.318-319]

    a. brown algae
    b. water molds
    c. green algae
    d. red algae

1. ____ Extracts of these are used commercially in foods and cosmetics.

2. ____ The group that is the most similar to plants.

3. ____ Heterotrophs that absorb nutrients that form during their life cycle.

4. ____ The diatoms are a member of this group.

5. ____ Chloroplasts in these organisms may have evolved from cyanobacteria.

6. ____ The deepest living of the marine algae.

7. ____ Chalk and limestone deposits are the remains of these organisms.

8. ____ Many of this group act as decomposers in aquatic habitats.

9. ____ Examples include *Chlamydomonas* and *Volvox*.

10. ____ The group that is the most similar to the fungi.

Complete the Table

For each of the species listed below, indicate the group of protists to which it belongs. Choose from green algae, red algae, brown algae, water molds. [pp.317-319]

Species	Group
Volvox	11.
Porphyra	12.
Chlamydomonas	13.
Chara	14.
Phytopthora	15.

20.9 AMOEBOZOANS [p.320-321]

Selected Terms
pseudopods [p.320], Amoeba proteus [p.320], amebic dysentery [p.320], slime molds [p.320], "social amoeba" [p.320], plasmodium [p.320], Dictyostelium discoideum [p.320], cyclic AMP [p.321]

Boldfaced Terms
[p.320] amoebozoans _____

_____

[p.320] amoebas _____

_____

[p.320] plasmodial slime mold _____

_____

[p.320] cellular slime molds _____

_____

Complete the Table
For each of the species listed below, indicate the group of protists to which it belongs. Choose from amoeboid protozoans, dinoflagellates, euglenoids, apicomplexans, flagellated protozoans, ciliated protozoans, brown algae, red algae, green algae. List a disease that is causes, if any. If the organism is not pathogenic, just put NP in the last column. [pp.313-321]

Species	Group	Disease
Paramecium	1.	11.
Plasmodium	2.	12.
Chlamydomonas	3.	13.
Giardia lamblia	4.	14.
Porphyra	5.	15.
Trypanosoma brucei	6.	16.
Volvox	7.	17.
Trichomonas vaginalis	8.	18.
Phytophthora	9.	19.
	10.	20.

SELF QUIZ

1. Red tide is caused by _____. [p.311]
   a. apicomplexans
   b. amoeboid protozoans
   c. dinoflagellates
   d. green algae
   e. ciliated protozoans

2. Which of the following is used agar and wraps for sushi [p.318]
   a. green algae
   b. euglenoids
   c. brown algae
   d. red algae
   e. diatoms

3. The malaria-causing parasite *Plasmodium* belongs to which of the following? [p. 316]
   a. flagellated protozoans
   b. ciliated protozoans
   c. dinoflagellates
   d. oomycota
   e. apicomplexans

4. Calcium carbonate and silica deposits at the bottom of ancient oceans are the remnants of _____. [p.314]
   a. eugleoids
   b. brown algae
   c. water molds
   d. alveolates
   e. mineral-shelled protozoans

5. Dinoflagellates, ciliates, and apicomplexans belong to the ____ group of protists. [p.314]
   a. mineral-shelled protozoans
   b. flagellated protozoans
   c. ciliates
   d. alveolates
   e. stramenopiles

6. A protist is a eukaryotic organism that is not _____. [p.311]
   a. a fungus
   b. a plant
   c. an animal
   d. a bacterium
   e. any of the above

7. Euglenoids have _____. [p.313]
   a. flagellum
   b. many mitochondria
   c. chloroplasts
   d. all of the above
   e. none of the above

8. Plasmodium cannot live _____. [p.316]
   a. at low temperatures
   b. in tropical countries
   c. in mammals
   d. in mosquitoes

9. These organisms have a two part silica shell. [p.317]
   a. water molds
   b. foraminiferans
   c. diatoms
   d. radiolarians
   e. dinoflagellates

10. Which of the following have a cell wall made of cellulose? [p.319]
    a. red and brown algae
    b. red algae and diatoms
    c. red algae and dinoflagellates
    d. red algae and the apicomplexans
    e. red and green algae

CHAPTER OBJECTIVES/ REVIEW QUESTIONS

1. List the general characteristics of a protist. [p.312]

2. Distinguish between flagellated protozoans, euglenoids, amoeboid protozoans, and ciliated protozoans. [pp.313-321]

3. Giardia is the protist responsible for _____. [p.313]

4. Understand the life cycle of *Plasmodium*. [p.316]

5. Understand the relationship between the dinoflagellates and red tide. [p.311 & 314]

6. Know the protist source of calcium carbonate and silica deposits. [p.314]

7. Understand the economic significance of the brown and red algae [pp.317-318]

8. Recognize the similarities of the green algae to the plants. [p.319]

10. Understand the life cycle of the two types of slime molds. [pp.320-321]

CHAPTER SUMMARY

Protists include many lineages of single-celled (1) _____ organisms and their closest (2) _____ relatives. Gene sequencing and other methods are clarifying how protist lineages are related to one another and to plants, (3) _____, and (4) _____.

Two of the earliest lineages of eukaryotic cells are known informally as flagellated (5) _____. The foraminiferans and (6) _____ are another ancient lineage.

Ciliated protozoans, (7) _____, and apicomplexans are single-celled (8) _____, predators, and parasites with a unique layer of tiny (9) _____ under their plasma membrane.

Water molds, diatoms, and brown algae are (10) _____. Red algae has chlorophyll (11) _____ and green algae has chlorophylls (12) _____ and (13) _____. One lineage of multicelled green algae, the charophyte algae, is the closest living relatives of land (14) _____.

A great variety of amoeboid species formally classified as separate lineages are now united as the (15) _____. They are the closest living protistan relatives of (16) _____ and animals.

INTEGRATING AND APPLYING KEY CONCEPTS

1. Some scientists do not believe that the protists should possess their own kingdom. Assume that you have been asked to reassign the major group of protists outlined in this chapter into the plant, animal and fungi kingdoms. Where would you place each group and why? What groups represent a special problem in reclassification? Why?

2. You read in the text that Malaria kills a child every 30 seconds. Because this disease is so prevalent and so devastating, many authorities are beginning to suggest the reintroduction of DDT use into malaria endemic regions. DDT was banned in the United States because of its environmental toxicity. Do you feel that DDT should be used to control the mosquitoes that carry malaria? Why or why not?

3. Prepare an evolutionary tree, similar to the one shown in Figure 20.2 of your text, that shows the relationship and organization of the protists in this chapter. Use only the following Terms in your tree: flagellates, brown algae, diatoms, *Garadia lamblia*, *Trypanosoma brucei*, *Leishmania donovani*, alveolates, shelled amoebas, foraminiferans, radiolarans, *Paramecium*, ciliates, dinoflagellates, *Karenia brevis*, apicomplexans, *Plasmodium*, water molds, green algae, *Chlamydomonas*, red algae, amoebozoans, *Amoeba proteus*, slime molds, *Physarum*, *Trichomonas vaginalis*.

21

PLANT EVOLUTION

INTRODUCTION

Plants are the reason we, as a species, survive. Almost all life on Earth, and certainly *Homo sapiens*, is dependent upon plants for oxygen and food. Plants have the unique and vital capacity to produce sugar from carbon found in the atmosphere and water in the environment. Thankfully, in the process they generate oxygen for us to breathe. Plants are also a tremendously successful and diverse group of organisms which has been on Earth for over 475 million years. During this time, the atmosphere and climate of the Earth have undergone drastic changes, and the plants have responded to these changes. This chapter discusses the four major groups of plants and their evolution. The life cycle of these four groups are described in detail, as are the structures that are vital to the life cycles of these plants.

FOCAL POINTS

- Figure 21.2 [p.326] is a diagram of a vascular plant leaf in cross section.
- Figure 21.3 [p. 326] introduces the generalized life cycle for plants and identifies the difference in sporophytes and gametophytes.
- Figure 21.4 [p. 327] compares the major plant groups.
- Figure 21.5 [p. 328] illustrates the life cycle of a moss.
- Figure 21.11 [p. 331] illustrates the life cycle of a fern.
- Figure 21.13 [p.332] depicts the time line for events in plant evolution.
- Figure 21.19 [p. 335] shows the life cycle of a conifer – the ponderosa pine.
- Figure 21.20 [p.336] illustrates a cherry flower.
- Figure 21.21[p. 336] diagrams the life cycle of an angiosperm – a lily.

INTERACTIVE EXERCISES

21.1 SPEAKING FOR THE TREES [p.325]
21.2 ADAPTATIVE TRENDS AMONG PLANTS [pp.326-327]

Selected Terms: deforestation [p.325], Wangari Maahai [p.325], Green Belt Movement [p.325], gametophyte [p.326], sporophyte [p.326].spore [p.326], gamete [p.326], meiosis [p.326], fertilization [p.326], haploid [p.326], diploid [p.326]

Boldfaced Terms
[p.326] embryophytes _____

_____

[p.326] cuticle _____

_____

[p.326] stomata _____

_____

[p.326] vascular tissues_____

_____

[p.326] xylem _____

_____

[p.326] phloem_____

_____

[p.326] lignin _____

_____

[p.327] pollen grain _____

_____

[p.327] seed _____

_____

Matching

Match each term with its description. [pp.326-327]

1. ____ stomata
2. ____ lignin
3. ____ phloem
4. ____ xylem
5. ____ cuticle
6. ____ gametophyte
7. ____ sporophytes
8. ____ spores
9. ____ pollen grains
10. ____ seed

a. A waxy coat that conserves water inside shoots on hot, dry days
b. A haploid body that produces haploid gametes
c. Vascualr tissue that distributes water
d. Develops from a diploid zygote into multicelled vegetative bodies that produce spores by way of meiosis
e. Resting structures, typically walled, that help a new generation wait out harsh environmental conditions; germinate and develop into gametophytes
f. Embryo sporophyte packaged in nutritive tissues and a touch, waterproof coat
g. An opening in the cuticle used in to maintain the balance of CO_2 and water intake.
h. Vascular tissue that distributes sugars
i. An organic compound in the cell walls of many plants that lends support
j. Formed from small microspores in gymnosperms and angiosperms; a protective wall encloses a few cells that will develop into a mature, sperm-bearing male gametophyte

21.3 THE BRYOPHYTES [pp.328-329]

Selected Terms
liverwort [p.328], hornwort [p.328], gametangium [p.328], sporangium, [p.328], bisexual [p.328], asexual [p.329], sexual reproduction [p.329], fragmentation [p.329], zygote [p.329], *Polytrichum* [p.329], *Sphagum* [p. 329], *Marchantia* [p.329]

Boldfaced Terms
[p.328] bryophyte _____

_____

[p.329] mosses _____

_____

[p.329] rhizoids _____

_____

[p.329] peat _____

_____

True/False
If the statement is true, write a "T" in the blank. If the statement is false, correct it by writing the correct word(s) for the underlined word(s) in the answer blank. [pp.328-329]

1. _____ Mosses , liverworts and hornworts are all <u>nonvascular</u> bryophytes.

2. _____ Bryophytes <u>have</u> true leaves, stems, and roots.

3. _____ Most bryophytes have <u>rhizomes</u> that absorb and attach gametophytes to the soil.

4. _____ In mosses, gametes form by mitosis at <u>gametophyte</u> shoot tips.

5. _____ The moss <u>gametophyte</u> is a stalk and a jacketed structure in which haploid spores develop.

6. _____ When organic remains of peat mosses accumulate over time, they become compressed into exceedingly moist mats called peat bogs.

7. _____ <u>Algae</u> are the simplest land plants and sometimes they are used as a fuel source.

8. _____ Hornworts and liverworts are considered a <u>bryophyte</u>.

9. _____ Gametes of bryophytes from in a structure called a <u>sporangium.</u>

10. _____ The gametophyte generation of bryophytes is <u>haploid.</u>

Dichotomous Choice

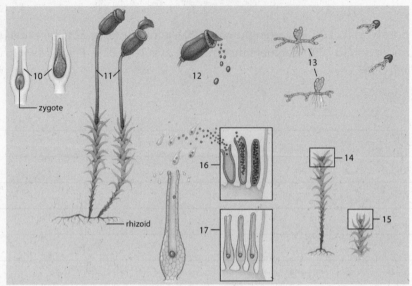

For each statement below, refer to the corresponding numbers on these sketches of the moss life cycle. You might also refer to Figure 21.5 in the textbook for addition information. Circle the Terms in parentheses that make each statement correct. [p.328]

10. The zygote shown is (haploid/diploid) and represents the future (gametophyte/sporophyte) plant.

11. The moss plants shown above the line represent (gametophyte/sporophyte) plants and their tissues are (haploid/diploid); the moss plants shown below the dotted line represent (gametophyte/sporophyte) plants and their tissues are (haploid/diploid).

12. (Mitosis/Meiosis) is occurring in the capsule-shaped structure and the cells being released from it are (haploid gametes/haploid spores).

13. The two structures shown represent young (gametophyte/sporophyte) plants and their tissues are (haploid/diploid).

14. The moss plant shown here represents a (gametophyte/sporophyte) plant.

15. The moss plant shown here represents a (gametophyte/sporophyte) plant.

16. The cells being released from this structure are (haploid sperms/diploid spores).

17. The cells at the base of the flask-shaped structures are (haploid spores/haploid eggs).

21.4 SEEDLESS VASCULAR PLANTS [pp.330-331]

Selected Terms

Lycopodium [p.330], *Equisetum* [p.330], "scouring rush" [p.330] flagellated sperm [p.330], club mosses, horsetails, & ferns [p.330], strobili [p.330], capsule [p.330], frond [p.330], fiddlehead [p.331]

Boldfaced Terms

[p.330] seedless vascular plant _____

_____

[p.330] rhizome _____

_____

[p.330] sori _____

_____

[p.330] epiphytes _____

_____

Choice
Choose from the following. [pp.330-331]

    a. club mosses
    b. horsetails
    c. ferns
    d. applies to a, b, and c

1. ____ *Equisetum*

2. ____ Club mosses, the best-known types, range from the tundra to the tropics.

3. ____ Seedless vascular plants

4. ____ Rust-colored patches, the sori, are on the lower surface of most fronds.

5. ____ Includes tropical tree types and species growing as epiphytes.

6. ____ American colonists and pioneers traveling west gathered these stems for use as disposable pot scrubbers.

7. ____ Their spores have been used as "flash powder."

8. ____ When chamber walls pop open, spores are dispersed by wind.

9. ____ Grow near many streams and have silica deposits in their tissues.

10. ____ Sporophytes usually have rhizomes and hollow stems.

11. ____ These are the most diverse seedless vascular plants.

12. ____ *Lycopodium*

13. ____ Young leaves develop in a coiled pattern that looks a bit like a fiddlehead and are uncoiled by maturity.

14. ____ Their "leaves" are called fronds.

15. ____ Depending on the species, strobili form at the tips of photosynthetic stems.

16. ____ After germination, each spore gives rise to a heart-shaped gametophyte, a few centimeters across.

Dichotomous Choice

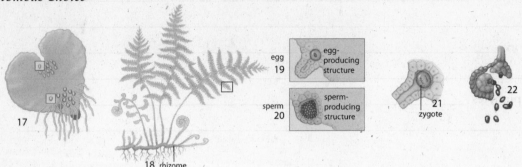

For each statement below, refer to the corresponding numbers on these sketches of the fern life cycle. You might also want to consult Figure 21.11 from the textbook for additional information. Circle the Terms in parentheses that make each statement correct. [p.331]

17. This structure represents the (gametophyte/sporophyte) plant; its tissues are (haploid/diploid).

18. The entire structure represents the (gametophyte/sporophyte) plant; its tissues are (haploid/diploid).

19. The egg shown is (haploid/diploid); (mitosis/meiosis) gave rise to the egg.

20. The sperm-producing structure is located on the (gametophyte/sporophyte) plant; its tissues are (haploid/diploid).

21. The zygote is (haploid/diploid) and is located within the egg-producing structure found on the (gametophyte/sporophyte) plant.

22. This structure functions in (sperm production/spore production) and the cells being released from it are (haploid/diploid); the structure itself is located on the (gametophyte/sporophyte) plant.

21.5 HISTORY OF THE VASCULAR PLANTS [pp.332-333]

Selected Terms

Ordovician [p.332], *Cooksonia* [p.332], Devonian [p.332], Carboniferous [p.332], decomposers [p.332], fossil fuel [p.332], gymnosperms [p.332], cycads [p.332], ginkgo [p.332], conifer [p.332], angiosperms [p.332], dinosaur [p.332], ovule [p.333], wood [p.333]

Boldfaced Terms

[p.332] coal _____

_____

[p.332] pollen sacs _____

_____

[p.332] microspores _____

_____

[p.332] megaspores _____

_____

[p.332] ovule _____

_____

[p.332] pollination_____

_____

Matching
Match each term with its description. [pp.332-333]

1. ____ carboniferous forests
2. ____ Late Devonian period
3. ____ Ordovician period
4. ____ coal
5. ____ nonrenewable source of energy
6. ____ two spore types in seed-bearing plants
7. ____ pollination
8. ____ ovules
9. ____ the rise of seed plants
10. ____ seed coat
11. ____ edible treasures from flowering plants

a. The arrival of pollen on female reproductive parts
b. energy source that cannot be replaced, such as coal
c. Coordinates with structural modifications, such as a water-conserving cuticle and two spore types
d. derived from the ovule tissues to protect the embryo
e. About 385 million years ago – oldest forest from what is now upstate New York
f. Formed by compression into great seams through heat and pressure on peat
g. Female reproductive structures that mature into seeds
h. Vast, represented by giant club mosses and giant horsetails
i. Leaves, fruits, seeds, stems, roots
j. About 450 million years ago – oldest fossils of vascular plants
k. Microspores and megaspores

21.6 GYMNOSPERMS—PLANTS WITH NAKED SEEDS [pp.334-335]

Selected Terms
gymnos [p. 334], *sperma* [p. 334], *Ginkgo biloba* [p.334], maidenhair tree [p.334], *Ephedra* [p.334], bristlecone pine [p.334], "pine nuts" [p.334], resin [p.334], "sago palms" [p.334], cone scales [p.334], seed cone [p.335], pollen cone [p.335], seed coat [p.335], nutritive tissue [p.335], embryo [p.335]

Boldfaced Terms
[p.334] gymnosperms_____

_____

[p.334] conifers _____

_____

[p.334] cycads _____

_____

[p.334] ginkgo _____

_____

[p.334] gnetophytes _____

_____

Choice

Choose from the following: [pp.334-335]

    a. cycads
    b. ginkgos
    c. gnetophytes
    d. conifers
    e. all gymnosperms (includes a, b, c, d)

1. ____ The thick, fleshy seeds of female trees smell terrible

2. ____ Includes pines, redwoods, and the bristlecone pine

3. ____ Only a single species survives, the maidenhair tree

4. ____ *Ephedra*

5. ____ Form pollen-bearing cones and seed-bearing cones on separate plants; many wild species face extinction

6. ____ Includes firs, spruces, and pines

7. ____ Ovules

8. ____ These trees are now widely planted; they have attractive, fan-shaped leaves and resist insects, disease, and air pollutants

9. ____ Ovules and seeds are exposed on a spore-producing structures

10. ____ Includes tropical trees, leathery vines, and desert shrubs

11. ____ Most species are trees and shrubs with needlelike or scalelike leaves

12. ____ Includes conifers, cycads, ginkgos, and gnetophytes

Dichotomous Choice

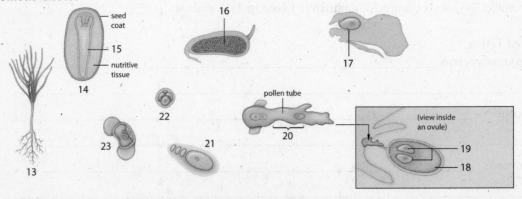

For each statement below, refer to the corresponding number on these sketches of the pine life cycle. You may also want to refer to Figure 21.19 from the textbook for additional information. Circle the Terms in parentheses that make each statement correct. [p.335]

13. This plant is a (gametophyte/sporophyte) and its tissues are (haploid/diploid).

14. The entire structure shown is a (seed/fruit).

15. This structure represents a young (gametophyte/sporophyte) plant; its tissues are (haploid/diploid).

16. The entire structure shown is a(n) (egg sac/pollen sac) and may be found as part of the pine (flower/cone).

17. This structure is the (egg/ovule) and it is (haploid/diploid).

18. The inside cellular mass of this structure represents the (male gametophyte/female gametophyte) plant; its cells are (haploid/diploid).

19. These structures are (eggs/sperms) and they are (haploid/diploid); they were produced by (mitosis/meiosis).

20. The cellular portion of this structure is known as the (male gametophyte/female gametophyte plant; it is (haploid/diploid).

21. The largest cell shown is a (microspore/megaspore); it was produced by (mitosis/meiosis) and is (haploid/diploid).

22. The quartet of cells shown are (microspores/megaspores); each is (haploid/diploid) and will develop into a(n) (ovule/pollen grain).

23. This structure is a (megaspore/pollen grain) and through pollination, it will be moved to a (male cone/female cone)

21.7 ANGIOSPERMS—THE FLOWERING PLANTS [pp.336-337]
21.8 ECOLOGICAL AND ECONOMIC IMPORTANCE OF ANGIOSPERMS [pp. 338-339]

Selected Terms
whorl [p.336], coevolve [p.336], herbaceous (nonwoody) [p.336], seed leaves (cotyledons) [p.336], filament [p.336], anther [p.336], petal [p.336], sepal [p.336], receptacle [p.336], calyx [p.336], corolla [p.336], microspores [p.336], adaptive radiation [p.336], nectar [p.336], stigma [p.337], triploid cell [p.337], style [p.337], megaspore [p.337], double fertilization [p.337], *Lilium* [p.337], legume [p.338], "hardwood" & "softwood" [p.339], aspirin [p.339], digitalis [p.339], caffeine [p.339], nicotine [p.339], pyrethrum [p.339]

Boldfaced Terms
[p.336] angiosperms _____

_____

[p.336] flower _____

_____

[p.336] stamens _____

_____

[p.336] carpel _____

_____

[p.336] ovary _____

_____

[p.336] pollinators _____

_____

[p.336] fruit _____

_____

[p.336] monocots _____

_____

[p.336] eudicots _____

_____

[p.337] endosperm _____

_____

[p.339] secondary metabolite _____

_____

Matching

Match each term with its description. [pp.336-337]

1. ____ flower
2. ____ coevolution
3. ____ pollinators
4. ____ ovary
5. ____ calyx
6. ____ examples of eudicots
7. ____ examples of monocots
8. ____ stamen
9. ____ carpel
10. ____ cotyledons

a. cabbages, daisies, most other herbaceous plants, most flowering shrubs and trees, and cacti
b. microspores form here
c. all of the sepals combined
d. seed leaves
e. agents that deliver pollen of one species to female parts of the same species
f. a specialized reproductive structure found only in angiosperms
g. megaspores form here
h. orchids, palms, lilies, and grasses such as rye, sugarcane, corn, rice, and wheat
i. where ovules develop
j. refers to two or more species evolving jointly because of their close ecological interactions

Dichotomous Choice

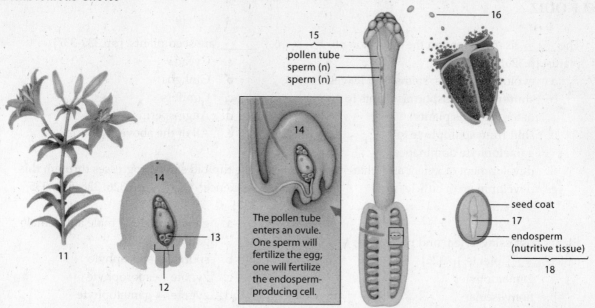

The pollen tube enters an ovule. One sperm will fertilize the egg; one will fertilize the endosperm-producing cell.

For each statement below, refer to the corresponding numbers on these sketches of the lily, a monocot. You may also ant to refer to Figure 21.21 in the textbook for more information. Circle the Terms in parentheses that make each statement correct. [p.337]

11. The plant is a (gametophyte/sporophyte); its tissues are (haploid/diploid).

12. The reduced plant is a (female gametophyte/female sporophyte); its cells are (haploid/diploid).

13. The single cell is a(n) (egg/spore); it is (haploid/diploid) and it was produced by (mitosis/meiosis).

14. This structure is called an (ovule/ovary); its cells are (haploid/diploid) and it later can develop into a(n) (fruit/seed).

15. The reduced plant is a (male gametophyte/male sporophyte); its cells are (haploid/diploid) and they were produced by (mitosis/meiosis).

16. The small cell is a (released spore/released pollen grain) and it is (haploid/diploid).

17. This structure is a(n) (embryo gametophyte plant/embryo sporophyte plant); its cells are (haploid/diploid) and it was produced by (mitosis/meiosis) of a (haploid zygote/diploid zygote).

18. The entire structure is called a (seed/fruit); it is a matured (ovary/ovule).

SELF QUIZ

1. The _____ is *not* a trend in the evolution of plants. [p.326]
    a. evolution of roots, stems, and leaves
    b. shift from homosporous plants to heterosporous plants
    c. shift from sporophyte to gametophyte dominance
    d. development of xylem and phloem
    e. development of cuticles and stomata

2. Plants possessing xylem and phloem are called _____ plants. [p.326]
    a. gametophytes
    b. nonvascular
    c. vascular
    d. sporophytes
    e. seedless

3. Existing nonvascular plants do *not* include _____. [p.328]
    a. horsetails
    b. mosses
    c. liverworts
    d. hornworts

4. _____ are *not* seedless vascular plants. [pp.330-333]
    a. Club mosses
    b. Gymnosperms
    c. Horsetails
    d. Ferns

5. In horsetails, club mosses, and ferns, _____. [pp.330-331]
    a. spores give rise to gametophytes
    b. the dominant plant is a gametophyte
    c. the sporophyte bears sperm- and egg-producing structures
    d. The gametophyte bears sperm- and egg-producing structures
    e. both a and d are correct

6. _____ are seed plants. [pp.332-337]
    a. Cycads
    b. Ginkgos
    c. Conifers
    d. Angiosperms
    e. All of the above

7. The diploid stage progresses through this sequence: _____. [pp.326, 328, 331, 335, 337]
    a. gametophyte → male and female gametes
    b. spores → sporophyte
    c. zygote → sporophyte
    d. zygote → gametophyte
    e. spores → gametes

8. In which of the following is the gametophyte the dominant plant? [pp.326, 328, 331, 335, 337]
    a. seedless vascular plants
    b. gymnosperms
    c. bryophytes
    d. angiosperms
    e. ferns

9. Microspores give rise to _____. [pp.336-337]
    a. megaspores
    b. female gametophytes
    c. male cones
    d. pollen grains
    e. embryos

10. All of the following (aspirin, pyrethrum, digitalis) are considered to be _____. [p.339]
    a. resins
    b. primary metabolites
    c. secondary metabolites

CHAPTER OBJECTIVES/REVIEW QUESTIONS

1. Define *deforestation* and cite the damage it has done to the world's forests. [p.325]

2. Be able to list the trends that have occurred in the evolution of plants. [pp.326-327]

3. List the groups included in the modern bryophytes and describe their major features. [pp. 328-329]

4. Be familiar with the major features of the moss life cycle. The _____ plant is the dominant form in this life cycle. [p.328]

5. The main lineages of seedless vascular plants are lycophytes, horsetails, and _____; describe the habitats and major features of each. [pp.330-331]

6. Be familiar with the major features of the fern life cycle. The _____ plant is the dominant form in this life cycle. [p.331]

7. The major carbon treasure left to us by vast Carboniferous forests was _____. [p.332]

8. Define the following as features that developed to allow the rise of seed-bearing plants: *microspores, pollination, megaspores,* and *ovules*. [pp.332-333]

9. Be familiar with representatives and the features of conifers, cycads, ginkgos, and gnetophytes. [p.334-335]

10. Be familiar with the major features of the pine life cycle. The _____ plant is the dominant form in this life cycle. [p.335]

11. The outstanding feature of the angiosperms is the _____, a specialized reproductive shoot. [p.336]

12. Define *coevolution, pollinator, eudicots,* and *monocots*. [pp.336-337]

13. Be familiar with the major features of the angiosperm life cycle. The _____ plant is the dominant form in this life cycle. [p.337]

CHAPTER SUMMARY

The earliest known plants date from (1) _____ million years ago. Ever since then, environmental changes have triggered (2) _____, adaptive (3) _____, and extinctions. Structural and (4) _____ adaptations of lineages are responses to some of the changes.

Bryophytes are (5) _____, with no internal pipelines to conduct water and solutes through the plant body. A (6) _____ -producing stage dominates their life cycle, and sperm reach the eggs by swimming through droplets or films of (7) _____.

Lycophytes, whisk ferns, horsetails, and ferns have (8) _____ tissues but do not produce (9) _____. A large spore-producing body that has (10) _____ vascular tissues dominates the life cycle. As with bryophytes, (11) _____ swim through water to reach eggs.

Gymnosperms and, later, angiosperms, radiated into higher and (12) _____ environments. The packaging of male gametes in (13) _____ grains and embryo sporophytes in (14) _____ contributed to the expansion of these groups into new habitats.

Angiosperms alone make (15) _____, which wind, water, and animals help pollinate. In distribution and (16) _____, angiosperms are the most successful group of plants. Nearly all plant species that we rely upon for food are (17) _____. Aspirin, nicotine, and digitalis are (18) _____ of plants.

INTEGRATING AND APPLYING KEY CONCEPTS

1. Prepare a table using the four column titles: "Plant group? Dominant plant? Vascular tissue present? and Seeds present?" List the following plant groups in the leftmost column: bryophytes, club mosses, horsetails, ferns, gymnosperms, and angiosperms. Then fill in the information indicated by the column titles for each plant group. Check your work against the text. List evolutionary trends you can discern after looking over the information.

2. Discuss why an angiosperm, such as an oak tree, can reach great heights, but a bryophyte never grows taller than 8 inches tall.

3. Describe various ways in which humans have exploited plants and plant parts for their own uses.

22

FUNGI

INTRODUCTION

Somewhere between the plants and the animals are the fungi. While this group of organisms shares characteristics with both, molecular studies have shown us that this diverse group of organisms is more closely related to animals than to plants. The fungi are worldwide in existence and important economically, ecologically and medically. Fungi are important in the food industry in the production of bread, beer and cheese for example. Fungi are the most numerous and adept decomposers. And fungi do cause many plant and animal diseases. This chapter describes these organisms and their life cycles, with proper emphasis on their importance and diversity.

FOCAL POINTS

- Figure 22.3 [p.345] animates the life cycle of a black bread mold, a representative fungus.
- Figure 22.8 [p.347] shows the life cycle of a typical club fungus.
- Figure 22.9 [p.348] illustrates a lichen growing on a birch tree.

INTERACTIVE EXERCISES

22.1 HIGH – FLYING FUNGI [p.343]
22.2 FUNGAL TRAITS AND DIVERSITY [p.344]

Selected Terms
Sugar cane rust [p.343], coffee rust fungus [p.343], wheat stem rust [p.343], Ug99 [p.343], pathogenic [p.343], decomposers [p.343], chitin [p.344], polysaccharide [p.344], cross-walls (septa) [p.344], mitosis (asexual spores) [p.344], sexual spores (meiosis) [p.344]

Boldfaced Terms
[p.344] fungi _____

_____

[p.344] saprobe _____

_____

[p.344] mycelium _____

_____

[p.344] hypha _____

_____

[p.344] dikaryotic _____

_____

Fill in the Blank [pp.344-345]

1. Spores for the sugar cane rust were introduced by a storm from Cameroom to the _____.

2. A new strain of wheat rust was discovered in Uganda in 1999 and named _____?

3. Most fungi are _____ with important ecological roles.

4. Fungi are spore-making eukaryotic _____.

5. A saprobe feeds on organic _____ and remains.

6. A single fungal filament is called a _____.

7. A mat composed of many (answer to number 6 above) is called a _____.

8. Dikaryotic means that a cell contains two genetically _____ nuclei.

22.3 CHYTRIDS, ZYGOTE FUNGI, AND RELATIVES [p.345]
22.4 SAC FUNGI [p.346]
22.5 CLUB FUNGI [p.347]

Selected Terms

Rhizopus stolonifer [p.345], flagellated spores [p.345], aquatic decomposers [p.345], cellulose [p.345], zygospores [p.345], mating strains [p.345], zygomycosis [p.345], "mycosis" [p.345], mutual beneficial [p.345], ascus [p.346], cytoplasmic fusion [p.346], ascocarp [p.346], morels [p.346], truffles [p.346], mold [p.346], *Penicillium* {p.346], blue cheese [p.346], antibiotic drug [p.346], penicillin [p.346], yeast [p.346] fermentation [p.346], roundworms [p.346], *Arthobotrys* [p.346], basidiomycete [p.347], basidocarp [p.347], fruiting bodies [p.347], gills [p.347], mushroom [p.347], cap [p.347], lignin [p.347], honey mushroom [p.347], smut [p.347], rust [p.347], coral fungus [p.347], chanterelle [p.347], shelf fungus [p.347], puffball [p.347], LSD [p.347], toxin [p.347], hallucination [p.347], psilocybin [p.347], pathogen [p.347]

Boldfaced Terms

[p.345] chytrids _____

_____

[p.345] zygote fungi _____

_____

[p.345] glomeromycetes _____

_____

[p.346] sac fungi _____

_____

[p.347] club fungi _____

_____

Completion

For each of the fungi listed, fill in the table and list what type of fungi they are (zygote, sac, club) and list any health, economical or ecological importance for that organism.

Fungi	Zygote, sac or club	Importance
Rhizopus stolonifer	1.	7.
yeast	2.	8.
Penicillium	3.	9.
Arthrobotrys	4.	10.
morel	5.	11.
puffball	6.	12.

22.6 FUNGI AS PARTNERS [p.348]
22.7 FUNGI AS PATHOGENS [p.349]

Selected Terms

Cyanobacteria [p.348], symbiotic [p.348], fragmentation [p.348], powdery mildew [p.349], chestnut blight & American chestnut tree [p.349], alkaloids [p.349], ergotism [p.349], "blind staggers" [p.349], "athlete's foot" [p349], 'ringworm" p.349], vaginitis [p.349], histoplasmosis [p.349], coccidioidomycosis or valley fever [p.349] *Claviceps purpurea* [p.349]

Boldfaced Terms

[p.348] lichens _____

_____

[p.348] mutualism _____

_____

[p.348] mycorrhiza _____

_____

Labeling

For each number choose the corresponding label from the diagram below. [p.347]

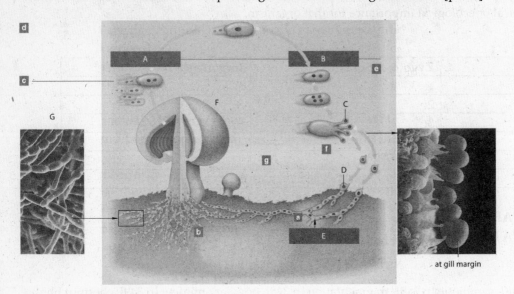

at gill margin

1. _____ meiosis

2. _____ a hypha

3. _____ cytoplasmic fusion

4. _____ a mushroom or fruiting body

5. _____ cells with two nuclei form on the gills

6. _____ a spore

7. _____ nuclear fusion

Choice [pp.345-349]

For each of the following statements, choose the group that is best described by the statement.

    a. lichens
    b. zygote fungi
    c. sac fungi
    d. mycorrhizae

8. _____ Most food-spoiling fungi belong to this group.

9. _____ This group has the ability to colonize hostile environments

10. _____ Baking yeast belongs to this group.

11. _____ Bread mold belongs to this group.

12. _____ A fungus and a photoautotroph in a symbionic relationship.

13. _____ Truffles are this type of fungi.

14. _____ A fungus that in a mutualistic relationship with the roots of a plant.

SELF QUIZ

1. A fungus that forms a mutualistic relationship with the root of a plant is called a _____. [p.348]
   a. lichen
   b. oomycota
   c. endophytic symbiont
   d. mycorrhizae
   e. none of the above

2. A single cellular filament of a fungi is called a _____. [p.344]
   a. hypha
   b. mycelium
   c. spore
   d. fruiting body
   e. mycorrhiza

3. A multicelled, eukaryotic, heterotrophic saprobe would belong to which of the following classifications? [p.44]
   a. animal
   b. plant
   c. protist
   d. fungi
   e. none of the above

4. Which of the following may form a symbiotic relationship with a fungus to form a lichen? [p.348]
   a. red algae
   b. brown algae
   c. green algae
   d. diatoms
   e. euglenoids

5. A mat of absorptive filaments of a fungus is called a _____. [p.344]
   a. hypha
   b. mycelium
   c. spore
   d. fruiting body
   e. mycorrhiza

6. LSD is produced by some _____.
   a. sac fungi
   b. club fungi
   c. zygote fungi

7. Morels and truffles are examples of a ____.
   a. sac fungi
   b. club fungi
   c. zygote fungi

8. *Penicillium* and yeast are examples of a ____.
   a. sac fungi
   b. club fungi
   c. zygote fungi

9. Black bread mold is an example of a ____.
   a. sac fungi
   b. club fungi
   c. zygote fungi

10. Chitin is found in the cell walls composed of chitin. Chitin is a ____.
    a. protein
    b. fat
    c. polysaccharide

CHAPTER OBJECTIVES/ REVIEW QUESTIONS

1. Identify the main characteristics of fungi. [p.344]

2. Understand the basic anatomy of fungi. [p.344]

3. Understand the life cycle of fungi. [p.344]

4. Distinguish between the various groups of fungi. [pp.345-349]

5. Understand the symbiotic relationships of lichens and mycorrhizae. [p.348]

Fungi **219**

6. Understand the pathogenic types of fungi and some of the plant and animal diseases they cause. [p.349]

CHAPTER SUMMARY

Fungi are single-celled and multicelled (1) _____ more closely related to (2) _____ than to plants. They feed by secreting digestive enzymes into (3) _____ matter in their surroundings, then absorbing the released nutrients. Multicelled species forms a mesh of absorptive (4) _____, some of which intertwine as spore-producing structures.

(5) _____ fungi, (6) _____ fungi, and (7) _____ fungi are three major groups. Others include chytrids, and ancient lineage with flagellated spores, and the (8) _____ parasites called microsporidians.

(9) _____ formation dominates fungal life cycles. Zygote fungi make thick-walled zygotes that give rise to (10) _____ spores. Many sac fungi and club fungi make complex (11) _____ -bearing structures. In sac fungi, (12) _____ spores form in a sac. In club fungi, they form at the tip of a club-shaped (13) _____.

Many fungi live on, in, or with other (14) _____. Some live inside plant leaves, stems, or roots. Others form (15) _____ by living with algae or cyanobacteria.

A minority of fungi are (16) _____. Certain species cause diseases in crop plants and (17) _____.

INTEGRATING AND APPLYING CONCEPTS

1. Why do some species of fungi produce toxins?
2. Why can some animals like deer eat some poisonous mushrooms and humans cannot?
3. Why do scientists place fungi near the animals rather than plants?

23

ANIMALS I: MAJOR INVERTEBRATE GROUPS

INTRODUCTION

The organisms that make up the world of animals are defined by several shared characteristics, including multicellularity, heterotrophic feeding in all, and sexual reproduction and tissue formation in most. While all animals share these common characteristics, they have evolved many different and sometimes unique adaptations as well that allow them to survive in various environments and conditions. This chapter begins our exploration of the animal world with a look at the invertebrates, the animals with no backbone. From the sponges to the insects and on to the spiny skinned echinoderms you will learn about the characteristics of these diverse animals that allow them able to survive, make them unique, make them important to us and sometimes give them the ability to cause disease.

FOCAL POINTS

- Figure 23.1 [p.354] shows the major animal groups covered in chapters 23 and 24.
- Figure 23.3 [p.355] animates the differences between body cavity arrangements in animals.
- Figure 23.6 [p.357] illustrates a sponge body.
- Figure 23.8 [p.358] shows a diagram of the polyp and medusa body forms.
- Figure 23.9 [p.358] shows nematocyst action.
- Figure 23.10 [p.359] is an animated look at the life cycle of a hydroid.
- Figure 23.13 [p.360] shows the organs systems of a planaria.
- Figure 23.14 [p.361] shows the life cycle of a beef tapeworm.
- Figure 23.17 [p.363] depicts the body plan of an earthworm..
- Figure 23.19 [p.364] illustrates the aquatic snail body plan.
- Figure 23.23 [p.366] illustrates the roundworm body plan.
- Figure 23.26 [p.368] shows a spider's external organization.
- Figure 23.28 [p.369] illustrates the clawed lobster body plan.
- Figure 23.30 [p.370] gives examples of insect appendages and body plan
- Figure 23.31 [p.370] shows different patterns of insect development.
- Figure 23.36 [p.373] shows anatomical features of a sea star.

INTERACTIVE EXERCISES

23.1 OLD GENES, NEW DRUGS [p.353]
23.2 ANIMAL TRAITS AND TRENDS [pp.354-355]
23.3 ANIMAL ORIGINS AND EARLY RADIATIONS [p.356]

Selected Terms
cone snail [p.353], *Conus geographus* [p.353], venom [p.353], ziconotide [p.353], haploid gamete [p.354], flagellated sperm [p.353], zygote [p.354], Metazoa [p.354], asymmetrical [p.354], anterior [p.355], yubular gut [p.355], complete digestive system [p.355], colonial origin [p.356], "collared flagellate" [p.356],

Trichoplax adhaerens [p.356], *Spriggina* [p.356], Ediacarans [p.356], Ediacara Hills [p.356], Cambrian explosion [p.356], supercontinent [p.356], speciation [p.356], homeotic genes [p.356], mutations [p.356]

Boldfaced Terms

[p.353] vertebrate _____

_____

[p.353] invertebrate _____

_____

[p.354] animals _____

[p.354] ectoderm _____

[p.354] endoderm _____

[p.354] mesoderm _____

[p.354] radial symmetry _____

_____

[p.354] bilateral symmetry _____

_____

[p.355] cephalization _____

_____

[p.355] protostomes _____

[p.355] deuterostomes _____

_____

[p.355] pseudocoelom _____

_____

[p.355] coelom _____

_____

[p.355] segmentation _____

_____

[p.356] choanoflagellates _____

_____

[p.356] placozoan _____

_____

True-False

If the statement is true, place a "T" in the space provided. If false, correct the underlined word so that the statement is correct. [p.354]

1. _____ All animals are multicelled and <u>do not</u> possess cell walls.

2. _____ Animals are <u>autotrophs</u>.

3. _____ Animals typically reproduce sexually, and <u>many</u> reproduce asexually as well.

4. _____ Most animals are <u>motile</u> during all or part of the life cycle.

Matching

Choose the most appropriate description for each of the following terms. [pp.354-355]

5. ____ vertebrates

6. ____ bilateral symmetry

7. ____ cephalization

8. ____ endoderm

9. ____ invertebrates

10. ____ ectoderm

11. ____ mesoderm

12. ____ asymmetrical

13. ____ gut

14. ____ radial symmetry

15. ____ coelum

16. ____ anterior

17. ____ segmentation

18. ____ pseudocoelom

a. the tissue layer that is the source of muscles and other organs; it is between ecto and endoderm.

b. the site of food digestion

c. front end of the body

d. the division of the body into repeated interconnecting units.

e. animals that possess a backbone.

f. a body arrangement in which many appendages and organs are paired.

g. a body cavity between the gut and body wall.

h. the tissue layer that develops into the gut lining.

i. the most common form of animal, does not have a backbone.

j. a body form in which the parts are arranged around a central axis.

l. a reduced, unlined coelum, present in some early animals.

m. the development of the anterior end into a distinct head.

n. having no symmetry

o. the tissue layer that lines the body surfaces

Fill-in-the-Blanks [p. 356]

How did animals arise? Perhaps the ancestor of the earliest animals was a (19_____. Each

of these organisms has a ring of absorptive structures (microvilli) around the base of a (20) _____.

Over time, some of the colonial cells have shown (21) _____ of labor. This is evidenced by the fact

that cells on the outside of the colony are involved with (22) _____ while cells on the insides are specialized for (23) _____. Comparisons of (24) _____ sequences also point to choanoflagellates as the group most likely to have been involved in the early stages of (25) _____ evolution. Animals first lived in the (26) _____, and in the (27) _____ there was an explosion in animal diversity such that all major animal lineages evolved within 40 million years.

Labeling
Choose the most appropriate label for each of the numbers in the diagram below. You may also want to consult Figure 23.1 in the textbook for more information. [p.354]

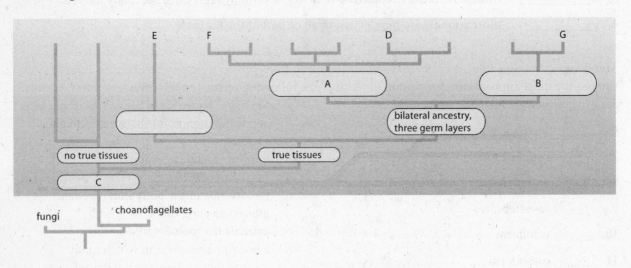

28. ____
29. ____
30. ____
31. ____
32. ____
33. ____
34. ____

a. protostomes
b. deuterostomes
c. multicelled body
d. roundworms
e. cnidarians
f. flatworms
g. chordates

23.4 SPONGES [p.357]
23.5 CNIDARIANS [pp.358-359]
23.6 FLATWORMS [pp.360-361]

Selected Terms
Phagocytosis [p.357], intracellular [p.357], ciliated larva [p.357], gemmules [p.357], bath sponge [p.357], coral, sea anemones, & jellyfish [p.358], medusa [p.358], polyp [p.358], gastrovascular cavity [p.358], gastrodermis [p.358], mesoglea [p.358], *cnidos* [p.358], stinging cells [p.358], cnidocytes [p.358], gland cells [p.358], bilateral ciliated larvae [p.358], *Obelia* [p.359], *Physalia* [p.359], budding [p.359], "coral bleaching" [p.358], siphonophores [p.359], Portuguese man-of-war [p.358], reef-building coral [p.359], flukes, tapeworms, & turbellarian [p.360], eyespot [p.360], flame cell [p.360], "flickering" cilia [p.360],

pharynx [p.360], intermediate host [p.361], blood fluke [p.361], *Schistosoma* [p.361], *Taenia saginata* [p.361], scolex [p.361], proglottid [p.369], cyst [p.361]

Boldfaced Terms

[p.357] sponges _____

_____

[p.357] hermaphrodite _____

_____

[p.357] larva _____

_____

[p.358] cnidarians _____

_____

[p.358] nematocysts _____

_____

[p.358] nerve net _____

_____

[p.358] hydrostatic skeleton _____

_____

[p.360] flatworms _____

_____

[p.360] planarian _____

_____

Choice

For each statement, choose the correct group of animals to which the statement best applies. Answers may be used more than once.

      a.   sponges [p.357]
      b.   cnidarians [pp.358-359]
      c.   flatworms [pp.360-361]
      d.   all of the above

1.   ____   Possess nematocysts

2.   ____   Possess a primitive nervous system called a nerve net.

3.   ____   The first animal with organs.

4.   ____   Have no symmetry, tissues, or organs.

5.   ____   All members of this group have ectoderm, endoderm and mesoderm.

6.   ____   May reproduce asexually by fragmentation.

7.   ____   The majority are bilateral.

8. ____ Body forms are called polyps and medusas.

9. ____ Many of this group has radial symmetry.

10. ____ Members of this group may utilize definitive and intermediate hosts.

11. ____ Lack a body coelum.

Labeling

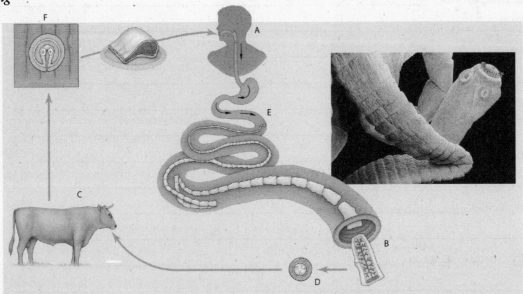

This is the life cycle of the beef tapeworm. Please match the following statement with the proper label from the picture above. You may also want to consult Figure 23.14 from the textbook for additional information. [p.361]

12. ____ the definitive host

13. ____ the larvae, with the inverted scolex of future tapeworm

14. ____ the intermediate host

15. ____ scolex

16. ____ fertilized egg

17. ____ sexually mature proglottid

23.7 ANNELIDS [pp.362-363]
23.8 MOLLUSKS [pp.364-365]
23.9 ROUNDWORMS [p.366]

Selected Terms
Oligo- [p.362], poly- [p.362], oligochaete [p.362], polychaete [p.362], sandworm [p.362], fan worms [p.362], feather-duster worms [p.362], leeches [p.362], earthworms [p.362], *Hirudo medicinalis* [p.362], palp [p.362], parapod [p.362], "castings" [p.363], longitudinal & circular muscles [p.363], clitellum [p.363], gizzard [p.363], mantle [p.364], radula [p.364], sensory tentacles [p.364], "belly foot" [p.364], inhalant siphon [p.364], snails & slugs [p.364], sea slugs [p.364], Spanish shawl nudibranch [p.364], *Flabellina*

iodinea [p.365], mussels, clams, oysters, & scallops [p.364], shell [p.364], squid, octopus, nautilus, & cuttlefish [p.365], jet propulsion [p.365], "head footed" [p.365], slime trail [p.365], nematodes [p.366], Nematoda [p.366], phylum [p.366], cuticle-covered [p.366], collagen [p.366], *Caenorhabditis elegans* [p.366], *Trichinella spiralis* [p.366], trichinosis [p.366], lymphatic filariasis [p.366], elephantiasis [p.366], hookworms [p.366], pinworms [p.366], *Ascaris lumbricoides* [p.366], *Wucheria bancrofti* [p. 366]

Boldfaced Terms

[p.362] annelids _____

_____

[p.362] closed circulatory system _____

_____

[p.364] mollusks _____

_____

[p.364] gastropods _____

_____

[p.364] open circulatory system _____

_____

[p.364] bivalves _____

_____

[p.365] cephalopods _____

_____

[p.366] roundworms _____

_____

[p.366] molting _____

_____

Choice
For each statement, choose the correct group of animals to which the statement best applies. Answers may be used more than once.
- a. mollusks [pp.364-365]
- b. annelids [pp.362-363]
- c. roundworms [p.366]
- d. none of the above

1. ____ Members of this group have the ability of jet propulsion.

2. ____ Has a false coelum packed with reproductive organs.

3. ____ Many of this group have a radula.

4. ____ Cephalopods belong to this group.

5. ____ Are responsible for diseases such as trichinosis and elephantitis.

Animals I: Major Invertebrate Groups **227**

6. _____ Possess multiple hearts and a closed circulatory system.

7. _____ Possess an extension of their body called a mantle.

8. _____ Leeches belong to this group.

9. _____ Also known as the nematodes.

10. _____ The most segmented of all the animals.

Matching

Match each of the following body parts to its correct function. [pp.362-365]

11. _____ nerve cords

12. _____ chaetae

13. _____ mantle

14. _____ radula

15. _____ brain

16. _____ flame cell

17. _____ gill

a. clusters of nerve cell bodies that control activity.
b. regulates volume of body fluids.
c. a cloak-like extension of the body mass.
d. lines of communication that help the brain coordinate activities
e. chitin-reinforced bristles that assist in movement, present in most annelids
f. a tongue-like organ that assists in feeding
g. respiratory organ found in most mollusks

Labeling

Provide the name of each indicated body part in the diagram below. [p.364]

18. _____

19. _____

20. _____

21. _____

22. _____

23. _____

24. _____

25. _____

Complete-the-Table [p. 374]

Species	Disease
Trichinella spiralis	26.
Ascaris lumbricoides	27.
Wuchereria bancrofti	28.

23.10 KEYS TO ARTHROPOD DIVERSITY [p.367]

23.11 SPIDERS AND THEIR RELATIVES [p.368]

23.12 CRUSTACEANS [p.369]

23.13 INSECT TRAITS AND DIVERSITY [pp.370-371]

23.14 THE IMPORTANCE OF INSECTS [p.372]
23.15 ECHINODERMS [p.373]

Selected Terms

chelicerates [p.367], spiders, lobsters, shrimp, crabs, millipedes, & centipedes [p.367], jointed leg [p.367], chitin [p.367], cephalothorax [p.368], pedipalps [p.368], horseshoe crabs [p.368], telson [p.368], scorpions, ticks, & mites [p.368], spinners [p.368], tarantulas & jumping spiders [p.368], Lyme disease [p.368], krill [p.369], euphausiids [p.369], copepod [p.369], barnacles [p.369], swimmerets [p.369], walking legs [p.369], abdomen [p.369], tracheal tubes [p.370], silverfish [p.370], hymenopterans [p.371], Med fly [p.372], malaria [p.372], African sleeping sickness [p.372], Chagas disease [p.372], bubonic plague [p.372], typhus [p.372], "spiny skinned" [p.373], calcium carbonate [p.373], tube feet [p.373], sea star [p.373], gonad [p.373], ossicle [p.373], sea urchin & sea cucumbers [p.373], ampullae [p.373]

Boldfaced Terms

[p.367] arthropods _____

_____

[p.367] exoskeleton _____

_____

[p.367] antennae _____

_____

[p.367] metamorphosis _____

_____

[p.368] chelicerates _____

_____

[p.368] arachnids _____

_____

[p.369] crustaceans _____

_____

[p.370] insects _____

_____

[p.373] echinoderm _____

_____

[p.373] water-vascular system _____

_____

True-False

Mark a "T" in the space provided if the statement is true. If the statement is false, correct the underlined word so that the statement is correct. [pp.367-371]

1. _____ The arthropods are <u>bilateral</u> animals with a <u>closed</u> circulatory system.

2. _____ In many arthropods, there is a <u>metamorphosis</u> of body form between the embryo and adult stages.

3. _____ Arthropods utilize a hardened <u>endoskeleton</u> for support.

4. _____ The <u>jointed</u> appendages of the arthropods allows for the diverse evolution of this group.

5. _____ Arthropods have a complete digestive system and a <u>pseudocoel</u>.

Choice

For each of the following statements, identify the group(s) of arthropods that is best described. Some statements may have more than one answer.

    a.   crustaceans [p.369]
    b.   spiders and their relatives [p.368]
    c.   insects [pp.370-371]
    d.   all arthropods

6. _____ These are marine organisms that are frequently consumed by humans.

7. _____ Possess special appendages called chelicerae

8. _____ Body is divided into three distinct parts – head, thorax and abdomen.

9. _____ Undergo periodic molts to shed their exoskeleton.

10. _____ In terms of number of species, the largest group on the planet.

11. _____ Gills and trachea are the respiratory structures.

12. _____ Specialize in predatory or parasitic life styles.

13. _____ Includes the copepods.

14. _____ These have incomplete or complete metamorphosis

15. _____ Includes the horseshoe crabs and mites.

16. _____ Winged invertebrates.

Fill-in-the-Blanks [pp.370-371]

In most adult insects, the (17) _____ body is divided into three distinct parts: a head, (18) _____ , and (19) _____ [p.370]. The head has paired sensory (20) _____ and paired mouthparts that are specialized for biting, chewing and other tasks. The (21) _____ has three pairs of legs and, usually, (22) _____ pairs of wings. Insects are the only winged (23) _____ .

Insects have a (24) _____ digestive system divided into a foregut, a midgut, and hindgut.

Insects that undergo incomplete metamorphosis go from eggs to (25) _____ and then to (26) _____. Insects that undergo complete metamorphosis go from eggs to (27) _____, then (28) _____ and then to adults.

True-False
Mark a "T" in the space provided if the statement is true. If the statement is false, correct the underlined word so that the statement is correct. [p.373]

29. _____ The water-vascular system of Echinoderms is used for <u>locomotion</u>.

30. _____ Most Echinoderms are <u>bilateral</u> as adults.

31. _____ Echinoderms have interlocking <u>calcium carbonate</u> plates on their surface.

32. _____ Echinoderms have a <u>centralized</u> nervous system.

Completion
Complete the table below by filling in the group of animal and the effect on humans for each of the individuals listed [pp.368-372].

Animal	Group	Impact on Humans
Med flies	33.	38.
ticks	34.	39.
biting flies	35.	40.
fleas	36.	41.
mosquitoes	37.	42.

List [p.367]
List the six key adaptations that contributed to the evolutionary success of the arthropods:

43. _____   46. _____

44. _____   47. _____

45. _____   48. _____

SELF QUIZ

1. Nematocysts are present in the _____.
    [pp.358-359]
    a. Cnidarians
    b. Flatworms
    c. Roundworms
    d. Echinoderms

2. Which of the following is a characteristic of an animal? [pp.354-355]
    a. multicelled
    b. aerobic heterotrophs
    c. most are motile
    d. most reproduce sexually
    e. all of the above are correct

3. A water-vascular system is a characteristic of the _____. [p.373]
   a. Echinoderms
   b. Crustaceans
   c. Poriferans
   d. Cnidarians

4. Which of the following is the first group to develop organ systems? [pp.360-361]
   a. Roundworms
   b. Flatworms
   c. Cnidarians
   d. Arthropods

5. Collar cells are found in _____. [p.357]
   a. flatworms
   b. cnidarians
   c. sponges
   d. roundworms
   e. arthropods

6. _____ symmetry allows for the process of cephalization. [pp.354-355]
   a. radial
   b. axial
   c. perpendicular
   d. bilateral
   e. none of the above

7. Which of the following represents the earliest known form of animals? [p.356]
   a. Poriferans
   b. Cnidarians
   c. Ediacarans
   d. Arthropods

8. A complete lack of symmetry is a characteristic of the _____. [p.357]
   a. Cnidarians
   b. Echinoderms
   c. Poriferans
   d. Arthropods
   e. Flatworms

9. You have identified an organism with true tissues and an organ with no segments or coelom. To which of the following groups could this organism belong? [pp.360-361]
   a. Cnidarians
   b. Flatworms
   c. Arthropods
   d. Chordates
   e. Annelids

10. These are the most segmented organisms in the animal kingdom. [pp.362-363]
    a. Annelids
    b. Roundworms
    c. Flatworms
    d. Echinoderms

11. Which of the following tissue layers is correctly matched to its correct function? [p.354]
    a. ectoderm – source of muscles and other organs.
    b. mesoderm – lines the body surfaces
    c. endoderm – forms the lining of the gut
    d. none of the above are matched correctly
    e. all of the above are matched correctly

12. The only winged invertebrates are the _____. [pp.370-372]
    a. spiders
    b. crustaceans
    c. cephalopods
    d. insects
    e. Echinoderms

13. Cephalopods belong to which of the following groups? [pp.364-365]
    a. Annelids
    b. Mollusks
    c. Roundworms
    d. Arthropods

14. The body cavity of an animal is called the
    _____. [p.355]
    a. peritoneum
    b. gut
    c. protostome
    d. coelum
    e. medoderm

15. Jointed appendages are a characteristic of
    the _____. [p.367]
    a. Echinoderms
    b. Arthropods
    c. Poriferans
    d. Mollusks

16. Examples of these are pinworms,
    hookworms, and *Ascaris*. They are members
    of the _____. [p.366]
    a. flatworms
    b. annelids
    c. roundworms
    d. all of these

CHAPTER OBJECTIVES/ REVIEW QUESTIONS

1. List the general characteristics of all animals. [pp.354-355]

2. The majority of animals are _____ (vertebrates, invertebrates). [pp.354-355]

3. Explain how bilateral symmetry leads to cephalization. [pp.354-355]

4. Distinguish between a gut and coelum. [p.355]

5. List the primary difference between a protostome and a deuterostome. [p.355]

6. Define segmentation. [p.355]

7. Recognize the general organization of the animal kingdom, and the key events in the evolution of the animal body form. [p.356]

8. Identify the key characteristics of a poriferan. [p.357]

9. Identify the key characteristics of a cnidarian. [pp.358-359]

10. Explain how organs can be combined into an organ system. [pp.360-361]

11. Distinguish between an intermediate and definitive host. [p.361]

12. Explain the purpose of the flame ells [p.360]

13. List the advantages of segmentation. [pp.362-363]

14. Identify the key characteristics of a mollusk. [pp.364-365]

15. Understand the relationship of roundworms to human diseases. [p.366]

16. Identify the key characteristics of all arthropods. [p.367]

17. What is metamorphosis and how does it relate to insect development? [p.370]

18. Distinguish between crustaceans, cheicerates and insects. [pp.368-372]

19. Identify the key characteristics of an Echinoderm. [p.373]

20. Give the purpose of the water vascular system. [p.373]

CHAPTER SUMMARY

Animals are multicelled (1) _____ that ingest other organisms, grow and (2) _____ through a series of stages, and actively move about during all or part of the (3) _____ cycle. Cells of most animals form (4) _____ and extracellular matrixes.

The earliest animals were small and structurally (5) _____. Their descendents evolved larger bodies with a more (6) _____ structure and greater (7) _____ among specialized parts.

Animals' body plans vary. Bodies may or may not show (8) _____. There may or may not be an internal (9) _____ cavity, a head, or division into (10) _____. An early divergence gave rise to two major branches: protostomes and (11) _____.

Sponges and placozoans have no body (12) _____ or true (13) _____. Cnidarians are (14) _____ symmetrical, with two tissue layers and a gelatinous matrix between the two.

Most animals show (15) _____ symmetry. Bilateral animals have tissues, organs, and organ (16) _____. All adult tissues arise from two or three simple layers that form in early (17) _____.

In diversity, numbers, and distribution, (18) _____ are the most successful animals. In the seas, (19) _____ are the dominant arthropod lineage; on land, (20) _____ rule.

Echinoderms are on the same branch of the animal family tree as the (21) _____. They are (22) _____ with bilateral ancestors, but adults now have a decidedly (23) _____ body plan.

INTEGRATING AND APPLYING KEY CONCEPTS

1. Juvenile hormone disruptors are becoming an important weapon in the war against pest insects. These chemicals stop the insect from molting. How could this be used to control pest populations? If a farmer finds a pest insect on his crop and applies these chemicals, will he immediately decrease the damage to his crops? Why or why not?

2. Consider the diversity of arthropods. Make a list of all of the habitats that arthropods have been successful in.

3. Roundworms have numerous parasitic members. What does it take to evolve into a parasite? What would be an advantage and a disadvantage of evolving into for example an *Ascaris*?

24

ANIMALS II: THE CHORDATES

INTRODUCTION

The diversity of life on Earth is further reflected in the animal world, and in this chapter that diversity is illustrated through a study of the vertebrate animals. This chapter discusses in a sequential, organized way, the evolution of vertebrate animals from the seas in which they began to the present day humans. The relationships between such varied animals as the fish and the human being are discussed and the innovations that took place at each evolutionary branch are pointed out. The varied animals all share some physical features that have been maintained throughout the ages. Again, the ability of species to adapt to various environments is illustrated, but what happens when a group of animals does not respond quickly enough to environmental changes is also discussed. Such was the case, we believe, with the dinosaurs.

FOCAL POINTS

- Figure 24.1 [p.378] body plan of a lancelet.
- Figure 24.2 [p.378] diagram of a tunicate.
- Figure 24.3 [p379] evolutionary tree diagram for the vertebrates with key traits for each clade.
- Figure 24.5 [p.380] compares the gill supporting structures in fishes.
- Figure 24.8 [p.391] diagram of the body plan of a perch, a bony fish.
- Figure 24.9 [p.382] transition to tetrapods.
- Figure 24.14 [p.384] diagram of an amniote.
- Figure 24.15 [p.384] is a family tree of the amniotes.
- Figure 24.18 [p.386] shows a bird skeleton.
- Figure 24.20 [p.387] generalized mammalian skull.
- Figure 24.23 [p.388] proposed evolutionary tree for the primates.
- Table 24.1 [p.388] organizes primate classification.
- Figure 24.24 [p.389] presents samples of primate skulls.
- Figure 24.25 [p.389] some skeletal differences associated with differences in posture and locomotion between a human and a gorilla.
- Figure 24.26 [p.290] shows a sampling of fossilized hominid skulls from Africa.

INTERACTIVE EXERCISES

24.1 WINDOWS ON THE PAST [p.377]
24.2 THE CHORDATE HERITAGE [pp.378-379]

Selected Terms
Sinosauropteryx [p.377], *Archaeopteryx* [p.377], *Confuciusornis sanctus* [p.377], "missing links" [p.377], Darwin [p.377], *On the Origin of Species* [p.377], radiometric dating [p.377], feathers [p.377], "theory" [p.377], vertebrate evolution [p.377], gill slits [p.378], anus [p.378], metamorphosis [p.378], "tunic" [p.378], braincase [p.378], cranium [p.378], cartilage [p.378], bone [p.378], hagfish [p.378], jaws [p.379] gills[p.379], lungs [p.379]

Boldfaced Terms

[p.378] chordates _____

_____

[p.378] notochord _____

_____

[p.378] lancelets _____

_____

[p.378] tunicates _____

_____

[p.378] craniates _____

_____

[p.378] vertebrates _____

_____

[p.378] endoskeleton _____

_____

[p.379] tetrapods _____

_____

[p.379] amniotes _____

_____

Labeling

Fill in the matching letter of the following structures. [p.378]

1. _____ eyespot

2. _____ anus

3. _____ pharynx with gill slits

4. _____ notochord

5. _____ tail extending past the anus

6. _____ dorsal nerve cord

7. _____ Which four letters represent the four
defining traits of a chordate?

Matching

Choose one of the following for each of the statements. [pp.377-379]

8. ____ another name for cranium

9. ____ first bird with a beak

10. ____ evolution

11. ____ larval forms has chordate traits, adult only retains pharynx with gills

12. ____ these assist in stabilizing and guiding the body through the water.

13. ____ these have a large surface area to exchange gases with the environment.

14. ____ this means ancient wing

15. ____ this led to the development of more efficient circulatory systems in land animals.

16. ____ vertebrae and other skeletal components

17. ____ these are internally moistened sacs for gas exchange.

a. *Archaeopteryx*
b. vertebrae
c. fins
d. gills
e. lungs
f. tunicate
g. braincase
h. endoskeleton
i. scientific theory
j. *Confuciusornis*

Labeling

Label each of the indicated groups on the diagram below. [p.339]

18. _____

19. _____

20. _____

21. _____

22. _____

24.3 THE FISHES [pp.380-381]
24.4 AMPHIBIANS – THE FIRST TETRAPODS [pp.382-383]

Selected Terms

Jawless [p.380], fins [p.380], scales [p.380], horny teeth [p.380], keratin[p.380], dentin[p.380], enamel [p.380], spiracle [p.380], plankton [p.381], venom gland [p.381], swim bladder [p.381], cloaca [p.381], coelacanths [p.381], lungfish [p.381], pelvic & pectoral fins [p.381], lobe-finned fishes [p.382], three chambers [p.382], Devonian [p.382], axolotls [p.382], salamander [p.382], *Acanthostega* [p.382], *Ichthyostega* [p. 382], tadpoles [p.383]

Boldfaced Terms

[p.380] hagfish _____

_____

[p.380] lampreys _____

_____

[p.380] cartilaginous fishes _____

_____

[p.381] bony fishes _____

_____

[p.381] ray-finned fishes_____

_____

[p.381] lobed-finned fishes_____

_____

[p.382] amphibians _____

_____

Matching

Match each of the following groups to its correct description. [pp.380-383]

1. _____ tetrapods

2. _____ cartilaginous fish

3. _____ lungfishes

4. _____ lobe-finned fish

5. _____ ray-finned fish

a. the most diverse fish in terms of body color, size and shape.

b. sharks and rays are the major representatives of this group.

c. the coelacanth is the last remaining type of this group.

d. possess both gills and lungs.

e. four-legged walkers

Labeling [p.381]

Fill in the matching letter of the following structures: You may also want to consult the textbook Figure 24.7 for additional information.

6. ____   swim bladder

7. ____   ovary

8. ____   gills

9. ____   kidney

10. ____   stomach

11. ____   nerve cord

12. ____   cloaca

13. ____   liver

14. ____   brain

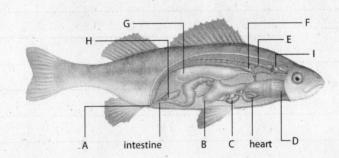

Matching [pp.382-382]

15. ____   gills are retained into adulthood

16. ____   any amphibian has this feature

17. ____   this group has the most species

18. ____   another name for a larval frog or toad

19. ____   these are adapted to drier conditions

20. ____   resemble the early tetrapods

a. heart with three chambers
b. salamanders and newts
c. tadpole
d. toads
e. axolotls
f. frogs and toads

24.5 EVOLUTION OF THE AMNIOTES [p.384]
24.6 NONBIRD REPTILES [p.385]
24.7 BIRDS- REPTILES WITH FEATHERS [p.386]

Selected Terms
Asteroid impact [p.384], yolk sac [p.384], embryo [p.384], amnion [p.384], chorion [p.384], allantois [p.384], albumin[p.384], Cretaceous [p.384], "heated from outside" [p.385], "cold-blooded [p.385], Komodo dragon [p.385], "beak" [p.385], caiman [p.385], sternum [p.386], penis [p.386], cloacal kiss [p.386], pelvic girdle & pectoral girdle [p.386]

Boldfaced Terms
[p.384] amniote eggs _____

_____

[p.384] reptiles _____

_____

[p.384] dinosaurs_____

_____

[p.385] ectotherms _____

_____

[p.385] endotherms _____

_____

[p.386] bird _____

_____

Matching

Match each of the following descriptions to the correct term. [pp.384-386]

1. _____ Produce eggs that have four membranes, have scaly skins and kidneys.

2. _____ The extinct members of the reptile clade.

3. _____ The outstanding feature of birds.

4. _____ This group has basic amniote features, but is not a bird or mammal.

5. _____ A protein that keeps reptiles waterproof

6. _____ The event at the end of the Cretaceous that ended the reign of the dinosaurs.

a. keratin
b. asteroid impact
c. dinosaurs
d. feathers
e. reptiles
f. amniotes

Complete the Table

Provide the missing information for each of the following groups of reptiles. [pp.384-385]

Group/Animal	Key characteristics
7.	Most advanced reptiles; closest relatives of the birds
8.	A herbivorous lizard
9.	The largest lizard
10.	These can weight over 2000 pounds
turtles	11.

Labeling [p.386]

Fill in the matching letter of the following structures. You may want to consult your textbook Figure 24.18B for additional information.

12. ____ skull

13. ____ pelvic girdle

14. ____ sternum

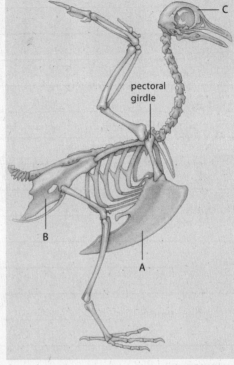

24.8 MAMMALS-THE MILK MAKERS [p.387]
24.9 PRIMATES TRAITS AND EVOLUTIONARY TRENDS [p.388-389]
24.10 EMERGENCE OF EARLY HUMANS [pp.390-391]
24.11 EMERGENCE OF MODERN HUMANS [pp.392-393]

Selected Terms

Milk [p.387], anthropoid lineage [p.388], depth perception [p. 388], upright walking [p. 389], better grips [p. 389], opposable movements [p.389], modified jaws and teeth [p.389], brain, behavior and culture [p.389], *Sahelanthropus tchadensis* [p.390], *Ardipithecus ramidus* [p.389], *Australpithecus afarensis* [p.389], *Paranthropus boisei* [p.390], *Homo habilus* [p.390], *H. erectus* [p.390], East African Rift Valley [p.391], Olduvai Gorge [p.391], *H. neanderthalensis* [p.392], *H. floresiensis* [p.392], *Homo sapiens* [p.392], "stone age" [p.393, "high tech" [p.393]

Boldfaced Terms

[p.387] mammals_____

_____

[p.387] monotremes _____

_____

[p.387] marsupials _____

_____

[p.387] placental mammals _____

_____

[p.388] primates_____

_____

[p.389] hominids _____

_____

[p.389] bipedalism _____

_____

[p.389] culture _____

_____

[p.390] australopiths_____

_____

[p.391] humans _____

_____

[p.392] multiregional model _____

_____

[p.392] replacement model _____

_____

True-False

If the statement is true, place a T in the space provided. If false, correct the underlined word so that the statement is correct. [p.387]

1. _____ Two of the unique characteristics of <u>mammals</u> are their ability to make hair and excrete milk.

2. _____ <u>Marsupials</u> are the only mammals that live in the oceans.

3. _____ <u>Monotremes</u> are the oldest mammalian lineage.

4. _____ About 50% of all mammals are rats and <u>bats.</u>

5. _____ <u>Placental</u> mammals had a competitive edge over marsupials due to a higher metabolic rate, precise control over body temperature, and a more efficient method of nourishing embryos.

6. _____ Molars and premolars in the jaws of mammals are close to the <u>front</u> of the mouth.

Fill-in-the-Blanks

The five trends that define the lineage leading to humans were set in motion among the first (7)

_____ [pp.388-389]. First, a reliance on the sense of smell became less important than (8)

_____ [pp.388-389] vision. Second, skeletal changes promoted (9) _____ [pp.388-389] –

upright walking – which freed (10) _____ [pp.388-389] for novel tasks. Third, bone and muscle

changes led to diverse (11) _____ [pp.388-389] motions. Fourth, (12) _____ [pp.388-389]

became less specialized to accommodate an omnivorous diet. Fifth, the evolution of the (13) _____

[pp.388-389], behavior, and (14) _____ [pp.388-389] became interlocked. Culture is the sum of a

social group's (15) _____ [pp.388-389] patterns, passed on through the generations by (16)

_____ [pp.388-389] and symbolic behavior. In brief, uniquely human traits emerged by

modification of (17) _____ [pp.388-389] that had evolved earlier, in ancestral forms.

Matching
Match each of the following groups to its correct description. [pp.390-393]

18. ____ *Homo habilis*

19. ____ *Australopithecus aferensis*

20. ____ *Paranthropus*

21. ____ *Homo erectus*

22. ____ *Homo sapiens*

a. The "handy-man"
b. Indisputably, the first bipedal hominid
c. The multiregional and replacement models attempt to explain the origin of this species.
d. "upright man"
e. one of the australopiths; wide face and large molars; died out 1.2 million years ago.

SELF QUIZ

1. Which of the following groups is immediate ancestor of the tetrapods? [pp.378-379]
   a. cartilaginous fish
   b. ray-finned fish
   c. lungfish
   d. coelacanths
   e. none of the above

2. Bipedalism and culture are found in all _____. [p.389]
   a. synapsids
   b. vertebrates
   c. dinosaurs
   d. hominids
   e. primates

3. Which of the following is a not a characteristic of all chordates? [pp.378-379]
   a. gills slits in a pharynx
   b. vertebrae made of calcium
   c. a tail that extends past the anus
   d. a notochord
   e. a nerve cord parallel to the gut and notochord

4. Reptiles, mammals and birds are all _____. [p.379]
   a. amniotes
   b. examples of organisms with four-chambered hearts in the entire group
   c. warm-blooded
   d. examples of sauropsids

5. The evolution of the _____ in fish set the stage for other key innovations leading to tetrapods. [p.381]
   a. brain
   b. lungs
   c. jaws
   d. fins
   e. legs

6. The term "handy-man" refers to _____. [p.391]
   a. all hominids
   b. *Homo habilus*
   c. *Australopithecus aferensis*
   d. *Homo sapiens*
   e. all primates.

7. Tunicates and lancelets are _____. [p.378]
   a. fish
   b. vertebrates
   c. amphibians
   d. invertebrate chordates
   e. hominids

8. Which of the following are components of an amniote egg? [p.384]
   a. allantois
   b. chorion
   c. amnion
   d. albumin
   e. all of the above

9. An asteroid impact on Earth attempts to explain the extinction of the _____ [p.384]
   a. Australopithecus species
   b. Dinosaurs
   c. Tuataras
   d. Monotremes

10. *Archaeopteryx* represents the first of the _____. [p.377]
   a. tetrapods
   b. lungfish
   c. vertebrate chordates
   d. feathered reptiles
   e. hominids

CHAPTER OBJECTIVES/ REVIEW QUESTIONS

1. List the four key characteristics of all chordates. [p.378]

2. Understand the difference between the invertebrate and the vertebrate chordates. [pp.378-379]

3. List the key innovations in the development of the vertebrate body plan. Be able to place these innovations in correct chronological order on an evolutionary diagram. [pp.378-379]

4. Distinguish between the cartilaginous and the bony fishes. [pp.380-381]

5. List the key characteristics of an amphibian. [pp.382-383]

6. Explain why amphibian species are on the decline. [p.383]

7. Explain the significance of an amniote. [p.384]

8. List one hypothesis that have been advanced to account for the extinction of the dinosaurs. [p.384]

9. Give the key characteristics of the major reptile groups. [p.385]

10. Explain how birds are adapted for flight. [p.386]

11. List the key characteristics of mammals. [p.387]

12. List the three major groups of mammals. [p.387]

13. Distinguish between a primate and a hominoid. [pp.388-389]

14. List the five trends in primate development that define the lineage to humans. [pp.388-389].

15. Give the significance of the australopiths. [p.390]

16. Recognize the key species in the evolution of the hominoids. [p.390]

17. Understand the two major hypotheses on the origins of *Homo sapiens*. [pp.392-393]

CHAPTER SUMMARY

A unique set of four traits characterizes the (1) _____: a supporting rod (notochord), a dorsal (2) _____ cord, a pharynx with (3) _____ in the wall, and a (4) _____ extending past an anus. Certain invertebrates and all (5) _____ belong to this group.

In some vertebrate lineages, a (6) _____ replaced the notochord as the partner of (7) _____ used in motion. Jaws evolved, sparking the evolution of novel (8) _____ organs and (9) _____ expansions. On land, lungs replaced (10) _____, and more efficient blood circulation enhanced (11) _____ exchange. Fleshy fins with skeletal supports evolved into (12) _____, which are now typical of vertebrates on land.

Vertebrates first evolved in the (13) _____, where lineages of cartilaginous and bony (14) _____ persist. Of all vertebrates, modern bony fishes show the most (15) _____. Mutations in master genes that control body plans were pivotal in the rise of aquatic (16) _____ and their move onto dry land.

As a group, the (17) _____ - known informally as the reptiles, birds, and (18) _____ are vertebrate lineages that radiated into nearly all habitats on (19) _____.

Primates that were ancestral to the (20) _____ lineage became physically and behaviorally adapted to change in global (21) _____ and available (22) _____. Behavioral and cultural flexibility helped humans disperse from (23) _____ throughout the world.

INTEGRATING AND APPLYING KEY CONCEPTS

1. There are two hypotheses regarding the flight of birds. The "tree down" hypothesis in which early bird-like lizards glided down from trees and the "ground up" in which early bird-like lizards ran on the ground and glided upward. Which one of the hypotheses do you think it was and why? Think about what is necessary for flight to occur.

2. Sharks and their relatives have a cartilaginous skeleton whereas most fish species have a bony skeleton. What advantage does one have over the other? Is there an advantage?

3. Fish, amphibians, and reptiles have their internal body temperatures controlled by the environment. What are the advantages and the disadvantages? Think expenses of staying warm and amounts of food needed as a path for your answer.

4. Why were many dinosaurs able to grow to be so big? Could this be one of the multitude of reason that dinosaurs went extinct? Why?

5. Knowing what you do about human evolution, what changes might the human species have in the next million years. Identify areas of the body that might change. Are those changes beginning now?

25

PLANT TISSUES

INTRODUCTION

This chapter investigates the different types of tissues in plants. These tissue types are very different than those found in the animals, thus special care should be taken to learn the function and location of the key tissue types. The plant body is organized into two sections – below ground and above ground. Below ground structures are used for absorbing nutrients and water and for storage and support. Above ground structures conduct photosynthesis, produce flowers, fruits, and seeds.

FOCAL POINTS

- Figure 25.1 [p.397] illustrates wood's porous structure.
- Figure 25.2 [p.398] is a diagram of a tomato plants with all of its parts.
- Figure 25.3 [p.398] shows the location of three plant tissue types in cross section.
- Figure 25.4 [p.399] diagrams the differences between a monocot and eudicot.
- Figure 25.6 [p.400] illustrates the locations complex tissues in the stem of a sunflower.
- Table 25.1[p.401] shows the overview of flowering plant tissues.
- Figure 25.8 [p.402] illustrates the primary structures of a eudicot and a monocot stem.
- Figure 25.9 [p.403] shows the growth in the stem of a eudicot.
- Figure 25.10 [p.404] illustrates the structure of a leaf and leaf forms.
- Figure 25.12 [p.405] shows the internal anatomy of a eudicot leaf.
- Figure 25.14 [p.406] illustrates the two types of root systems.
- Figure 25.15 [p.406] shows tissue formation and organization in a typical root.
- Figure 25.16 [p.407] diagrams the internal structure of roots.
- Figure 25.17 [p.408] illustrates the secondary growth of a plant.
- Figure 25.18 [p.409] diagrams the structure of a woody stem.

INTERACTIVE EXERCISES

25.1 SEQUESTERING CARBON IN FORESTS [p.397]
25.2 ORGANIZATION OF THE PLANT BODY [pp.398-399]

Selected Terms
carbon [p.397], carbon dioxide [p.397], Earth's climate [p.397], global emissions [p.397], photosynthesis [p.397], cellulose [p.397], forestry [p.397], 260,000 species [p.398], flowering plants [p.398], roots & shoots [p.398], water & nutrients [p.398], xylem [p.399], phloem [p.399], epidermis [p.399], monocot [p.399], eudicot [p.399], cotyledon [p.399], embryo [p.399], vascular tissue [p.399]

Boldfaced Terms
[p.398] ground tissue system _____

_____

[p.398] vascular tissue system_____

_____

[p.399] dermal tissue system _____

_____

[p.399] cotyledons _____

_____

[p.399] meristems _____

_____

[p.399] primary growth _____

_____

[p.399] secondary growth _____

_____

Short Answer

1. Explain why carbon is so important to sequester. [p.397]

_____

_____

_____

Identification

Use the terms in exercises 2-6 to identify each part of this illustration. Then complete each exercise by noting the letter of each term's description in the parentheses. [p.398]

2. _____ ground tissues ( )

3. _____ vascular tissues ( )

4. _____ dermal tissues ( )

5. _____ shoot system ( )

6. _____ root system ( )

a. Above ground, includes stems, leaves, and flowers (reproductive shoots)

b. Typically grows below ground, anchors aboveground parts, absorbs water and dissolved minerals, cells store and release food, and anchors the aboveground parts

c. Serves basic functions, such as food and water storage

d. Protects all exposed plant surfaces

e. Threads through ground tissue and delivers water and solutes through the plant

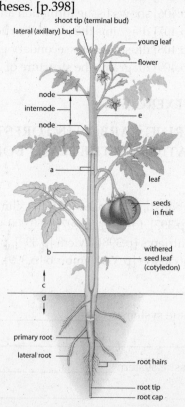

shoot tip (terminal bud)
lateral (axillary) bud
young leaf
flower
node
internode
node
e
a
leaf
seeds in fruit
b
withered seed leaf (cotyledon)
c
d
primary root
lateral root
root hairs
root tip
root cap

7. Explain the difference between primary and secondary growth with regards to meristems. [p.399]

_____

_____

_____

Complete the Table

8. There are two classes of flowering plants, monocots and eudicots. Complete the table by supplying the information requested. [p.399]

Class	Number of Cotyledons	Number of Floral Parts	Leaf Venation	Pollen Grains	Vascular Bundles
a. Monocots					
b. Eudicots					

25.3 COMPONENTS OF PLANT TISSUES [pp.400-401]

Selected Terms
radial sections [p.400], tangential sections [p.400], transverse sections [p.400], sclereids [p.400], fibers [p.400], cuticle [p.400], chloroplast [p.401], stomata [p.401]

Boldfaced Terms
[p.400] parenchyma _____

_____

[p.400] collenchyma _____

_____

[p.400] sclerenchyma _____

_____

[p.400] epidermis _____

_____

[p.401] mesophyll _____

_____

[p.401] xylem _____

_____

[p.401] tracheids _____

_____

[p.401] vessel members _____

_____

[p.401] phloem_____

_____

[p.401] sieve tubes _____

_____

[p.401] companion cells _____

_____

Labeling
Label the indicated components of the diagram below. [p.401]

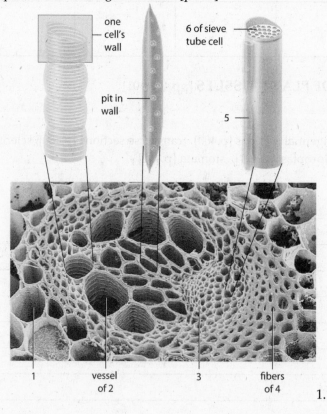

2. _____   5. _____

3. _____   6. _____

4. _____

Choice

For each of the following statements, choose the appropriate form of plant tissue from the list below. [pp.400-401]

    a. xylem
    b. phloem
    c. epidermis
    d. cuticle

7. _____ conserves water and protects against pathogens

8. _____ the main cells are sieve-tube members

9. _____ conducts water and dissolved mineral ions

10. _____ secretes the cuticle

11. _____ consists of vessel members and tracheids

12. _____ rich in a waxy substance called cutin

13. _____ first dermal tissue to form on a plant

14. _____ conducts sugars and other organic solutes

15. _____ companion cells assist in its function

25.4 PRIMARY SHOOTS [pp.402-403]
25.5 A CLOSER LOOK AT LEAVES [pp.404-405]

Selected Terms

terminal buds [p.403], lateral buds [p.403], vascular cylinder [p.402], epidermis [p.402], pith [p.402], cell division [p.403], node [p.403], node [p.403], internode [p.403], axillary or lateral buds [p.403], epidermis [p.404] petiole [p.404], sheath [p.404], blade [p.404], *palisade* mesophyll, [p.405] *spongy* mesophyll [p.405], plasmodesmata [p.405]

Boldfaced Terms

[p.402] vascular bundles _____

_____

[p.405] veins _____

_____

Labeling

Label each numbered structure in these illustrations of a eudicot stem. [p.403]

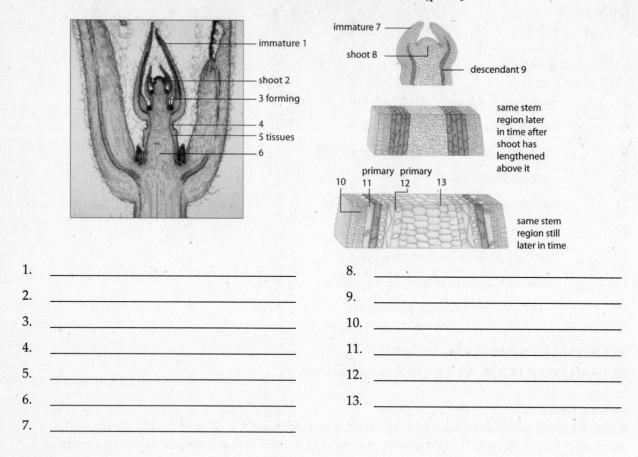

1. _____
2. _____
3. _____
4. _____
5. _____
6. _____
7. _____

8. _____
9. _____
10. _____
11. _____
12. _____
13. _____

Labeling

Name each of the genetically dictated, internal, cross-sectional stem patterns as either eudicot or monocot. [p.402]

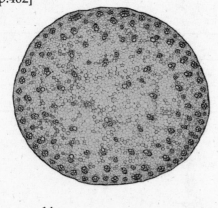

14. _____

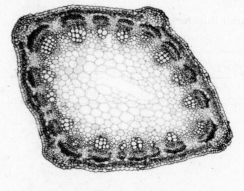

15. _____

Labeling

Name each numbered structure in these illustrations of leaf structure and organization. [p.404]

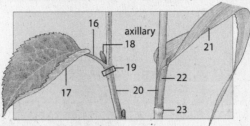

16. _____   20. _____

17. _____   21. _____

18. _____   22. _____

19. _____   23. _____

25.6 PRIMARY ROOTS [pp.406-407]

Selected Terms

pericycle [p.406], root tip [p.406], root cap [p.406], root hair [p.406], apical meristem [p.406], vessel element [p.406]

Boldfaced Terms

[p.406] taproot system _____

_____

[p.406] fibrous root system _____

_____

[p.407] roots hairs _____

_____

[p.407] vascular cylinder _____

_____

[p.407] endodermis _____

_____

Labeling

Label the numbered parts of these illustrations. (*Note:* Some numbers are used more than once; they refer to the same structure.) [pp.406-407]

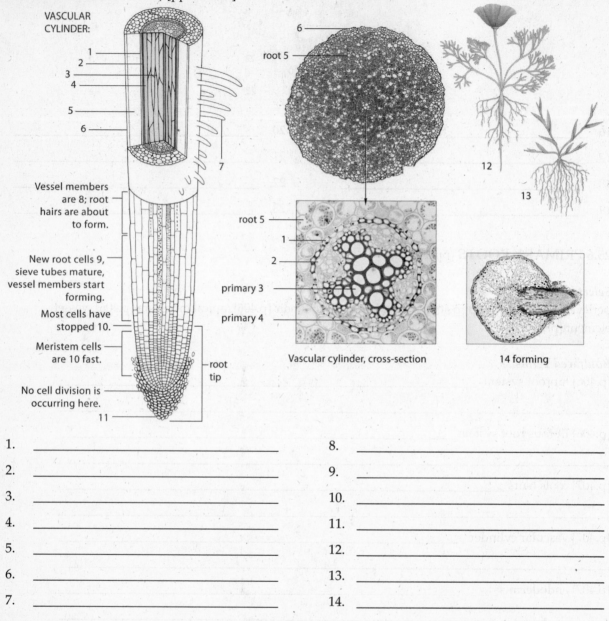

VASCULAR CYLINDER:

1
2
3
4
5
6
7

Vessel members are 8; root hairs are about to form.

New root cells 9, sieve tubes mature, vessel members start forming.

Most cells have stopped 10.

Meristem cells are 10 fast.

No cell division is occurring here.

11

root tip

6
root 5

12

13

root 5
1
2
primary 3
primary 4

Vascular cylinder, cross-section

14 forming

1. _____   8. _____

2. _____   9. _____

3. _____   10. _____

4. _____   11. _____

5. _____   12. _____

6. _____   13. _____

7. _____   14. _____

25.7 SECONDARY GROWTH [pp.408-409]

Selected Terms

pith [p.408], cortex [p.408], primary phloem [p.408], primary xylem [p.408], secondary phloem [p.408], secondary [p.408], primary growth [p.408], secondary growth [p.408], "tree ring" [p.409], tracheids [p.409], parenchyma rays [p.409]

Boldfaced Terms

[p.408] lateral meristem _____

_____

[p.408] vascular cambium _____

_____

[p.408] wood _____

_____

[p.409] cork cambium_____

_____

[p.409] bark _____

_____

[p.409] cork _____

_____

[p.409] heartwood _____

_____

[p.409] sapwood _____

_____

Labeling

Label each part of these illustrations, which show woody stems and primary and secondary growth [pp.408-409]

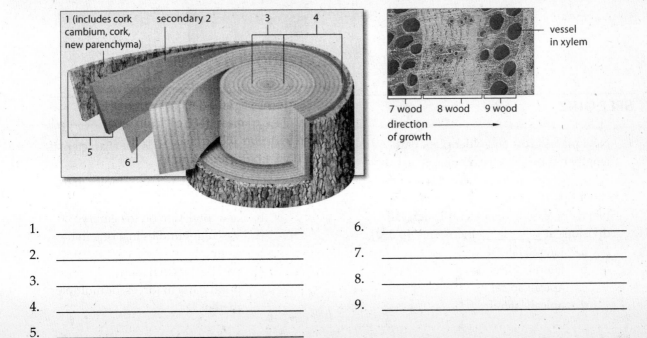

1. _____   6. _____

2. _____   7. _____

3. _____   8. _____

4. _____   9. _____

5. _____

25.8 VARIATION ON A STEM [p.410]
25.9 TREE RINGS AND OLD SECRETS [p.411]

Matching

Match each of the following forms of asexual reproduction to the correct description. [p.410]

1. ____ plants arise from buds on short underground stems

2. ____ plants arise from above ground horizontal stem

3. ____ plants arise from auxiliary buds on underground carbohydrate-storing stems

4. ____ plants arise from auxiliary buds on underground stems

5. ____ flattened, photosynthetic stems that store water

6. ____ plants arise from underground horizontal stem

a. rhizome
b. cladodes
c. corm
d. runner
e. tuber
f. bulb

Identify

Identify the structures that are marked with 7 in the figure below. [p.411]

7. _____

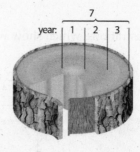

SELF QUIZ

1. New plants grow and older plant parts lengthen through cell divisions at _____ meristems present at root and shoot tips; older roots and stems of woody plants increase in diameter through cell divisions at _____ meristems. [p.398]
   a. lateral; lateral
   b. lateral; apical
   c. apical; apical
   d. apical; lateral

2. Which of the following is *not* characteristic of monocots? [p.399]
   a. scattered vascular bundles in the stem
   b. one cotyledon on the embryo
   c. vascular bundles in a ring in the stem
   d. parallel leaf venation
   e. floral parts in threes, or multiples thereof

3. Which of the following is *not* considered a type of simple tissue? [pp.400-401]
   a. xylem
   b. parenchyma
   c. collenchyma
   d. sclerenchyma
   e. a and d

4. Of the following cell types, which one does not appear in vascular tissues? [p.401]
   a. vessel members
   b. cork cells
   c. tracheids
   d. sieve-tube members
   e. companion cells

5. New stems and leaves form on plants by the activity of _____. [p.403]
   a. internodes
   b. apical meristems
   c. lateral meristems
   d. cork cambium
   e. cotyledons of the embryo

6. The major photosynthetic cell layer of many leaves is found in the _____. [p.405]
   a. lower epidermis
   b. leaf vein
   c. palisade mesophyll
   d. upper epidermis
   e. spongy mesophyll

7. A primary root and its lateral branchings represent a _____ system. [p.406]
   a. lateral root
   b. adventitious root
   c. taproot
   d. branch root
   e. fibrous root

8. A root vascular cylinder is composed of _____ [p.406]
   a. epidermis, phloem, and xylem
   b. pericycle and epidermis
   c. pericycle and endodermis
   d. pericycle, phloem, and xylem
   e. epidermis, cortex, and pith

9. Plants whose vegetative growth and seed formation continue year after year are _____ plants. [p.408]
   a. annual
   b. perennial
   c. nonwoody
   d. herbaceous

10. "Wood formed at the start of the growing season and has large-diameter, thin-walled xylem cells," describes _____. [p.408]
    a. hard wood
    b. late wood
    c. soft wood
    d. heartwood
    e. early wood

CHAPTER OBJECTIVES/REVIEW QUESTIONS

1. Distinguish between the ground tissue system, the vascular tissue system, and the dermal tissue system. [p.398-399]

2. Distinguish between primary and secondary growth. [p.399]

3. Distinguish between monocots and dicots by listing their characteristics and citing examples of each group. [p.399]

4. Be able to visually identify and generally describe the simple tissues and cells called parenchyma, collenchyma, and sclerenchyma. [pp.400-401]

5. Name and describe the functions of the conducting cells in xylem and phloem. [p.401]

6. Understand the importance of the epidermis. [p.401]

7. Be able to visually distinguish between monocot stems and dicot stems, as seen in cross section; identify the cells present by name and function. [p.402]

8. Understand the difference between a terminal and lateral bud. [p.403]

9. Describe the structure (cells and layers) and major functions of tissues composing a leaf.[p.405]

10. Distinguish a taproot system from a fibrous root system [p.406]

11. Describe the origin and function of root hairs. [p.407]

12. Describe the formation of cork and bark occurs by cork cambium activity. [p.408]

13. Distinguish early wood from late wood; heartwood from sapwood. [p.409]

14. Be able to distinguish the 6 different types of stems. [p.410]

15. Explain the origin of the annual growth rings (tree rings) seen in a cross section of a tree trunk. [p.411]

CHAPTER SUMMARY

Seed-bearing (1) _____ plants have a (2) _____ system, which includes stems, leaves, and reproductive parts. Most also have a (3) _____ system. Such plants consist mostly of ground tissues. Their vascular tissues distribute (4) _____, nutrients, and products of (5) _____. Their dermal tissues cover all surfaces exposed to the (6) _____.

Plants lengthen or thicken only at active (7) _____: zones where undifferentiated cells are dividing rapidly. Meristems near the tips of young shoots and roots drive (8) _____ growth, or the lengthening of plant parts.

Ground, vascular, and (9) _____ tissue systems of monocot and eudicot stems and leaves show patterns of (10) _____. Patterns of leaf growth and internal leaf structure maximize (11) _____ interception, (12) _____ conservation, and (13) _____ exchange.

Ground vascular and dermal tissue systems of monocot and (14) _____ roots show different patterns of organization. Root systems absorb water and (15) _____ ions, and may (16) _____ the plant.

In many plants, older branches or roots put on (17) _____ growth; they thicken during successive growing seasons. Extensive secondary growth is known as (18) _____.

INTEGRATING AND APPLYING KEY CONCEPTS

1. Many plants can be asexually reproduced using stem cuttings. What tissues of the stem allow the stem to produce an entirely new plant?

2. Why is it is it important for plants to be able to reproduce asexually?

3. What is an advantage to plants of reproducing sexually?

26

PLANT NUTRITION AND TRANSPORT

INTRODUCTION

This chapter explains the processes by which plants move water and nutrients throughout the tissues of the plant. Plant nutrients and symptoms of nutrient deficiencies are discussed. Adaptations of plant roots for absorbing and storing nutrients are explored. The chapter introduces several important principles of nutrient and water movement, including the cohesion-tension theory and the translocation of organic nutrients.

FOCAL POINTS

- Figure 26.2 [p.416] shows soil horizons.
- Table 26.1 [p.416] [p.416] shows plant nutrients and symptoms of deficiency.
- Figure 26.4 [p.418] shows examples of root specializations that help take up nutrients.
- Figure 26.6 [p.420] tracheids and vessel members from xylem.
- The importance of symbiotic relationships to plant nutrition. [pp.444-445]
- Figure 26.7 [p.421] demonstrates the cohesion-tension theory of water transport.
- Figure 26.8 [p.422] shows water conserving structures in plants.
- Figure 26.12 [p.425] illustrates the translocation of organic compounds in phloem.

INTERACTIVE EXERCISES

26.1 MEAN GREEN CLEANING MACHINES [p.415]
26.2 PLANT NUTRIENTS AND SOIL [pp.416-417]

Selected Terms
phytoremediation [p.415], toxic metals [p.415], popular trees [p.415], *Populus trichocarpa* X *deltoids* [p.415], TCE [p.415], Alpine pennycross [p.415], macronutrients [p.416], micronutrients [p.416], O horizon, A horizon, B horizon, C horizon, bedrock [p.416], sand, silt, clay [p.416], mineral ions [p.417]

Boldfaced Terms
[p.416] soil _____

_____

[p.416] humus _____

_____

[p.417] loams _____

_____

[p.417] topsoil _____

_____

[p.417] leaching _____

_____

[p.417] soil erosion _____

_____

Matching
Match each of the following terms with the correct description.

1. ____ micronutrients [p.416]
2. ____ leaching [p.417]
3. ____ soil erosion [p.417]
4. ____ macronutrients [p.416]
5. ____ soil [p.416]
6. ____ humus [p.416]
7. ____ loams [p.417]
8. ____ topsoil [p.417]
9. ____ sand, silt, clay [p.416]
10. ____ phytoremediation [p.440]

a. the different types of soil particles
b. nutrients that are required in amounts above 0.5 percent of the plant's dry weight
c. trace nutrients
d. mineral particles mixed with variable amounts of decomposing organic material
e. decomposing organic material
f. soil that is equal part sand, silt and clay
g. the layer of soil most important for plant growth
h. the use of plants to remove toxic compounds from soil
i. loss of soil due to the action of wind or water
j. the removal of nutrients from soil by water

26.3 HOW DO ROOTS ABSORB WATER AND MINERALS? [pp.418-419]

Selected Terms
root hairs [p.418], hyphae, [p.418], mutually beneficial relationships [p.418], nitrogen gas [p.418], ammonia [p.418], nitrogen-fixing bacteria [p.418], cytoplasm [p.419], permeable [p.419], vascular cylinder [p.419], root cortex [p.419], epidermis [p.419], endodermis [p.419], primary phloem [p.419], primary xylem [p.419], cortex [p.419], tracheids [p.419]

Boldfaced Terms
[p.418] mycorrhiza _____

_____

[p.418] nitrogen fixation _____

_____

[p.418] root nodules _____

_____

[p.419] Casparian strip _____

_____

Dichotomous Choice
Circle the word in parentheses that makes each statement correct. [pp.418-419]
1. Mycorrhizae and root nodules are two examples of (mutualism/parasitism).

2. Plants lack the ability to use (ammonia/ gaseous nitrogen) directly.

3. Nitrogen-fixing bacteria reside in localized swellings on legume plants known as (root hairs/root nodules).

4. In the mycorrhizae, the fungus receives (sugars/ minerals) from the plant.

5. In a mycorrhizal interaction between a young root and a fungus, the plant benefits by receiving (nitrogen/ minerals) from the fungus.

Labeling
Identify each numbered part of the following illustration. Use Figure 26.5 from the textbook for information. [p.419]

6. _____

7. _____

8. _____

9. _____

10. _____

11. _____

12. _____

13. _____

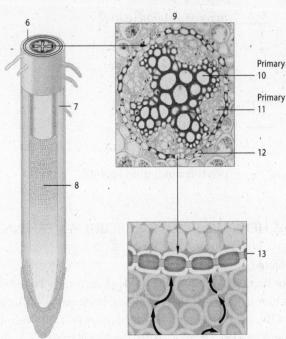

Matching
Match each of the following structures of the plant root with its correct function. [p.419]

14. _____ Casparian strip

15. _____ vascular cylinder

16. _____ exodermis

17. _____ endodermal cells

a. separates the vascular tissue from the root cotex; secrete the Casparian strip.
b. a waxy band that forces water to move through the endodermal cells.
c. contains the xylem and phloem
d. located just below the surface, these cells have a Casparian strip the prevents water and solute flow.

26.4 WATER MOVEMENT INSIDE PLANTS? [pp.420-421]
26.5 WATER-CONSERVING ADAPTATIONS OF STEMS AND LEAVES [pp.422-423]

Selected Terms
xylem [p.420], lignin-impregnated walls [p.420], Henry Dixon [p.420], evaporation [p.420], hydrogen bonds [p.420], mesophyll [p.421], vein [p.421], upper edidermis [p.421], vascular cambium [p.421], phloem [p.421], cuticle [p.422], stomata [p.422], cutin [p.422], osmosis [p.423], abscisic acid [p.423]

Boldfaced Terms

[p.420] cohesion-tension theory _____

_____

[p.420] transpiration _____

_____

[p.422] guard cells _____

_____

Concept Map

Provide the missing terms in the following concept map of the cohesion-tension theory. [p.420-421]

1. _____

2. _____

3. _____

4. _____

5. _____

Matching

Match each of the following with its function. [pp.422-423]

6. ____ stomata

7. ____ cuticle

8. ____ stoma

9. ____ abscisic acid (ABA)

10. ____ guard cells

a. a waxy covering that prevents water loss
b. the opening between two guard cells
c. the cells that are responsible for regulating the movement of gas into leaves
d. signals stomata to close during water shortages
e. the combination of guard cells and stoma

26.6 MOVEMENT OF ORGANIC COMPOUNDS IN PLANTS? [pp.424-425]

Selected Terms

phloem [p.424], vascular tissue [p.424], conducting tubes [p.424], fibers [p.424], parenchyma cells [p.424], carbohydrates [p.424], starch [p.424], sucrose [p.424], aphids [p.424], pressure gradient [p.424], sieve tube [p.424], source [p.425], sink [p.425], tugor [p.425]

Boldfaced Terms

[p.424] translocation _____

_____

[p.425] pressure flow theory _____

_____

Matching

Choose the most appropriate definition for each term. [pp.424-425]

1. ____ translocation
2. ____ sieve tube
3. ____ companion cells
4. ____ source
5. ____ sink
6. ____ pressure flow theory

a. cells found in phloem that are responsible for the movement of organic compounds
b. process that moves sucrose and organic compounds through phloem

c. process by which internal pressure moves a solute-rich solution from a source to a sink
d. a region of a plant where organic compounds are being unloaded from the sieve-tube system
e. a region of a plant where organic compounds are being loaded into the sieve-tube system
f. cells adjacent to the sieve-tube system whose purpose is to supply energy for active transport

Labeling

Provide the indicated terms in the diagram of the translocation process below. [p.425]

7. _____
8. _____
9. _____
10. _____
11. _____
12. _____

interconnected sieve tubes

7
(e.g., mature
leaf cells)

a 8 transport
mechanisms move
solutes into the
sieve tube, against
concentration
gradients.

10

b As a result of
the increased solute
concentration, the
water potential is
decreased in the
sieve tube, and
water moves in,
increasing 11
pressure.

c The pressure
then pushes solutes
by bulk flow between
a source and a sink,
with water moving
into and out of the
system all along the
way.

flow

d Both pressure and
solute concentrations
gradually decrease
between the source
and the sink.

e 9 are
unloaded into sink
cells, and the water
potential in those
cells is lowered.
Water moves out of
the sieve tube and
into sink cells.

12
(e.g., developing
root cells)

SELF QUIZ

1. Which of the following is not a macronutrient? [p.416]
   a. carbon
   b. manganese
   c. calcium
   d. nitrogen
   e. potassium

2. Gaseous nitrogen is converted to a plant usable form in the _____ of the plant. [p.418]
   a. root nodules
   b. root hairs
   c. mycorrhizae
   d. leaves
   e. stems

3. The removal of toxic compounds by plants is called _____. [p.415]
   a. photosynthesis
   b. transmutation
   c. translocation
   d. phytoremediation
   e. nitrification

4. The Casparian strip is located within the _____ of the plant. [p.419]
   a. roots
   b. leaves
   c. stems
   d. stomata

5. The cohesion-tension theory explains the movement of _____ in plants. [pp.420-421]
   a. minerals
   b. carbohydrates
   c. water
   d. bacteria

6. The stomata of a plant are involved in the exchange of _____. [pp.422-423]
   a. water
   b. ammonia
   c. gases
   d. sugars

7. Sieve tubes are located in the _____ of the plant. [p.424]
   a. xylem
   b. phloem
   c. mycorrhizae
   d. root hairs

8. In the pressure flow theory, solutes are moved into the system by _____. [pp.424-425]
   a. turgor pressure
   b. translocation
   c. transpiration
   d. active transport

9. A soil that has equal parts sand, silt and clay is called a _____. [pp.416-417]
   a. topsoil
   b. loam
   c. horizon
   d. humus

10. Which of the following processes represent the loss of nutrients in a soil? [p.417]
    a. translocation
    b. transpiration
    c. erosion
    d. leaching

CHAPTER OBJECTIVES/REVIEW QUESTIONS

1. Understand the concept of phytoremediation. [p.415]

2. Distinguish between a macro and micronutrient. [p. 416]

3. Understand the components of soil. [pp.416-417]

4. Distinguish between soil erosion and leaching. [p.417]

5. Understand the importance of root nodules, mycorrhizae and root hairs in the processing of water and minerals by a plant. [pp.418-419]

6. Understand the structure of a root hair and how the different components are involved in water and solute absorption. [pp.418-419]

7. Define transpiration. [p.420]

8. Outline the steps of the water-cohesion theory. [pp.420-421]

9. Understand the importance of the cuticle and stomata in water conservation. [pp.422-423]

10. Understand how organic materials move in a plant. [pp.424-425]

11. Define translocation. [p.424]

12. Understand the pressure-flow theory. [p.425]

CHAPTER SUMMARY

Many plant structures and functions are adaptations to limited amounts of (1) _____ and dissolved (2) _____ ions. The (3) _____ systems of vascular plants absorb water and mine the soil for nutrients, and many have symbionts that help them do so.

The amount of water and minerals available to plants to take up depends on the composition of the (4) _____. Soil is vulnerable to leaching and (5) _____.

Xylem distributes absorbed water and (6) _____ from roots to leaves. (7) _____ from leaves pulls up water molecules that are hydrogen-bonded to one another in long columns inside xylem. New molecules entering leaves replace the ones evaporating away.

A cuticle and (8) _____ help plants conserve water, a limited resource in most land habitats. Closed stomata stop water loss but also stop (9) _____ exchange. Some plant adaptations are trade-offs between water conservation and (10) _____.

(11) _____ distributes sucrose and other organic compounds from photosynthetic cells in leaves to living cells throughout the plant. (12) _____ compounds are actively loaded into conducting cells, then unloaded in growing tissues or storage tissues.

INTEGRATING AND APPLYING KEY CONCEPTS

1. How do you think that maple syrup is made from maple trees? Which specific systems of the plant are involved, and why are maple trees tapped only at certain times of the year?

2. Under stressful conditions, plant roots may expel mutualistic bacteria and fungi. Explain the possible effects of this event and how the process can become cyclical.

3. Fungi are often pathogens of plants. How would you propose that by the process of natural selection mycorrhizae may have evolved?

4. Considering your answer to question 3, the fungal-plant association was an important adaptation for the successful conquest of land. Why?

27

PLANT REPRODUCTION AND DEVELOPMENT

INTRODUCTION

This chapter begins with the plight of the honey bees, extremely important pollinators for fruit and nut crops across the United States. Chapter 27 introduces a typical plant life cycle from sporophyte to gametophyte; the structure of a flower with all of the parts; the formation of the fruit and the various types of fruits; and ultimately the formation of seeds. Asexual reproduction of plants is presented. Plants use tropisms to respond to various environmental stimuli as well as hormones that control aspects of growth and development. Plant defenses and senescence are presented in the remaining pages of this chapter.

FOCAL POINTS

- Figure 27.2a [p.430] illustrates the structures of a typical flower.
- Figure 27.2b [p.430] shows variations in floral structures.
- Figure 27.3 [p.430] outlines the life cycle of a typical flowering plant.
- Figure 27.5 [p.432] illustrates the life cycle of a cherry tree, a eudicot.
- Figure 27.6 [p.434] shows the seed of a shepherd's purse, a eudicot.
- Figure 27.9 [p.436] shows the anatomy of a corn seed, a monocot.
- Figure 27.10 a&b [p.436] illustrates the development of a corn seed into a corn plant.
- Figure 27.11 a&b [p.437] illustrates the development of a bean seed into a bean plant.
- Table 27.1 [p.438] lists the major plant hormones and their effects on plants.
- Figure 27.19 [p.442] illustrates the phytochrome activation pathway.

INTERACTIVE EXERCISES

27.1 PLIGHT OF THE HONEYBEE [p.429]
27.2 REPRODUCTIVE STRUCTURES OF FLOWERING PLANTS [pp.456-457]

Selected Terms
beekeepers [p.429], pollen [p.429], parasitic mites [p.429], pesticides [p.429], neonicotinoids [p.429], nectar [.p429], colony collapse [p.429], coevolved [p.429], sporophyte [p.430], diploid spore-producing plant [p.430], haploid gametophyte [p.430], lateral buds [p.430], receptacle [p.430], whorl [p.430], sepal [p.430], calyx [p.430], petal [p.430], corolla [p.430], filament [p.430], anther [p.430], pollen sacs [p.430], meiosis [p.430], *Prunus* [p.430], stigma [p.430], style [p.430], pistil [p.431], female gametophyte [.431], haploid female gametophyte [p.431], fertilization [p.431], seed germinates [p.431], sexual reproduction [p.431], wind pollination [p.431], fragrance [p.431], honey [p.431]

Boldfaced Terms
[p.429] pollinator _____

_____

[p.430] flower _____

_____

[p.430] stamens _____

_____

[p.431] carpels _____

_____

[p.431] ovary _____

_____

[p.431] ovule _____

_____

Identification

Identify each of the numbered components in the below diagram. [p.430]

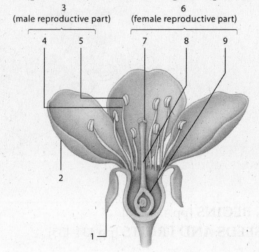

1. _____   6. _____
2. _____   7. _____
3. _____   8. _____
4. _____   9. _____
5. _____

Identification

Identify each of the numbered components in the below diagram. [p.430]

10. _____

11. _____

12. _____

13. _____

14. _____

15. _____

16. _____

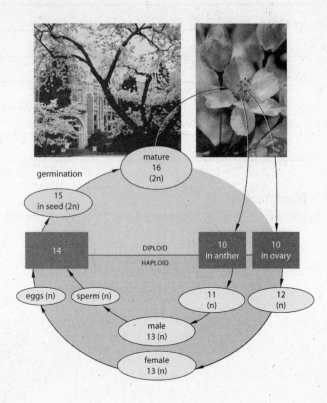

27.3 A NEW GENERATION BEGINS [pp.432-433]
27.4 FROM ZYGOTES TO SEEDS AND FRUITS [pp.434-435]

Selected Terms

pollen tube [p.432], sperm cells [p.432], endosperm mother cell [p.433], seedling [p.433], zygote [p.434], triploid (3n) [p.434], mitotic [p.434], embryo sporophyte [p.434], coyyledon [p.434], "vegetables" [p.434], simple fruits [p.434], aggregate fruit [p.434], multiple fruits [p.434], true fruits [p.434], accessory fruits [p.434], dry or juicy (fleshy) [p.434], *Rubus* [p.434], berry [p.434], pome [p.434], tufted fruits [p.435]

Boldfaced Terms

[p.432] megaspores _____

_____

[p.432] microspores _____

_____

[p.432] dormancy _____

_____

[p.432] pollination _____

_____

[p.458] double fertilization _____

_____

[p.432] endosperm _____

_____

[p.434] seed _____

_____

[p.434] fruit _____

_____

Identification

Identify the numbered parts of the eudicot life cycle using the figure on the following page. [pp.432-433]

1. _____   16. _____

2. _____   17. _____

3. _____   18. _____

4. _____   19. _____

5. _____   20. _____

6. _____   21. _____

7. _____   22. _____

8. _____   23. _____

9. _____   24. _____

10. _____   25. _____

11. _____   26. _____

12. _____   27. _____

13. _____   28. _____

14. _____   29. _____

15. _____   30. _____

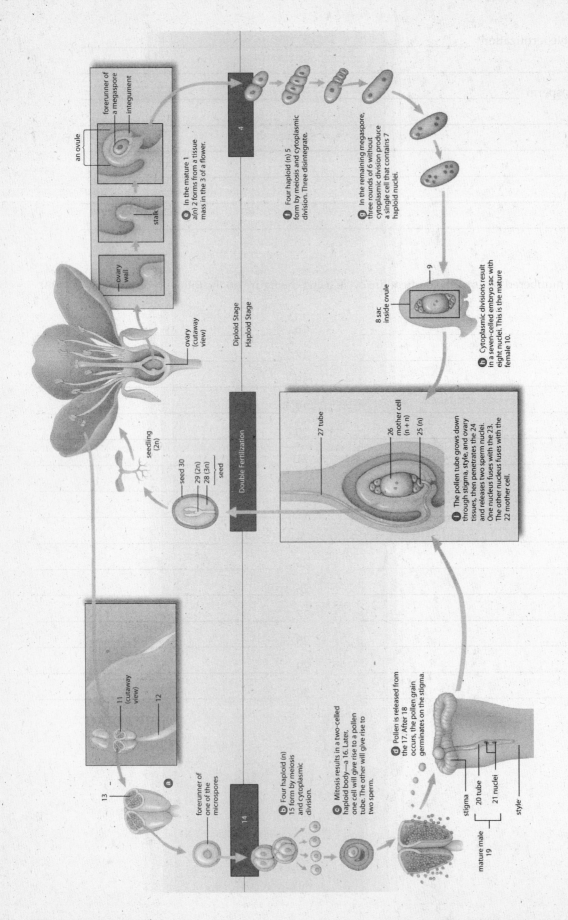

an ovule

forerunner of a megaspore

integument

a In the mature 1 a(n) 2 forms from a tissue mass in the 3 of a flower.

stalk

ovary wall

ovary (cutaway view)

Diploid Stage
Haploid Stage

4

e Four haploid (n) 5 form by meiosis and cytoplasmic division. Three disintegrate.

f In the remaining megaspore, three rounds of 6 without cytoplasmic division produce a single cell that contains 7 haploid nuclei.

8 sac inside ovule

9

g Cytoplasmic divisions result in a seven-celled embryo sac with eight nuclei. This is the mature female 10.

seedling (2n)

seed 30

29 (2n)

28 (3n)

seed

Double Fertilization

27 tube

26 mother cell (n + n)

25 (n)

i The pollen tube grows down through stigma, style, and ovary tissues, then penetrates the 24 and releases two sperm nuclei. One nucleus fuses with the 23. The other nucleus fuses with the 22 mother cell.

11 (cutaway view)

12

13

a

forerunner of one of the microspores

14

b Four haploid (n) 15 form by meiosis and cytoplasmic division.

c Mitosis results in a two-celled haploid body—a 16. Later, one cell will give rise to a pollen tube. The other will give rise to two sperm.

d Pollen is released from the 17. After 18 occurs, the pollen grain germinates on the stigma.

stigma

20 tube

21 nuclei

style

mature male 19

Matching

Match each of the following descriptions to the correct term. [pp.434-435]

31. ____ fruits that contain floral parts

32. ____ a pineapple is an example

33. ____ contain a pit

34. ____ has one to many seeds, no pit, and a fleshy fruit

35. ____ fruits that contain only the ovary wall and its contents

36. ____ has seeds in an elastic core

37. ____ a strawberry is an example

a. aggregate fruit
b. multiple fruit
c. drupes
d. berry
e. pome
f. true fruit
g. accessory fruit

27.5 ASEXUAL REPRODUCTION IN PLANTS [p.435]

Selected Terms

asexual reproduction [p.435], clone [p.435], King's holly [p.435], *Lomatia tasmanica* [p.435], grafted [p.435], Runner [p.463], rhizome [p.463], corm [p.463], tuber [p.463], bulb [p.463]

Boldfaced Terms

[p.435] vegetative reproduction _____

_____

[p.435] tissue culture propagation _____

_____

Matching

Match each of the following forms of asexual reproduction to the correct description. [p.435]

1. ____ plant derived from a parent plant cell that is not reversibly differentiated

2. ____ new plant arises from a structure dropped from the parent plant

a. vegetative reproduction
b. tissue culture propagation

27.6 Patterns of Development in Plants [pp.436-437]

Selected Terms

embryonic [p.436], monocot [p.436], dormancy [p.436], eudicot [p.436], cotyledon [p.436], hypocotyl [p.436], plumule [p.436], radicle [p.436], seed coat [p.436], primary root [p.436], branch root [p.436], adventitious foot [p.436], primary leaf [p.436], coleoptile [p.436], hydrolysis [p.436], rood nodule [p.437], withered cotyledon [p.437], evolutionary adaptation [p.437]

Boldfaced Terms

[p.436] germination_____

_____

Identification

Identify each of the numbered parts of the plant diagrams below. [pp.436-437]

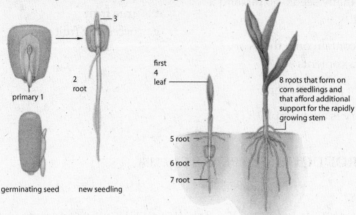

primary 1

2 root

3

germinating seed   new seedling

first 4 leaf

8 roots that form on corn seedlings and that afford additional support for the rapidly growing stem

5 root

6 root

7 root

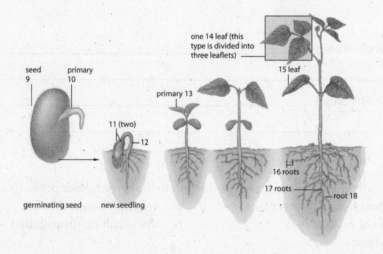

seed 9   primary 10

11 (two)

12

germinating seed   new seedling

one 14 leaf (this type is divided into three leaflets)

15 leaf

primary 13

16 roots

17 roots

root 18

1. _____
2. _____
3. _____
4. _____
5. _____
6. _____
7. _____
8. _____
9. _____

10. _____
11. _____
12. _____
13. _____
14. _____
15. _____
16. _____
17. _____
18. _____

27.7 PLANT HORMONES AND OTHER SIGNALING MOLECULES [pp.438-439]
27.8 ADJUSTING THE DIRECTION AND RATES OF GROWTH [pp.440-441]

Selected Terms

stimulus [p.438], inhibit [p.438], abscission [p.438], brassinosteroids [p.439], salicylic acid [p.439], nitric oxide[p.439], systemin [p.439]

Boldfaced Terms

[p.438] hormones _____

_____

[p.438] gibberellins _____

_____

[p.438] auxins _____

_____

[p.439] abscisic acid (ABA) _____

_____

[p.439] cytokinins _____

_____

[p.439] ethylene _____

_____

[p.440] tropism _____

_____

[p.440] gravitropism _____

_____

[p.440] phototropism _____

_____

[p.441] solar tracking _____

_____

[p.441] thigmotropism _____

_____

Choice

For each of the following, choose the appropriate plant hormone from the list below. Some answers can be used more than once. You may also want to consult Table 27.1 in the textbook for assistance. [pp.438-439]

    a. gibberellins
    b. cytokinins
    c. auxins
    d. ethylene
    e. abscisic acid

1. _____ compound that triggers stem elongation

2. _____ stimulates rapid cell division in root and shoot meristems and in maturing fruit

3. _____ prevents the growth of lateral buds along the lengthening stem

4. _____ the only gaseous plant hormone

5. _____ helps seeds germinate by stimulating cell division and elongation

6. _____ causes stomata to close in water-stressed plants

7. _____ helps control the growth of most tissues, fruit ripening, leaf dropping and other aging responses

8. _____ applied to nursery stock before shipping to induce dormancy

9. _____ induces and maintains dormancy in buds and seeds

10. _____ opposes the effects of auxins; makes lateral buds grow

11. _____ stimulates transport of some photosynthetic produces to seeds; in seeds it triggers protein synthesis and embryo formation

12. _____ keeps leaves from aging; used to prolong shelf life in cut flowers

13. _____ used to brighten the rids of citrus fruits

27.9 SENSING RECURRING ENVIRONMENTAL CHANGES [pp.442-443]
27.10 PLANT DEFENSES [pp.444-445]

Selected Terms

wavelength [p.442], rubisco [p.442], photosystem II [p.442], ATP synthase [p.442], phototropism [p.442], short-day plants [p.442], long-day plants [p.442], day-neutral plants [p.442], vernalis [p.443], dormancy [p.443], jasmonates [p.444], herbivores [p.444], antifungal & antiviral proteins [p.444], abscission zone [p.445]

Boldfaced Terms

[p.442] biological clock _____

_____

[p.442] circadian rhythm _____

_____

[p.442] phytochromes_____

_____

[p.442] photoperiodism _____

_____

[p.443] vernalization_____

_____

[p.444] systemic acquired resistance _____

_____

[p.444] abscission _____

_____

[p.444] senescence_____

_____

Dichotomous Choice

Circle the Terms in parentheses that make each statement correct. [pp.442-443]

1. Summer blooming irises are (short-day/ long-day) plants.

2. Chrysanthemums are examples of (short-day/ long-day) plants.

3. Sunflowers are (short-day/ day-neutral) plants that form flowers when they mature, without regard to the season.

4. "Short-day" plants flower only when the nights are (shorter/longer) than the critical value.

5. "Long-day" plants flower only when nights are (shorter/ longer) than a critical value.

6. Phytochrome is activated by (red/ far-red) wavelengths.

7. When a dark interval is interrupted by a pulse of (red/ far-red) light, phytochrome becomes inactivated.

8. If red light is followed by a pulse of far-red light, phytochrome is (activated/ inactivated) and plants behave as though the dark period was uninterrupted.

Matching

Match each of the following terms its correct description. [pp.442-445]

9. ____ abscission

10. ____ senescence

11. ____ dormancy

12. ____ vernalization

a. the period from full-maturity to death of the plant

b. the shedding of plant parts

c. a period of arrested growth

d. flowering only after exposure to low winter temperatures

SELF QUIZ

1. The portion of the carpel that contains the ovule is the _____. [p.430]
   a. stigma
   b. anther
   c. style
   d. ovary
   e. filament

2. The phase of the life cycle of a plant that gives rise to spores in the _____. [p.430]
   a. female gametophyte
   b. male gametophyte
   c. seed
   d. sporophyte
   e. vegetative growth

3. An immature fruit is a(n) _____ and an immature seed is a(n) _____. [pp.432-433]
   a. ovary; megaspore
   b. ovary; ovule
   c. megaspore; ovule
   d. ovule; ovary

4. In flowering plants, one sperm nucleus fuses with that of an egg, while the sperm fuses with the endosperm. This is called _____. [p.432]
   a. vernalization
   b. double fertilization
   c. vegetative propogation
   d. meiosis

5. Which of the following is not a fruit? [pp.434-435]
   a. drupes
   b. pomes
   c. berry
   d. seed

6. An example of asexual reproduction in flowering plants is _____. [p.435]
   a. vegetative reproduction
   b. tissue culture propagation
   c. a and b

7. In dry climates, germination depends on _____. [p.436]
   a. number of daylight hours
   b. temperature
   c. water
   d. soil oxygen level
   e. all of the above

8. Which of the following controls the growth of plant tissues, fruit ripening, leaf drop and other aging responses? [pp.438-439]
   a. auxins
   b. cytkinins
   c. gibberellins
   d. ethylene
   e. abscisic acid

9. The sum of all processes that lead to the death of a plant or some of its parts is called _____. [p.444]
   a. dormancy
   b. vernalization
   c. abscission
   d. senescence
   e. none of the above

10. Phytochrome is activated by _____. [pp.442-443]
    a. red wavelengths of light
    b. water
    c. auxins
    d. far-red wavelengths of light
    e. none of the above

CHAPTER OBJECTIVES/REVIEW QUESTIONS

1. Understand the life cycle of a flowering plant. [p.430]

2. Be able to label the parts of a flower. [p.430]

3. Distinguish between microspore and megaspores. [p.432-433]

4. Distinguish between pollination and fertilization. [p.432-433]

5. Understand the process of double fertilization, the structure involved, and the chromosome state (ploidy) of the resulting tissues. [pp.432-433]

6. Understand the terminology associated with seeds and fruits. [pp.434-435]

7. Understand the basic mechanisms of seed dispersal. [p.434]

8. Understand the major methods of asexual reproduction in plants. [p.435]

9. Understand the conditions under which germination occurs. [p.436]

10. For each of the plant hormones, recognize its influence on plant growth and development. [pp.4438-439]

11. Understand the major forces of tropism in plants. [pp.440-441]

12. Understand the concept of a biological clock. [p.442]

13. Understand photoperiodism. Be able to distinguish between short-day, long-day, and day-neutral plants. [p.442]

14. Understand the activation/deactivation pathway of phytochrome. [pp.442-443]

15. Understand the difference between dormancy, senescence and abscission. [pp.444-445]

CHAPTER SUMMARY

(1) _____ reproduction is the dominant reproductive mode of flowering plant life cycles, which typically depend on (2) _____. Spores and gametes form in the specialized reproductive shoots called (3) _____.

Sperm-bearing male gametophytes (pollen grains) and female (4) _____ that bear eggs inside ovules form in reproductive parts of flowers. After (5) _____, ovules mature into seeds, each an embryo (6) _____ and tissues hat nourish and protect it.

As seeds develop, tissues of the (7) _____ and often other floral parts mature into fruits, which function in seed (8) _____. Air currents and (9) _____ are the main dispersal agents.

Many species of flowering plants also reproduce (10) _____ by vegetative growth and other mechanisms.

Interactions among (11) _____ and other signaling molecules control plant growth and development. Hormone synthesis starts at seed (12) _____ and guides all events of the life cycle, such as (13) _____ and shoot development, flowering, fruit formation, and dormancy.

Plant cell receptors that govern hormone (14) _____ respond to (15) _____ cues, such as gravity, sunlight, and seasonal shifts in (16) _____ length and temperature.

INTEGRATING AND APPLYING KEY CONCEPTS

1. A lumber company removes all but a single oak tree from a forest. What changes have occurred in the environment of the oak tree? Which hormones and tropisms may be involved in the response of the oak to the changing environment?

2. You have been hired by a company to bring Costa-Rican plants to the United States for sale. The plants require 12 hours of continuous darkness per night in order to flower and produce seeds. Explain what is happening within this plant.

3. As an agricultural consultant you have been assigned the task of maximizing profits by growing plants as quickly as possible and reducing the amount of loss due to damage or rotting on the way to market. Describe the hormones that you would most likely use in your work and when they would be applied to the plant.

4. Explain the concept that "there is no such thing as a vegetable with respect to fruits."

28

ANIMAL TISSUES AND ORGAN SYSTEMS

INTRODUCTION

All animals are comprised of a set of common tissue types. These tissue types are then organized into specialized organ systems that perform specific tasks for the organism. These organ systems work together to maintain homeostasis within the specific animal's body. In this chapter you will be introduced to the major tissue types and organ systems found in the animals, specifically humans.

FOCAL POINTS

- Figure 28.1 [p.449] illustrates how stem cells differentiate into other specialized cell types.
- Figure 28.2 [p.450] is a diagram of the levels of organizations in a human body.
- Figure 28.5 [p.452] illustrates the three types of simple epithelia.
- Figure 28.6 [p.453] is a diagram of glandular epithelium.
- Figure 28.8 [p.454] illustrates the types of connective tissues.
- Figure 28.9 [pp.455-456] the three types of muscle tissues.
- Figure 28.10 [p.457] illustrates a neuron.
- Figure 28.12 [p.458] illustrates the main body cavities of the human body.
- Figure 28.13 [pp.458-459] illustrates the human body organ systems and their functions.
- Figure 28.14 [p.459] illustrates some of the ways that organs interact.
- Figure 28.15 [p.460] illustrates skin structure.
- Figure 28.17 [462] shows an example of negative feedback.

INTERACTIVE EXERCISES

28.1 STEM CELLS [p.449]
28.2 ORGANIZATION OF ANIMAL BODIES [pp.450-451]
28.3 EPITHELIAL TISSUE [pp.452-453]
28.4 CONNECTIVE TISSUE [pp.454-455]

Selected Terms
embryonic [p.449], paralysis [p.449], stem cell research [p.449], development [p.450], vertebrate [p.450], epithelial tissue [p.450], connective tissue [p.450], muscle tissue [p.450], nervous [p.450], cardiac muscle [p.450], homeostasis [p.450], organs [p.450], anatomy [p.451], physiology [p.451], diffusion [p.451], circulatory system [p.4511], digestive tract [p.451]. respiratory tract [p.451], simple epithelium [p.452], squamous epithelium [p.452], cuboidal epithelium [p.452], columnar epithelium [p.452], cilia [p.452], oviduct [p.452], cancer & carcinoma [p.453], fibroblast [p.454], chondrocytes [p.454], collagen fibers [p.454], red blood cells, white blood cells & platelets [p.455], plasma [p.455]

Boldfaced Terms
[p.449] stem cell _____

_____

[p.450] extracellular fluid _____

_____

[p.450] interstitial fluid _____

_____

[p.450] plasma _____

_____

[p.452] epithelium _____

_____

[p.452] basement membrane _____

_____

[p.452] microvilli _____

_____

[p.453] exocrine glands _____

_____

[p.453] endocrine glands _____

_____

[p.454] connective tissues _____

_____

[p.454] loose connective tissue _____

_____

[p.454] dense, irregular connective tissue _____

_____

[p.454] dense, regular connective tissue _____

_____

[p.454] cartilage _____

_____

[p.455] adipose tissue _____

_____

[p.455] bone tissue _____

_____

[p.455] blood _____

_____

Matching

Match each of the following to its correct definition. [pp.449-455]

1. ____ endocrine glands

2. ____ loose connective tissue

3. ____ glands

4. ____ stem cells

5. ____ dense, irregular connective tissue

6. ____ exocrine glands

7. ____ cartilage

8. ____ dense, regular connective tissue

9. ____ homeostasis

10. ____ adipose tissue

11. ____ blood

12. ____ bone tissue

13. ____ plasma

a. glands that secrete products into interstitial fluid; do not possess ducts

b. contains fibroblasts and fibers dispersed widely throughout the matrix

c. contains fibroblasts and fibers in an orderly configuration

d. living cells within calcium hardened secretions

e. cells are derived from stem cells in bone

f. contain the body's main energy reservoir

g. contains ducts or tubes that deliver secretions to a free epithelial surface

h. a mixture of collagen fibers and glycoproteins

i. maintaining conditions of the internal environment

j. cells that can divide to produce different types of cells

k. the fluid portion of blood

l. these seal epithelial membranes together

m. the general term for organs that secrete substances onto the skin, or into a body cavity or interstitial fluid.

Choice

Choose the appropriate class of tissue for each statement.

  a. epithelial tissue [pp.452-453]
  b. connective tissue [pp.454-455]

14. ____ fibroblasts are usually the main cell type

15. ____ one free surface is exposed to the environment

16. ____ cells within an extracellular matrix

17. ____ arranged in a sheet-like configuration

18. ____ cells are usually squamous, cuboidal or columnar in shape

28.5 MUSCLE TISSUES [pp.456-457]
28.6 NERVOUS TISSUE [p.457]

Selected Terms

contract [p.456], muscle fibers [p.456], "voluntary muscle" [p.456], "involuntary" muscle [p.456], sphincters [p.456], electrochemical signals [p.457], neuroglial cells [p.457]

Boldfaced Terms

[p.456] skeletal muscle tissue _____

_____

[p.456] cardiac muscle tissue _____

_____

[p.456] smooth muscle tissue _____

_____

[p.457] nervous tissue _____

_____

[p.457] neurons _____

_____

True/False

If the statement is correct, place a "T" in the blank. If the statement is incorrect, correct the underlined word to make the statement true. [pp.456-457]

1. _____ <u>Neurons</u> are the primary cells of the nervous system.

2. _____ The <u>skeletal</u> muscle tissues contain the highest number of mitochondria and the lowest level of glycogen of all muscle tissues.

3. _____ Cells of the <u>nervous</u> system function by contraction.

4. _____ Smooth muscle tissue controls the operation of <u>voluntary</u> muscles.

5. _____ Cells of the <u>nervous</u> system respond to stimuli using electrical and chemical signals.

6. _____ <u>Skeletal</u> muscle tissue lines the walls of most soft internal organs.

28.7 ORGANS AND ORGAN SYSTEMS [pp.458-459]
28.8 CLOSER LOOK AT A ORGAN-HUMAN SKIN [pp.460-461]
28.9 INTEGRATED ACTIVITIES [p.462-463]

Selected Terms

cranial cavity [p.458], spinal cavity [p.458], thoracic cavity [p.458], abdominal cavity [p.458], pelvic cavity [p.458], organ system [p.458], integumentary, nervous, muscular, skeletal, circulatory, endocrine, lymphatic, respiratory, digestive, urinary, reproductive systems [pp.458-459], keratin [p.460], oil gland [p.460], hypodermis [p.460], hair follicle [p.460], sweat gland [p.460], adipose [p.460], stratified squamous epithelium [p.460], keratinocyte [p.460], melanin [p.461], UV radiation [p.461], Vitamin D [p.461], vitiligo [p.461], cultured skin [p.461], response [p.462], stimulus [p.462], gap junction [p.463]

Boldfaced Terms

[p.458] coelom _____

_____

[p.460] epidermis _____

_____

[p.460] dermis _____

_____

[p.462] sensory receptor _____

_____

[p.462] negative feedback _____

_____

[p.463] apoptosis _____

_____

Matching

Use the figure to the right and select the letter that represents the correct body cavity and place it next to its name. [p.458]

1. ____ pelvic cavity

2. ____ diaphragm

3. ____ cranial cavity

4. ____ thoracic cavity

5. ____ spinal cavity

6. ____ abdominal cavity

Matching

Match the correct choice with its description. [pp.456-463]

7. ____ a circular muscle

8. ____ a process of cellular suicide

9. ____ response that reverses a change in homeostasis

10. ____ a cell that detects a specific stimulus

11. ____ the outer most skin layer

a. sensory receptor
b. apoptosis
c. dermis
d. sphincter
e. negative feedback

Labeling and Matching

Write the name of each organ system described by the statement. Then match the letter from the illustration to the correct statement. You may also want to consult Figure 28.13 in the textbook for additional information. [pp.458-459]

12. _____ ( ) Rapidly transports many nutrients to and from cells; helps stabilize internal pH and temperature.

13. _____ ( ) Rapidly delivers oxygen to the tissue that bathes all living cells; removes carbon dioxide waste of the cells; helps regulate pH.

14. _____ ( ) Maintains the volume and composition of the internal environment; excretes excess fluid and blood-borne wastes.

15. _____ ( ) Supports and protects body parts; provides muscle attachment sites; produces red blood cells; stores calcium, phosphorous.

16. _____ ( ) Hormonally controls body function; works with nervous system to integrate short- and long-term activities.

17. _____ ( ) *Female*: produces eggs; after fertilization, affords a protected, nutritive environment for the development of new individuals. *Male*: produces and transfers sperm to the female. Hormones of both systems also influence other organ systems.

18. _____ ( ) Ingests food and water; mechanically and chemically breaks down food and absorbs small molecules into the internal environment; eliminates food residues.

19. _____ ( ) Moves body and its internal parts; maintains posture; generates heat.

20. _____ ( ) Detects both internal and external stimuli; controls and coordinates responses to stimuli; integrates all organ system activities.

21. _____ ( ) Protects body from injury, dehydration, and some pathogens; controls its temperature; excretes some wastes; receives some external stimuli.

22. _____ ( ) Collects and returns some tissue fluid to the bloodstream; defends the body against infection and tissue damage

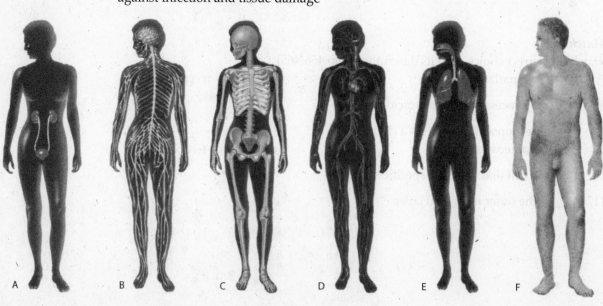

A   B   C   D   E   F

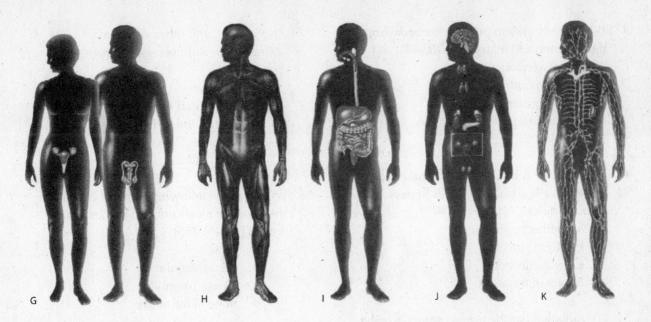

G   H   I   J   K

Labeling

Label each of the numbered components of the diagram below. [p.460]

23. _____

24. _____

25. _____

26. _____

27. _____

28. _____

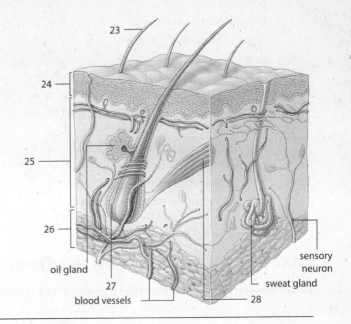

SELF QUIZ

1. An animal's structural traits are its _____ and its functional traits are its _____. [pp.450-451]
    a. anatomy/physiology
    b. physiology/anatomy

2. Which body system maintains the pH and temperature of the internal environment? [pp.458-459]
    a. nervous
    b. urinary
    c. reproductive
    d. circulatory

Animal Tissues and Organ Systems **287**

3. Which body system protects the body from dehydration and injury? [pp.458-459]
   a. respiratory
   b. circulatory
   c. integumentary
   d. urninary
   e. skeletal

4. Which of the following types of junctions between cells allow substances to pass through quickly? [p.479]
   a. adhering junctions
   b. tight junctions
   c. gap junctions
   d. all of the above

5. UV radiation has the biggest negative effect on the _____. [p.461]
   a. nervous system
   b. skin
   c. adipose tissue
   d. reproductive system

6. Select the correct choice from the selections below with respect of going from the outer surface of the skin to the inner most part of the skin. [p.460]
   a. epidermis, hypodermis, dermis
   b. dermis, hypodermis, epidermis
   c. dermis, epidermis, hypodermis
   d. epidermis, dermis, hypodermis

7. In a human, and other vertebrates, extracellular fluid consists of mainly of. [pp.450-451]
   a. plasma
   b. interstitial fluid
   c. extracellular fluid
   d. a and b
   e. a, b, and c

8. Which of the following forms of connective tissue is the most common in the human body? [pp.454-455]
   a. loose
   b. dense, regular
   c. dense, irregular
   d. none of the above

9. Which of the following is under voluntary control? [p.456]
   a. skeletal muscle
   b. smooth muscle
   c. cardiac muscle
   d. all of the above

10. Which of the following is not a form of connective tissue? [pp.454-455]
    a. blood
    b. cartilage
    c. adipose tissue
    d. bone
    e. epithelium

CHAPTER OBJECTIVES/REVIEW QUESTIONS

1. Understand the basic forms and functions of epithelium tissue. [pp.452-453]

2. Distinguish between an endocrine and exocrine gland [p.453]

3. Understand the basic forms and functions of connective tissue. [pp.454-455]

4. Recognize the importance of cartilage, adipose tissue, blood and bone as a connective tissue. [pp.454-455]

5. Understand the role of muscle tissue and the three major forms of muscle tissue. [p.456]

6. Recognize the role of nervous tissue in the body. [p.457]

7. For each body system, understand its general role in the human body. [pp.458-459]

8. Understand the structure and function of human skin. [pp.460-461]

9. Understand what happens with sun exposure and the skin.[pp.460-461]

10. Be able to give an example of negative feedback. [p.462]

11. Understand and be able to give an example of apoptosis. [p.463]

CHAPTER SUMMARY

Epithelial, (1) _____, muscle, and (2) _____ tissues are the basic categories of tissues in nearly all animals.

(3) _____ line the body surface and its internal cavities and tubes. They have protective and (4) _____ functions.

Connective tissues bind, support, strengthen, protect, and (5) _____ other tissues. They include soft connective tissues, (6) _____, bone, blood, and (7) _____ tissue.

(8) _____ tissues help move the body and its parts. The three kinds are skeletal, (9) _____, and smooth muscle tissue.

Nervous tissue provides local and long-distance lines of (10) _____ among cells. Its cellular components are (11) _____ and neuroglia.

Vertebrate organ systems (12) _____ the tasks of survival and reproduction for the body as a whole; they show a division of labor.

Human (13) _____ is an example of an organ system. It has epithelial layers, connective tissue, adipose tissue, glands, blood vessels, and (14) _____ receptors. It helps protect the body from injury and some pathogens, conserve (15) _____ and control body temperature, excrete (16) _____ and detect some external (17) _____.

(18) _____ feedback assists the body in reversing a process. Gap (19) _____ allow substances to pass quickly between adjoining cells. How the human fingers form is an example of (20) _____.

INTEGRATING AND APPLYING KEY CONCEPTS

1. As a stem cell researcher you have a line of cells that you wish to turn into nervous tissue. What type of characteristics would those cells have to develop? How about if you decided to make connective tissues from these cells? What would be the major differences in how the nervous cells and connective cells developed?

2. What are some of the uses of adult and embryonic stem cells?

29
NEURAL CONTROL

INTRODUCTION

While we share many characteristics with other animals, it is really our ability to feel emotions that sets the human apart from other species. The human brain is the organ responsible for this characteristic, and for many other traits that we possess. This chapter describes the vertebrate nervous system, with special attention paid to the organization and structure of the nervous systems of humans. The chapter briefly discusses the evolution of the nervous system, then goes on to an in-depth look at how nerve cells actually transmit messages to other cells. One main example is how the nerve cell "talks" to a muscle cell to initiate movement. Much information on the chemicals involved in these messages is given. From this point, the chapter then moves on to a discussion of drugs that can interfere with the function of the nervous system. Finally the chapter concludes with a detailed look at the human brain, and to the organization of the human nervous system. The various parts of the brain are discussed in detail, including the most recently evolved cerebral cortex and the limbic system; the parts of the brain that allow humans to experience emotion.

FOCAL POINTS

- Figure 29.2 [p.468] shows a gallery of invertebrate nervous systems..
- Figure 29.3 [p.468] shows the functional divisions of vertebrate nervous systems.
- Figure 29.4 [p.469] illustrates some of the major nerves of the human nervous system.
- Figure 29.6 [p.470] shows a diagram of a neuron.
- Figure 29.7 [p.471] illustrates the protein channels and pumps that span a neuron membrane.
- Figure 29.9 [p.473] shows a graph of changes in membrane potential changes during an action potential.
- Figure 29.10 [p.474] illustrates communication at a synapse. .
- Figure 29.13 [p.478] diagrams the structure of a vertebrate nerve.
- Figure 29.14 [p.479] shows the sympathetic and parasympathetic nerves.
- Figure 29.15 [p.480] shows the location and organization of the spinal cord.
- Figure 29.16 [p.481] illustrates a stretch reflex.
- Figure 29.17 [p.482] illustrates the development of the human brain.
- Figure 29.18 [p.482] shows other vertebrate brains.
- Figure 29.20 [p.483] illustrates the human brain.
- Figure 29.23 [p.485] shows the limbic system.

INTERACTIVE EXERCISES

29.1 IN PURSUIT OF ECSTASY [p.467]

29.2 EVOLUTION OF NERVOUS SYSTEMS [pp.468-469]

29.3 NEURONS – THE COMMUNICATORS [p.470]

29.4 MEMBRANE POTENTIALS [p.471]

29.5 A CLOSER LOOK AT ACTION POTENTIALS [p.472]

Selected Terms

ecstacy [p.467], MDMA [p.467], psychosis [p.467], illegal drug [p.467], psychoactive drug [p.467], multicelled organism [p.468], electrical signal [p.468], chemical messengers [p.468], cnidarians (hydra, jellyfish) [p.468], bilateral [p.468], anterior end [p.468], planarian [p.468], neuron cell bodies [p.468], cytoplasmic extensions [p.468], annelids [p.468], arthropods [p.468], dorsal nerve cord [p.469], autonomic nerves [p.469], sciatic nerve [p.469], spinal cord [p.469], cenral nervous system [p.469], somatic nerves [p.469], sympathetic division [p.469], parasympathetic system [p.469], neuron's nucleus [p.470], receptor [p.470], input zones [p.470], trigger zone [p.470], conducting zone [p.470], output zone [p.470], lipid bilayer [p.471], potassium ions & sodium ions [p.471], sodium-potassium pump [p.471], graded [p. 472], all or nothing spike [p.472], axon terminal [p.473]

Boldfaced Terms

[p.468] neurons _____

_____

[p.468] neuroglia cells _____

_____

[p.468] nerve net _____

_____

[p.468] sensory neurons _____

_____

[p.468] interneurons _____

_____

[p.468] motor neurons _____

_____

[p.468] ganglion _____

_____

[p.468] nerve cord _____

_____

[p.468] nerve _____

_____

[p.469] central nervous system _____

_____

[p.469] peripheral nervous system _____

_____

[p.470] dendrites _____

_____

[p.470] axons _____

_____

[p.471] membrane potential_____

_____

[p.471] resting potential _____

_____

[p.471] action potential_____

_____

[p.472] threshold potential_____

_____

[p.472] positive feedback _____

_____

Completion

Use the table below to compare dendrites and axons. [p.470]

Dendrites	Axons
1.	Carry information away from the cell body
One per cell in most sensory neurons; many per cell in most other neurons	2.
Branch close to cell body	3.
4.	Insulating sheath for most

Labeling

Use the following diagram to label the parts of a neuron involved in the generation of an action potential. Use page 471 from the textbook to assist you in completing this diagram.

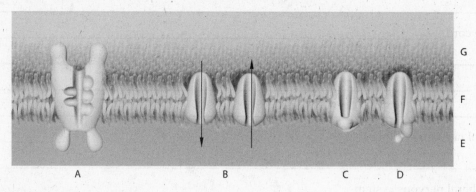

5. ____   cytoplasm of a neuron

6. ____   plasma membrane

7. ____   interstitial fluid

8. ____   gated channels that are open during an action potential; Na+ and K+ can pass through.

9. ____ sodium-potassium pump actively transports Na+ and K+ aginst the concentration gradient.

10. ____ passive transporters with open channels allow K+ to leak back in the cell, down the concentration gradient.

11. ____ gated channels that are shut while the cell is at rest.

True or False [pp.471-473]

13. ____ Without ion gradients, an action potential cannot be generated.

12. ____ In a resting cell, for every 3 Na+ inside the cell, there are 30 outside. Likewise, for every K+ inside, there is one outside the cell.

Labeling

Label the indicated components of a neuron in the diagram below. [p.470]

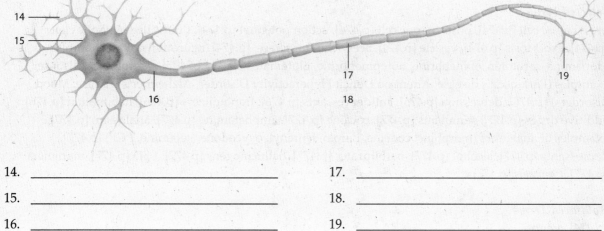

14. _____   17. _____

15. _____   18. _____

16. _____   19. _____

Matching [pp. 468-473]
Match each of the following terms with its correct description.

20. ____ sodium-potassium pumps

21. ____ resting membrane potential

22. ____ neurons

23. ____ excitable cells

24. ____ neuroglia

25. ____ action potential

26. ____ axons

27. ____ dendrites

a. The mechanism that maintains the correct ion concentration across the membrane.
b. The conducting and output zones of the neuron are located here.
c. Cells that support the metabolic activity of a neuron.
d. Along with neurons, these types of cells may generate an action potential.
e. The communication units of the nervous system.
f. This is normally –70 millivolts.
g. A brief, abrupt reversal of the electrical charge across a membrane.
h. Input zones for information.

True-False [pp.471-473]
The following statements pertain to the propagation of an action potential. If the statement is true, mark a T in the space provided. If the statement is false, correct the underlined word or Terms to make the statement true.

28. _____ Once a threshold level is reached there is an <u>all-or-nothing</u> membrane response.

29. _____ In positive feedback, the flow of sodium ions <u>decreases</u> as an outcome of its own occurrence.

30. _____ <u>Calcium</u> pumps are responsible for restoring the resting membrane potential.

31. _____ Action potentials move along the axon as a result of <u>self-propagating</u> events.

32. _____ Action potentials spread <u>away</u> from the membrane area where it started.

29.6 CHEMICAL COMMUNICATION AT SYNAPSES [pp.474-475]
29.7 DISRUPTING SIGNALING – DISORDERS AND DRUGS [pp.476-477]

Selected Terms
presynaptic cell [p.474], postsynaptic cell [p.474], action potential [p.474], fluid-filled synaptic cleft [p.474], exocytosis [p.474], vesicle [p.474], acetylcholinesterase [p.474], neurotransmitter examples (dopamine, serotonin, epinephrine, norepinephrine, glutamate, GABA) [p.475], neurological disorder examples (Parkinson's disease, Attention Deficit Hyperactivity Disorder, Alzheimer's Disease, Mood disorders) [p.476], depression [p.476], antidepressants [p.476], tranquilizers [p.476], habituation [p.476], additive drugs [p.477], stimulants [p.477], cocaine [p.477], amphetamine [p.477], analgesics [p.477], examples of analgesics (morphine, codeine, heroin, fentanyl, oxycodone, ketamine, PCP) [p.477], depressants [p.477], alcohol [p.477], barbiturates [p.477], hallucinogens [p.477], LSD [p.477], marijuana [p.477], *Cannabis* [p.477]

Boldfaced Terms
[p.474] synapse _____

_____

[p.474] neuromuscular junction _____

_____

[p.474] neurotransmitters_____

_____

[p.474] acetylcholine (ACh) _____

_____

[p.475] synaptic integration_____

_____

[p.477] endorphins_____

_____

Matching

Match each of the following definitions with the most appropriate term. [pp.474-475]

1. ____ The name given to the region between the output zone of one neuron and the input zone of a second neuron.

2. ____ The summing of the incoming signals at the postsynaptic neuron.

3. ____ Receptors on this cell respond to neurotransmitters released into the chemical synapse.

4. ____ The result of signals that nudge a cell membrane in the direction of an action potential.

5. ____ This cell releases neurotransmitter in response to an action potential.

6. ____ The result of signals that move the cell membrane away from an action potential.

a. postsynaptic neuron
b. synaptic integration
c. chemical synapse
d. excitatory effect
e. inhibitory effect
f. presynaptic neuron

Matching

For each of the following, indicate the type of neurotransmitter that may be responsible for the response indicated. Some answers may be used more than once. [p.475]

7. ____ Influences cells involved with sleeping.

8. ____ Influences emotions, dreaming and alertness.

9. ____ Initiates muscle contraction.

10. ____ A common inhibitory neurotransmitter

11. ____ Inhibits the response of cells involved in memory

12. ____ Is involved in award-based learning.

13. ____ Influences sensory perception.

a. serotonin
b. norepinephrine
c. acetylcholine
d. GABA
e. dopamine

Labeling [p. 474]

Label the indicated components of the chemical synapse in the diagram below. Look in the textbook on page 474 for more information to assist you in the labeling of this diagram.

14. _____ plasma membrane of the postsynaptic cell

15. _____ plasma membrane of an axon ending in a presynaptic cell.

16. _____ synaptic vesicle

17. _____ membrane receptor for neurotransmitter

18. _____ synaptic cleft

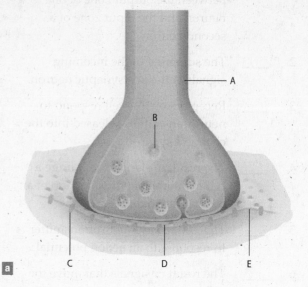

Completion [p.476]

Fill in the following table.

Disease	Neurotransmitter	Pathology associated with this neurotransmitter
ADHD	19.	20.
21.	Acetylcholine	Low levels in the brain
22.	Dopamine	Dopamine secreting neurons in the brain are destroyed

List [p.477]

List the 8 warning signs of drug addiction:

23. _____

24. _____

25. _____

26. _____

27. _____

28. _____

29. _____

30. _____

Choice [p.476-478]

Match the class of drug with the effects it has on the nervous system or an example of each.

    a. stimulants
    b. depressants
    c. analgesics
    d. hallucinogens

31. _____ marijuana is an example.

32. _____ distort sensory perception

33. _____ LSD

34. ____ make users feel alert; can interfere with fine motor skills.

35. ____ slow motor responses by inhibiting Ach output.

36. ____ heroin is an example

37. ____ cocaine is an example

38. ____ mimic the body's natural painkillers

39. ____ Ecstasy is an example.

40. ____ overdoses can cause strokes and heart attacks

41. ____ can be strongly addictive.

29.8 THE PERIPHERAL NERVOUS SYSTEMS [pp.478-479]
29.9 THE SPINAL CORD [pp.480-481]

Selected Terms
sheath of connective tissue [p.478], node [p.478], voluntary control [p.478], synaptic intergration [p.478], "fight or flight" [p.479], paralysis [p.480], spinal nerve [p.480], intervertebral disk [p.480], vertebra [p.480], stimulus [p.481], response [p.481], multiple sclerosis [p.481], autoimmune disorder [p.481], oligodendrocytes [p.481], neuroglial cells [p.481], Schwann cells [p.481]

Boldfaced Terms
[p.478] myelin _____

_____

[p.478] somatic nervous system _____

_____

[p.478] autonomic nervous system _____

_____

[p.479] sympathetic neurons _____

_____

[p.479] parasympathetic neurons _____

_____

[p.480] meninges _____

_____

[p.480] cerebrospinal fluid _____

_____

[p.480] white matter _____

_____

[p.480] gray matter _____

_____

[p.480] reflexes _____

_____

Labeling

Use page 481 in the textbook to assist you in labeling the diagram below. Match each of the following descriptions to its correct label in the diagram of a stretch reflex. [p.481]

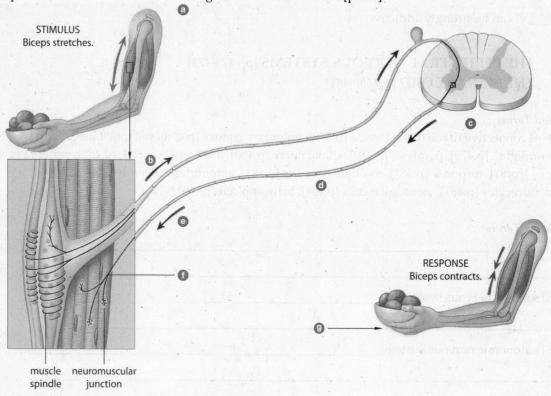

STIMULUS
Biceps stretches.

RESPONSE
Biceps contracts.

muscle   neuromuscular
spindle   junction

1. _____   Weight on an arm muscle stretches muscle spindles in the muscle sheath.

2. _____   Axon endings of the sensory neuron release a neurotransmitter that stimulate a motor neuron.

3. _____   Axon endings of the motor neuron synapse with muscle fibers in the stretched muscle.

4. _____   Stimulations makes the stretched muscle contract.

5. _____   ACh released from the motor neuron's axon endings stimulates cells making up the muscle fibers.

6. _____   Action potentials self-propagate along the motor neuron axon.

7. _____   Stretching stimulates sensory receptor endings in the muscle spindle.

True-False

If the statement is true, mark a T in the space provided. If false, correct the underlined word so that the statement reads is correct. [pp.478-481]

8. _____ All interneurons are in the <u>central</u> nervous system.

9. _____ Rapid signal transmission occurs in the <u>gray</u> matter.

10. _____ The brain and spinal cord are part of the <u>autonomic</u> nervous system.

11. _____ The <u>peripheral</u> nervous system carries signals in and out of the central nervous system.

12. _____ Mylenated axons are present in the white matter.

Matching

Match each of the following terms to its correct definition. [pp.504-505]

13. ____ cranial nerves

14. ____ parasympathetic nerves

15. ____ autonomic nerves

16. ____ fight-flight response

17. ____ spinal nerves

18. ____ somatic nerves

19. ____ sympathetic nerves

a. These nerves tend to slow down activity and begin housekeeping tasks.
b. There are thirty-one pairs of these in the body.
c. These carry signals about moving the head, trunk and limbs.
d. These deal with internal organs and structures.
e. This is caused by release of the neurotransmitter epinephrine (adrenaline).
f. These nerves are activated in times of stress or danger.
g. Twelve pairs of these connect directly to the brain.

Labeling

Provide the correct label for each of the indicated nerves in the diagram below. Consult your textbook page 479 for additional help. [p.479]

20. _____

21. _____

22. _____

23. _____

24. _____

25. _____

26. _____

Neural Control **299**

Labeling

Complete the following using the diagram of the spinal cord and vertebra. Consult your textbook page 480 for additional assistance. [p.480]

27. ____ spinal cord

28. ____ location of intervertebral disk

29. ____ meninges

30. ____ spinal nerve

31. ____ vertebra

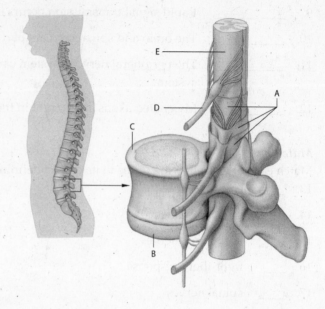

29.10 THE VERTEBRATE BRAIN [pp.482-483]
29.11 THE HUMAN CEREBRUM [pp.484-485]

Selected Terms
forebrain [p.482], midbrain [p.482], hindbrain [p.482], alcohol [p.482], caffeine [p.482], mercury [p.482], fissure [p.483], endocrine system [p.483], lobes of the brain (frontal, parietal, temporal, occipital) [p.484], association areas [p.484], Broca's area [p.484], "gut reactions" [p.485], amygdala [p.485], panic disorder [p.485], memories [p.485], short-term memory [p.485], long-term memory [p.485], declarative memory [p.485]

Boldfaced Terms
[p.482] blood-brain barrier _____

_____

[p.482] medulla oblongata _____

_____

[p.482] cerebellum _____

_____

[p.482] pons _____

_____

[p.483] cerebrum _____

_____

[p.483] thalamus _____

_____

[p.483] hypothalamus _____

_____

[p.484] cerebral cortex _____

_____

[p.484] primary motor cortex _____

_____

[p.484] limbic system _____

_____

Matching

Match each of the following subdivisions of the brain to its correct function. [p.506]

1. ____ cerebellum

2. ____ hypothalamus

3. ____ medulla oblongata

4. ____ pons

5. ____ brain stem

6. ____ cerebrum

7. ____ thalamus

a. This area helps to control motor skills and posture.

b. This is a coordinating center for sensory input.

c. This exists as two large hemispheres and is where most signal integration occurs.

d. This provides homeostatic control for the body.

e. This ancient tissue is found in the midbrain, forebrain and hindbrain.

f. This acts as a traffic center between the cerebellum and forebrain.

g. Contains reflex centers for respiration and circulation.

Fill-in-the-Blanks [pp.482-485]

A blood-brain barrier protects the brain and (8) _____ of vertebrates from harmful substances. It exerts some control over which solutes enter (9) _____ fluid.

The barrier works at the wall of (10) _____ that service the brain. In most brain regions (11) _____ junctions fuse all abutting cells that make up the wall of capillaries, so water-soluble substances must pass through the cells to reach the brain. Transport (12) _____ in the plasma membrane allow (13) _____, other vital nutrients, and some (14) _____ across. They bar many toxins and wastes, such as (15) _____. The barrier is not perfect; some toxins such as (16) _____, caffeine or mercury can pass through. Inflammation, toxins, and traumatic blows can destroy this barrier and compromise (17) _____ function.

Labeling

Label each of the indicated components of the human brain. You may want to consult your textbook for additional assistance. [p.483]

18. _____

19. _____

20. _____

21. _____

22. _____

23. _____

24. _____

25. _____

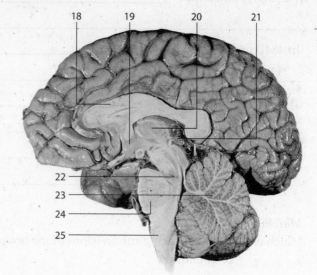

Labeling [pp.484-485]

Indicate whether each of the following statements is associated with the motor area (M), sensory area (S) or association area (A) of the cerebral cortex.

26. ____ This contains the Broca's area.

27. ____ Input from the skins and joints are processed here.

28. ____ The integrations of many inputs occur in this area.

29. ____ Controls the coordinated movement of skeletal muscles.

30. ____ Intellect and personality are the result of activity in this area.

31. ____ Sound and color are perceived here.

True-False [pp.484-485]

If the statement is true, mark a T in the space provided. If false, correct the underlined word so that the statement reads is correct.

32. _____ The control of <u>emotion</u> is provided by the limbic system.

33. _____ The limbic system is evolutionarily related to the <u>vision</u> systems.

34. _____ <u>Short-term</u> memory can store a tremendous amount of information regarding sensory input.

35. _____ Alzheimer's disease specifically effects to operation of the <u>hypothalamus</u>.

36. _____ The memory system processes facts and skills <u>together</u>.

SELF QUIZ

1. Which of the following group of animals is the first to display a simple nervous system? [p.468]
   a. vertebrates
   b. flatworms
   c. earthworms
   d. Cnidarians

2. Which of the following is active in a fight-flight response? [p.479]
   a. somatic nerves
   b. autonomic nerves
   c. sympathetic nerves
   d. parasympathetic nerves

3. The input zones of a neuron are located in which of the following structures? [p.470]
   a. axon
   b. dendrite
   c. cell body
   d. nucleus

4. The brain and spinal cord comprise what part of the nervous system? [p.469]
   a. central nervous system
   b. peripheral nervous system
   c. limbic system
   d. autonomic system

5. Which of the following is responsible for control of the motor skills? [pp.482-483]
   a. brain stem
   b. pons
   c. hypothalamus
   d. cerebellum

6. Caffeine and nicotine belong to what group of drugs? [p.477]
   a. depressants
   b. hallucinogens
   c. analgesics
   d. psychedelics
   e. stimulants

7. Which of the following process sensory input? [p.468]
   a. sensory neurons
   b. motor neurons
   c. interneurons
   d. transneurons

8. Which of the following is responsible for establishing the resting membrane potential in a neuron? [p.471]
   a. calcium pump
   b. sodium-potassium pump
   c. osmosis
   d. positive feedback

9. Homeostatic control of the internal environment occurs in the _____. [p.483]
   a. thalamus
   b. hypothalamus
   c. cerebrum
   d. medulla oblongata

10. Emotions are controlled by which of the following areas of the brain? [pp.484-485]
    a. hypothalamus
    b. limbic system
    c. motor areas
    d. association areas

CHAPTER OBJECTIVES/ REVIEW QUESTIONS

1. Describe the evolution of nervous systems from cnidarians to earthworms. [pp.468-469]

2. Describe the basic divisions of the human nervous system. [p.469]

3. Label the structure and zones of a neuron. [p.470]

4. Understand the concept of membrane potentials, including action potentials and resting membrane potentials. [pp.471-473]

5. How does a neuron restore a membrane potential following the generation of an action potential? [pp.471-473]

6. List the steps in the propagation of an action potential. [pp.471-473]

7. Indicate the role of the presynaptic and postsynaptic neurons. [pp.474-475]

8. How does synaptic integration process information from the neurons? [pp.474-475]

9. Understand the mechanism by which various drug classes interact with the nervous system. [pp.476-477]

10. What is the difference between a somatic and autonomic nerve? [p.469;478-479]

11. Describe the role of the sympathetic and parasympathetic nervous systems. [pp.478-479]

12. What is the fight-flight response and how is it controlled? [p.479]

13. What is the difference between gray and white matter? [p.480]

14. How does multiple sclerosis affect the structure of a nerve? [pp.480-481]

15. What is the role of the Schwann cell in a nerve? [p.481]

16. What is a reflex? [pp.480-481]

17. For each of the major areas of the brain, be able to identify its location and general function. [pp.482-485]

18. What substances can freely pass through the blood-brain barrier? [p.482]

19. Describe the general function of the motor, sensory and association areas of the brain. [pp.484-485]

20. What is the purpose of the limbic system? [pp.484-485]

21. Describe the differences between short and long-term memory. [p.485]

CHAPTER SUMMARY

Excitable cells called (1) _____ interconnect and form communication lines of animal nervous systems. In (2) _____ symmetrical animals, the neurons connect as a nerve net. In (3) _____ symmetrical animals, the concentration of neurons is at one end and one or more cords run the length of the body. Messages flow along a neuron's (4) _____. These messages are brief (5) _____ propagating reversals in the distribution of (6) _____ across the membrane. At an output zone, they are transduced to a (7) _____ signal that may stimulate or (8) _____ activity in another cell. (9) _____ drugs interfere with the information flow between cells. (10) _____ are the simples routes of information flow.

The central nervous system of vertebrates consists of the (11) _____ and the (12) _____. The (13) _____ nervous system consists of many pairs of nerves that connect the brain and spinal cord to the rest of the body. The(14) _____ is the most recently evolved part of the brain. Supporting cells in the human brain are called (15) _____ - the brain cannot function without them.

INTEGRATING AND APPLYING KEY CONCEPTS

1. Suppose that anger is eventually determined to be caused by excessive amounts of specific transmitter substances in the brains of angry people. Also suppose that an inexpensive antidote to anger that neutralizes these anger-producing transmitter substances is readily available. Can violent murderers now argue that they have been wrongfully punished because they were victimized by their brain's transmitter substances and could not have acted in any other way? Suppose an antidote is prescribed to curb violent tempers in an easily angered person. Suppose also that the person forgets to take the pill and subsequently murders a family member. Can the murderer still claim to be victimized by transmitter substances?

2. What is your hypothesis about how "truth drugs" work for individuals recalling memories?

30

SENSORY PERCEPTION

INTRODUCTION

In this chapter you will examine the different types of receptors that are used to detect both external and internal stimuli. The chapter discusses somatic and visceral sensations and shows a diagram of the brain with those areas illustrated. Vision from the anatomical to the issues regarding corrected vision and visual disorders are presented. Chemical senses, taste and smell, are discussed with illustrations for clarity. The closing pages of this chapter present the ear as a hearing organ and parts of the ear as balancing organs.

FOCAL POINTS

- Figure 30.2 [p.490] illustrates sensory receptors in the skin.
- Figure 30.4 [p.491] shows a map of where different body regions are represented in the human primary somatosensory cortex.
- Figure 30.5 [p.492] shows invertebrate eyes.
- Figure 30.6 [p.493] diagrams the structure of the human eye.
- Figure 30.9 [p.494] illustrates the retina of a human.
- Figure 30.10 [p.495] illustrates focusing problems of the human eye.
- Figure 30.12 [p.496] shows olfactory structures.
- Figure 30.13 [p.496] diagrams the tongue and taste buds.
- Figure 30.14 [p.497] illustrates structures of the human ear.
- Figure 30.16 [p.498] illustrates structures of the middle and inner ear.

INTERACTIVE EXERCISES

30.1 A WHALE OF A DILEMMA [p.489]
30.2 DETECTING STIMULI AND FORMING PERCEPTIONS [p.490]
30.3 SOMATIC AND VISCERAL SENSATIONS [p.491]

Selected Terms

Sensory [p.490], peripheral [p.490], central nervous system [p.490], somatosensory [p.491], histamine [p.491], prostaglandin [p.491], spinal interneuron [p.491], endorphins [p.491], aspirin [p.491], morphine [p.491]

Boldfaced Terms

[p.490] stimulus_____

_____

[p.490] mechanoreceptors _____

_____

[p.490] pain receptors_____

_____

[p.490] thermoreceptors _____

_____

[p.490] chemoreceptors _____

_____

[p.490] photoreceptors _____

_____

[p.490] sensory adaptation _____

_____

[p.490] sensation _____

_____

[p.490] perception _____

_____

[p.491] somatic sensation _____

_____

[p.491] visceral sensation _____

_____

[p.491] pain _____

_____

Matching

Match each of the following stimuli to the correct receptor. [p.490]

1. ____ temperature changes

2. ____ changes in pressure, position, or acceleration

3. ____ tissue damage

4. ____ change in the concentration of solutes in a fluid

5. ____ specific substances dissolved in fluid

6. ____ light energy

a. pain receptors
b. photoreceptors
c. mechanoreceptors
d. thermoreceptors
e. chemoreceptors
f. osmoreceptors

If the statement is true place a 'T' in the blank. If the statement is incorrect, correct the underlined word so that the statement is true. [p.490]

7. _____ Activation of a pain receptor in the skin would be an example of <u>visceral</u> pain.

8. _____ <u>Sensory adaptation</u> occurs when a neuron stops paying attention to a constant stimulus.

9. _____ The brain assesses a stimulus based on the <u>amplitude</u> and number of action potentials.

10. _____ A <u>stimulus</u> is a form of energy that activates receptor endings of a sensory neuron.

30.4 DO YOU SEE WHAT I SEE? [pp.492-493]
30.5 THE HUMAN RETINA [p.494]
30.6 VISUAL DISORDERS [p.495]

Selected Terms

vision [p.492], photoreceptors [p.520], cephalopod [p.492], orbit [p.492], conjunctiva [p.492], conjunctivitis [p.492], melanin [p.492], ciliary body [p.492], aqueous humor [p.493], vitreous humor [p.493], curvature [p.493], fovea [p.493], optic disk [p.493], optic nerve [p.493], rhodopsin [p.493], sclera [p.493], cornea [p.493], pupil [p.493], iris [p.493], choroid [p.493], color blindness [p.495], X-linked trait [p.495], astigmatism [p.495], nearsightedness [p.495], farsightedness [p.495], LASIK [p.495], cataract [p.495], macular degeneration [p.495], glaucoma [p.495]

Boldfaced Terms

[p.492] lens _____

_____

[p.492] compound eyes _____

_____

[p.492] camera eyes _____

_____

[p.492] cornea _____

_____

[p.492] iris _____

_____

[p.492] pupil _____

_____

[p.493] visual accommodation _____

_____

[p.494] rod cells _____

_____

[p.494] cone cells _____

_____

[p.494] fovea _____

_____

Labeling

First, label the indicated structures of the human eye. Then match each structure to the correct function or description from the list provided. [p.493]

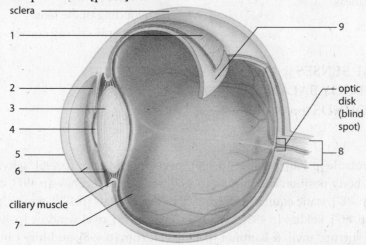

1. _____ ( )

2. _____ ( )

3. _____ ( )

4. _____ ( )

5. _____ ( )

6. _____ ( )

7. _____ ( )

8. _____ ( )

9. _____ ( )

a. The opening that lets light into the eye
b. Keeps the lens moist
c. Relays information from the eye to the brain for interpretation
d. The tissue that is densely packed with photoreceptors
e. Made of transparent crystalline proteins
f. A darkly pigmented tissue that absorbs wavelengths missed by the photoreceptors
g. Contracts to adjust the size of the pupil
h. Bends light in what is called visual accommodation
i. Fills the area in the eye behind the lens

True-False

If the statement is true place a 'T' in the blank. If the statement is incorrect, correct the underlined word so that the statement is true. [pp.492-494]

10. _____ Information from the right visual field is delivered to the <u>right</u> cerebral hemisphere.

11. _____ <u>Cone</u> cells are responsible for detecting dim light.

12. _____ There are <u>four</u> different types of cone cells.

13. _____ Rod cells and cone cells are derived from <u>cilia</u>.

14. _____ The visual pigment of <u>cone</u> cells is rhodopsin.

Matching

Match each of the following disorders to the correct description. [p.495]

15. ____ glaucoma

16. ____ farsightedness

17. ____ cataracts

18. ____ color blindness

19. ____ macular degeneration

20. ____ nearsightedness

21. ____ astigmatism

a. problems focusing on close objects
b. destruction of photoreceptors clouds straight-ahead vision
c. due to problems with one of the three types of cone cells
d. problems focusing on objects far away
e. excess accumulation of aqueous humor in the eye
f. clouding of the lens
g. unevenly curved cornea

30.7 THE CHEMICAL SENSES [p.496]
30.8 KEEPING THE BODY BALANCED [p.497]
30.9 DETECTING SOUNDS [pp.498-499]

Selected Terms

water soluble [p.496], volatile [p.496], papillae [p.496], sweet [p.496], sour [p.496], salty [p.496], bitter [p.496], umami [p.496], body position and motion [p.497], semicircular canals [p.497], cilia [p.497], dynamic equilibrium [p.497], static equilibrium [p.497], calcite crystals [p.497], stroke [p.497], vestibular nerve [p.497], saccule [p.497], koklias [p.498], organ of Corti [p.498], oval window [p.498], cochlea [p.498], auditory nerve [p.498], stirrup, anvil, & hammer [p.498], eardrum [p.498], auditory canal [p.498], round window [p.498], pinna [p.498], vibrating object [p.498], frequency [p.498], streptomycin [p.499], earwax [p.499]

Boldfaced Terms

[p.496] olfactory receptors _____

_____

[p.496] pheromones _____

_____

[p.496] vomeronasal organ _____

_____

[p.496] taste receptors_____

_____

[p.497] organs of equilibrium _____

_____

[p.497] vestibular apparatus _____

_____

[p.497] hair cells _____

_____

[p.497] outer ear _____

_____

[p.498] middle ear _____

_____

[p.498] inner ear _____

_____

[p.498] cochlea_____

_____

Choice

Choose the appropriate sense for each statement. Some statements may have more than one answer.
[pp.516-519]

      a. balance
      b. taste
      c. hearing
      d. smell

1. ____ involves the organ of Corti

2. ____ involves the vestibular apparatus

3. ____ involves the use of chemoreceptors

4. ____ conflicting sensory inputs may cause vertigo

5. ____ detects the frequency and amplitude of pressure variations

6. ____ involves structures found in the inner ear

7. ____ detects pheromones

8. ____ detects umami

True-False

If the statement is true place a 'T' in the blank. If the statement is incorrect, correct the underlined word so that the statement is true. [pp.496-499]

9. _____ Vomeronasal organs are involved in the detection of <u>taste</u> in mammals and reptiles.

10. _____ <u>Taste</u> receptors are primarily involved in the detection of volatile chemicals.

11. _____ Umami and plant toxins are detected by <u>taste</u> receptors.

12. _____ A <u>pheromone</u> is a chemical secreted by one individual to influence the social behavior of other members of the species.

13. _____ The saccule are involved in the process of <u>static</u> equilibrium.

14. _____ <u>Dynamic</u> equilibrium involves the use of a jellylike mass to sense changes in orientation of the head.

15. _____ The semicircular canals are part of the organs of equilibrium in mammals.

Labeling
Label each component of the human ear in the diagram below. [p.498]

16. _____

17. _____

18. _____

19. _____

20. _____

21. _____

22. _____

23. _____

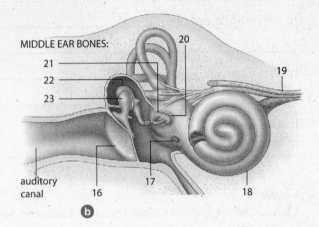

MIDDLE EAR BONES:

21

22

23

20

19

auditory
canal

16

17

18

b

Matching
Match each of the following terms with the correct definition or function. [pp.498-499]

24. ____ inner ear

25. ____ frequency

26. ____ middle ear

27. ____ amplitude

28. ____ outer ear

29. ____ basilar membrane

30. ____ round window

a. found in the middle ear enters cochlea
b. the cochlea is located here
c. this membrane responds to different frequencies
d. this contains the eardrum, anvil, stirrup, and hammer
e. this area is adapted for gathering sounds in the air
f. the number of wave cycles per second
g. the intensity, or loudness, or a sound

SELF QUIZ

1. In nearsightedness light rays from distant objects converge in front of the? [p.495]
   a. cornea
   b. choroid
   c. sclera
   d. retina

2. Amplification of sound waves occurs in the _____. [pp.498-499]
   a. inner ear
   b. outer ear
   c. middle ear
   d. vestibular apparatus

3. Which of the following is associated with the sense of balance? [p.497]
   a. vestibular apparatus
   b. cochlea
   c. basilar membrane
   d. eardrum

4. Which of the following is the location of the photoreceptors of the eye? [p.492]
   a. retina
   b. cornea
   c. pupil
   d. lens
   e. none of the above

5. The accumulation of excess aqueous humor behind the eye is called _____. [p.493]
   a. caratacts
   b. glaucoma
   c. cancer
   d. astigmatism
   e. none of the above

6. Mechanoreceptors are involved in which of the following senses? [p.490]
   a. taste
   b. smell
   c. balance
   d. vision

7. The hammer, anvil, and stapes are found in the [p.498]
   a. outer ear
   b. inner ear
   c. middle ear

8. Accommodation involves the ability to _____. [p.493]
   a. change the sensitivity of the rods and cones by means of neurotransmitters.
   b. change the shape of the lens by using special muscles
   c. change the shape of the cornea
   d. adapt to rapid changes in light intensity
   e. all of the above

9. Rods are cones are _____. [p.494]
   a. mechanoreceptors
   b. osmoreceptors
   c. chemoreceptors
   d. photoreceptors

10. Which of the following is due to the action of a chemoreceptor? [p.490]
    a. light
    b. smell
    c. taste
    d. all of the above
    e. only b or c

CHAPTER OBJECTIVES/REVIEW QUESTIONS

1. List the five kinds of receptors and identify the type of stimulus energy that each type detects. [p.490]

2. Understand the processing of somatic sensations by the brain. [p.491]

3. Explain the difference between somatic pain and visceral pain. [p.491]

4. Identify the structures of the human eye and give their function. [pp. 492-493]

5. Define the term visual accommodation. [p.493]

6. List the differences between rod cells and cone cells. [p.494]

7. Know the major vision disorders and their physiological causes. [p.495]

8. Explain the differences and similarities between olfactory receptors and taste receptors. [p.496]

9. Explain the process by which the body assesses balance. [p.497]

10. Explain the difference between static and dynamic equilibrium. [p.497]

11. Describe the functions of each part of the human ear. [pp.498-499]

CHAPTER SUMMARY

Sensory systems are front doors of the (1) _____ system. Each has sensory receptors, (2) _____ pathways to the brain, and brain regions that receive and (3) _____ the sensory input. A (4) _____ is a form of energy that activates a specific type of sensory receptor. Information becomes encoded in the (5) _____ and (6) _____ of action potentials sent to the brain along particular nerve pathways.

Touch, pressure, pain, temperature, and muscle sense are (7) _____ sensations. These sensations start at (8) _____ in skin, muscles, and the wall of internal organs.

The senses of smell and taste require (9) _____, which bind molecules of specific substances that have become dissolved in the fluid bathing them.

The sense of (10) _____ starts at mechanoreceptors, which detect (11) _____, velocity, acceleration, and other forces that influence the position and motion of the body or specific parts of it.

In vertebrates, the sense of (12) _____ starts with structures that collect, amplify, and sort out pressure variations caused by sound waves. The variations trigger (13) _____ in mechanoreceptors that send signals to the brain.

Most organisms have (14) _____ pigments, but vision requires eyes with a dense array of (15) _____ and image formation in the brain. Paired, camera-like eyes of cephalopoda and vertebrates collect and process information on distance, brightness, shape, position, and movement of visual (16) _____. A sensory pathway starts at the (17) _____ and ends in the (18) _____.

INTEGRATING AND APPLYING KEY CONCEPTS

1. Echolocation is common in some mammals (such as bats and whales), but not in primates. How would our special senses have to change in order for us to utilize echolocation? What types of receptors would we use?

2. Have you ever heard the saying –"Blind as a bat"? What does this mean?

3. The eyes of humans and squid are very similar. Explain why this may be an example of convergent evolution.

31

ENDOCRINE CONTROL

INTRODUCTION

This chapter begins with the discovery of hormones. The root meaning of the term "hormone" is presented. An exploration of the mechanisms by which the body uses hormones to control the activity of glands and organs follows. The chapter introduces the different types of hormones and then examines the function of the major endocrine glands in the human body. Abnormal conditions for each hormone are given within each section for a particular endocrine gland.

FOCAL POINTS

- Figure 31.2 [p.505] illustrates the major endocrine glands of the human body.
- Figure 32.3 [p.507] illustrates the differences between steroid and peptide hormones.
- Table 31.2 [p.508] displays the hormones secreted by the pituitary gland.
- Table 31.3 [p.511] lists the sources and functions of the other major human hormones.

INTERACTIVE EXERCISES

31.1 HORMONES IN THE BALANCE [p.503]
31.2 THE VERTEBRATE ENDOCRINE SYSTEM [pp.504-505]

Selected Terms
Synthetic chemicals [p.503], DDT [p.503], pesticide [p.503], atrazine [p.503], zebrafish [p.503], aromatase [p.503], sexual development [p.503], extinction [p.503], gap junction [p.504], Bayliss & Starling [p.504], prancreatic juices [p.504], "hormone" [p.504], *hormon* [p.528]

Boldfaced Terms
[p.503] endocrine disrupter _____

_____

[p.504] local signaling molecules _____

_____

[p.504] animal hormones _____

_____

[p.504] endocrine system _____

_____

Matching

Choose the most appropriate definition for each term. [p.504]

1. _____ neurotransmitters
2. _____ animal hormones
3. _____ local signaling molecules
4. _____ animal hormones

a. longer range communication molecules
b. secretory products of endocrine glands
c. change the chemical conditions of nearby tissues
d. released from the axons of nerve endings

Labeling

For each of the following, first provide the name of the endocrine gland. Second, select the hormone(s) secreted by the gland from the list provided. You might ant to consult Figure 31.2 for additional assistance. [p.505]

5. _____ ( )
6. _____ ( )
7. _____ ( )
8. _____ ( )
9. _____ ( )
10. _____ ( )
11. _____ ( )
12. _____ ( )
13. _____ ( )
14. _____ ( )

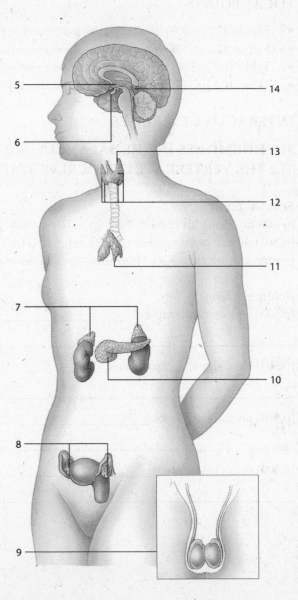

a. cortisol, aldosterone, epinephrine and norepinephrine
b. ACTH, TSH, FSH, and LH
c. Insulin
d. Sex releasing and inhibiting hormones, ADH and oxytocin
e. Melatonin
f. Estrogens and progesterone
g. Parathyroid hormone
h. Thymosins
i. Testosterone
j. Calcitonin and thyroid hormone

31.3 NATURE OF HORMONE ACTION [pp.506-507]

Selected Terms
cholesterol [p.506], amino acids [p.506], amine hormones [p.506], peptide hormones [p.506], protein hormones [p.506], bilayer [p.506], plasma membrane [p.506], cascade of reactions [p.506], cyclic AMP [p.506], androgen insensitivity [p.506]

Boldface Terms
[p.506] steroid hormones _____

_____

[p.506] second messenger _____

_____

Short Answer

1. List the four categories of hormones. [p.506]_____

   _____

   _____

   _____

2. List the factors that influence a cell's sensitivity to a hormone. [pp.506-507]_____

   _____

   _____

   _____

   _____

   _____

Choice
Indicate which class of hormones each of the following statements describes. [pp.506-507]
   a. peptide
   b. steroid

3. ____ Binds to receptors on the plasma membrane.

4. ____ Directly stimulate transcription into mRNA

5. ____ Lipid-soluble molecules that are derived from cholesterol.

6. ____ May involve the use of second messenger systems.

7. ____ Binds to receptors in the cytoplasm or nucleus.

Labeling

Provide the missing terms in the following diagrams of hormone action. [p.507]

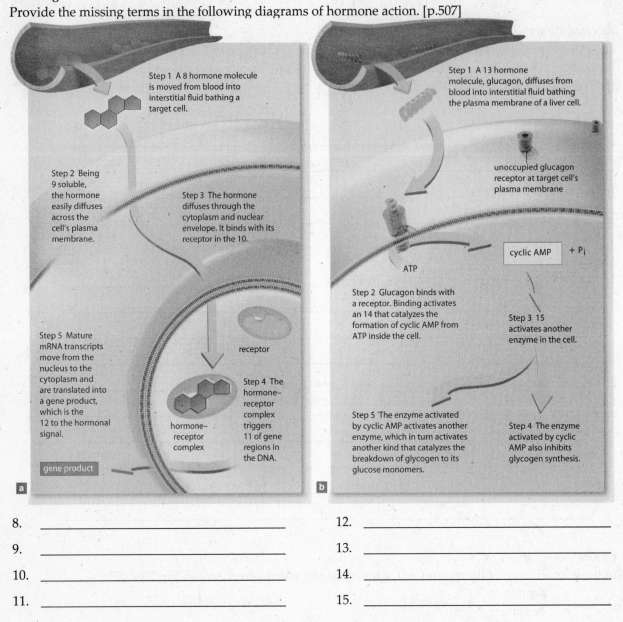

a Step 1 A 8 hormone molecule is moved from blood into interstitial fluid bathing a target cell.

Step 2 Being 9 soluble, the hormone easily diffuses across the cell's plasma membrane.

Step 3 The hormone diffuses through the cytoplasm and nuclear envelope. It binds with its receptor in the 10.

Step 5 Mature mRNA transcripts move from the nucleus to the cytoplasm and are translated into a gene product, which is the 12 to the hormonal signal.

receptor

hormone–receptor complex

Step 4 The hormone–receptor complex triggers 11 of gene regions in the DNA.

gene product

b Step 1 A 13 hormone molecule, glucagon, diffuses from blood into interstitial fluid bathing the plasma membrane of a liver cell.

unoccupied glucagon receptor at target cell's plasma membrane

cyclic AMP + P$_i$

ATP

Step 2 Glucagon binds with a receptor. Binding activates an 14 that catalyzes the formation of cyclic AMP from ATP inside the cell.

Step 3 15 activates another enzyme in the cell.

Step 5 The enzyme activated by cyclic AMP activates another enzyme, which in turn activates another kind that catalyzes the breakdown of glycogen to its glucose monomers.

Step 4 The enzyme activated by cyclic AMP also inhibits glycogen synthesis.

8. _____	12. _____
9. _____	13. _____
10. _____	14. _____
11. _____	15. _____

31.4 THE HYPOTHALAMUS AND PITUITARY GLAND [pp.508-509]
31.5 SOURCES AND EFFECTS OF OTHER VERTEBRATE HORMONES [p.510]
31.6 THYROID AND PARATHYROID GLANDS [p.511]

Selected Terms

mammary glands [p.508], uterus [p.508], adrenocorticotropic hormone [p.508], thyroid stimulating hormone [p.508], posterior lobe [p.508], anterior lobe [p.508],
luteinizing hormone [p.509], follicle stimulating hormone [p.509], prolactin [p.509], exocrine gland [p.509], pituitary gland [p.509], secretin [p.510], bicarbonate [p.510], erythropoietin [p.510], red blood cells [p.510], atrial natriuretic peptide [p.510], receptors [p.510], somatostatin [p.510], triiodothyronine [p.511],

thyroxine [p.511], iodine [p.511], hypothyroidism [p.511], thyroid enlargement (goiter) [p.511], dietary hypothyroidism [p.511], calcium [p.511], parathyroid hormone [p.511], osteoporosis [p.511]

Boldfaced Terms

[p.508] hypothalamus _____

_____

[p.508] pituitary gland _____

_____

[p.508] releasing hormones _____

_____

[p.508] inhibiting hormones _____

_____

[p.511] thyroid gland _____

_____

[p.511] parathyroid glands _____

_____

Choice

For each statement, choose the appropriate endocrine gland from the list below. [pp.508-511]

    a. anterior pituitary
    b. posterior pituitary
    c. hypothalamus
    d. thyroid
    e. parathyroid

1. ____ Hormones from this gland have the major control over blood calcium levels.

2. ____ Secretes calcitonin

3. ____ Secretes oxytocin and ADH

4. ____ Secretes growth hormones

5. ____ Abnormal secretions can lead to diabetes insipidus

6. ____ Secretes ACTH and LH

7. ____ Contains neurons that secrete hormones

8. ____ Goiter and Graves disease are associated with this gland.

9. ____ Requires iodine

10. ____ Directly connected to the pituitary glands

31.7 THE ADRENAL GLANDS [pp.512-513]
31.8 PANCREATIC HORMONES [p.514]

31.9 DIABETES [p.515]

Selected Terms

corticotrophin-releasing hormone [p.512], sympathetic division [p.512], hippocampus [p.512], Cushing's syndrome [p.513], "moon face" [p.513], Addison's disease [p.513], pancreatic islets [p.514], alpha cells [p.514], beta cells [p.514], delta cells [p.514], insulin [p.514], glucagon [p.514], diabetes mellitus [p.515], hyperglycemia [p.515], Type I diabetes [p.515], insulin pump [p.515], ketones & ketosis [p.515], Type II diabetes [p.515], obesity [p.515]

Boldfaced Terms

[p.512] adrenal glands _____

_____

[p.512] adrenal cortex _____

_____

[p.512] adrenal medulla _____

_____

[p.512] cortisol _____

_____

[p.514] pancreas _____

_____

Choice

For each of the following, choose the appropriate endocrine organ from the following list.

     a.   adrenal gland [p.512-513]
     b.   pancreatic islets [p.514]

1. _____ Is involved in the stress response.

2. _____ Diabetes mellitus is associated with this endocrine gland.

3. _____ Secretes somatostatin

4. _____ Has both endocrine and exocrine functions.

5. _____ Secretes glucagons and insulin.

6. _____ Cushing syndrome is associated with this endocrine gland.

7. _____ The fight-flight response is controlled from here.

True-False

If the statement is true, place a 'T' in the blank. If the statement is false, correct the underlined word so that the statement is correct. [pp.512-515]

8. _____ The adrenal <u>cortex</u> releases aldosterone and cotisol.

9. _____ Epinephrine and norepinephrine are involved in the <u>stress</u> response.

10. _____ The beta cells of the pancreatic islets secrete the <u>glucagon</u> hormone.

11. _____ Directly after a meal, when blood glucose levels are high, <u>alpha</u> cells are active.

12. _____ The <u>somatostatin</u> hormone is involved in regulating digestion and absorption.

13. _____ Somatostatin is secreted by the delta cells of the <u>adrenal medulla</u>.

14. _____ <u>Type I diabetes</u> is the result of an autoimmune response.

15. _____ In hypoglycemia, <u>low</u> blood glucose levels disrupt body functions.

16. _____ <u>Type II</u> diabetes may result in ketosis.

31.10 THE GONADS, PINEAL GLAND, AND THYMUS [p.516]
31.11 INVERTEBRATE HORMONES [p.517]

Selected Terms
gametes [p.516], testes [p.516], ovaries [p.516], sex hormones [p.516], eggs & sperm [p.516], estrogen & progesterone [p.516], testosterone [p.516], puberty [p.516], secondary sexual traits [p.516], melatonin [p.516], biological clock [p.516], thymosins [p.516], T cells [p.516], molting [p.517], cuticle [p.517], ecdysome [p.517]

Boldfaced Terms
[p.516] gonads _____

_____

[p.516] pineal gland _____

_____

[p.516] thymus _____

_____

Matching
Match each of the following terms to the correct statement. [pp.516-517]

1. ____ pineal gland

2. ____ ecdysone

3. ____ melatonin

4. ____ biological clock

5. ____ gonads

6. ____ puberty

7. ____ thymus

a. internal timing mechanism
b. hormones from this gland are associated with T cell maturity
c. endocrine gland that senses changes in light levels
d. invertebrate hormone that controls molting
e. primary reproductive glands
f. hormone associated with seasonal affective disorder
g. developmental stage when reproductive glands mature

SELF QUIZ

1. Goiter is associated with a problem in the _____ gland. [p.511]
    a. pineal
    b. adrenal
    c. pituitary
    d. thyroid
    e. parathyroid

2. Hormones from the parathyroid gland regulate the balance of _____ in the body. [p.511]
    a. iron
    b. calcium
    c. sex hormones
    d. potassium
    e. magnesium

3. The fight-flight response is regulated by the _____. [pp.512-513]
    a. adrenal gland
    b. pineal gland
    c. testes
    d. thyroid
    e. kidneys

4. This class of chemicals helps to integrate social behavior. [p.528]
    a. neurotransmitters
    b. animal hormones
    c. hormones
    d. local signaling molecules
    e. none of the above

5. This endocrine gland is responsible for the secretion of melantonin. [p.516]
    a. hypothalamus
    b. pituitary gland
    c. ovaries
    d. pineal gland
    e. pancreatic islets

6. The neurons in this gland secrete hormones instead of neurotransmitters. [p.508]
    a. pituitary
    b. adrenal
    c. pineal
    d. hypothalamus
    e. thyroid

7. Which of the following conditions can be due to a problem with the pituitary? [pp.508-509]
    a. Grave's disease
    b. Diabetes mellitus
    c. Hypoglycemia
    d. Seasonal affective disorder
    e. None of the above

8. Which of the following statements is true regarding the steroid hormones? [pp.506-507]
    a. they can diffuse directly across the plasma membrane.
    b. the bind to receptors in the cytoplasm or nucleus
    c. the stimulate gene transcription
    d. they are made from cholesterol
    e. all of the above are correct

9. The _____ cells of the pancreas produce insulin. [p.514]
    a. alpha
    b. beta
    c. gamma
    d. delta
    e. omega

10. Which of the following is associated with the activity of T cells in the immune system? [p.516]
    a. thymus
    b. thyroid
    c. adrenal gland
    d. pineal gland
    e. hypothalamus

CHAPTER OBJECTIVES/REVIEW QUESTIONS

1. Understand the categories of signaling molecules and the function of each in the body. [p.504]

2. Know the location and role of each gland in the endocrine system. [p.505 & p.510]

3. Know the four types of hormones. [p.506]

4. Compare the mode of action of a steroid and peptide hormone. [pp.506-507]

5. Understand the interaction of the hypothalamus and the pituitary gland. [pp.508-509]

6. List the major secretions of the anterior and posterior pituitary glands. [pp.508-509]

7. Understand the role and secretions of the thyroid and parathyroid glands. [p.511]

8. Understand the role and secretions of the adrenal gland. [pp.512-513]

9. Understand the role of the pancreatic islets, its secretions and the diseases associated with abnormal function of this gland. [pp.514-515]

10. Understand the role of the pineal gland and melantonin. [p.516]

11. Understand the role of the male and female gonads. [p.516]

12. Understand the relationship between ecdysone and invertebrate molting. [p.517]

CHAPTER SUMMARY

(1) _____ and other signaling molecules regulate the pathways that control metabolism, growth, (2) _____, and reproduction. Nearly all vertebrates have an (3) _____ system composed of the same hormone sources.

A hormone signal acts on any cell that has (4) _____ for it. Receptor activation leads to (5) _____ of the signals and a response in the targeted cell.

Invertebrates, the (6) _____ and pituitary gland are connected structurally and functionally. Together, they coordinate the activities of many other (7) _____.

(8) _____ feedback loops to the hypothalamus and pituitary control secretions from many glands. Other glands secrete hormones in response to local changes in the internal (9) _____.

Hormones control (10) _____ and other events in invertebrate life cycles. Vertebrate hormones and receptors for them first evolved in ancestral lineages of (11) _____.

(12) _____ organ systems are a focus of many chapters in this unit. This secretion is an overview of tissue-specific responses to hormones including the impact of variations in (13) _____.

INTEGRATING AND APPLYING KEY CONCEPTS

1. Many people are concerned about the presence of growth hormones in milk and other animal products. These hormones are given to animals to stimulate the production of milk or muscle mass in order to increase agricultural yield. To be harmful, these would have to mimic the action of a human growth-related hormone. Compare this to the material presented on ecdysone and molting in invertebrates and state why you think, or don't think, that this may be a problem.

2. What would be the possible effects of a tumor on the anterior pituitary that limited its function? The posterior?

32
STRUCTURAL SUPPORT AND MOVEMENT

INTRODUCTION

This chapter examines how muscles and bones interact in the process of movement. It begins with an introduction of animals skeletons in general and leads to the more complex. The human skeleton is details with diagram and description. A diagram of the knee is given as well as aliments that can occur to the skeleton. Next the anatomical features of joints with what they can and shouldn't do. The chapter introduces the basic principles of how skeletons and muscles function, as well as some of the diseases that are common with these two systems.

FOCAL POINTS

- Figure 32.4 [p.522] shows the skeleton of an early reptile.
- Figure 32.5 [p.523] illustrates some of the major bones of the human body.
- Figure 32.6 [p.524] illustrates a bone and an exploded view of the microscopic anatomy.
- Figure 32.8 [p.525] shows the anatomy of the knee.
- Figure 32.9 [p.526] shows opposing muscles of the upper arm.
- Figure 32.10 [p.527] illustrates some of the major muscles of the human body.
- Figure 32.12 [p.528] illustrates the action of a sarcomere.

INTERACTIVE EXERCISES

32.1 MUSCLES AND MYOSTATIN [p.521]
32.2 ANIMAL SKELETONS [pp.522-523]
32.3 BONES AND JOINTS [pp.524-525]

Selected Terms

testosterone [p.521], anabolic [p.521], myostatin [p.521], gastovascular cavity [p.522], earthworm [p.522], segment [p.522], sea anemone [p.522], shell [p.522], cuticle [p.522], clam [p.521], arthropod [p.522], shark [p.522], cartilaginous [p.522], pectoral [p.522], pelvic [p.522], pectoral [p.522], cranial bones [p.523], facial bones [p.523], sternum [p.523], sacrum [p.523], coccyx [p.523], scapula [p.523], clavicle [p.523], 206 bones [p.524], extracellular matrix mineralization [p.524], parathyroid hormone [p.524], calcium [p.524], osteoporosis [p.524],connective tissue [p.525], fibrous joint [p.525], cartilaginous joint [p.525], synovial joint [p.525], arthritis [p.525], osteoarthritis [p.525], rheumatoid arthritis [p.525], *spongy* bone [p.525], cruciate [p.525], menisci [p.525]

Boldfaced Terms

[p.522] hydrostatic skeleton _____

_____

[p.522] exoskeleton _____

_____

[p.522] endoskeleton_____

_____

[p.522] vertebral column _____

_____

[p.522] vertebrae _____

_____

[p.522] intervertebral disks _____

_____

[p.522] axial skeleton _____

_____

[p.522] appendicular skeleton _____

_____

Matching

For each of the following select the correct choice from the ones given below. Note some of the choices may NOT be used. [p.521]

1. ____ a regulatory protein that normally acts as a brake on the production of muscle proteins

2. ____ synthetic versions of natural-building hormones

3. ____ a sex hormone in males

4. ____ another hormone that stimulates synthesis of muscle proteins

a. testosterone
b. myostatin
c. muscular dystrophy
d. anabolic
e. estrogen
f. human growth

Choice

For each of the following, choose the form of skeleton that is best described by the statement. [p.522]

    a. endoskeleton
    b. exoskeleton
    c. hydrostatic skeleton

5. ____ The vertebrates possess this form of skeleton

6. ____ Arthropods possess this form of skeleton.

7. ____ Internal body parts accept the applied force of muscle contraction.

8. ____ Muscle contractions redistribute a volume of fluid in the body cavity.

9. ____ Earthworms and soft-bodied invertebrates have this form of skeleton.

10. ____ External body parts receive the applied force of muscle contraction.

Labeling

Match the statement with the appropriate label from the figure. You might want to consult Figure 32.5 for addition information. [p.523]

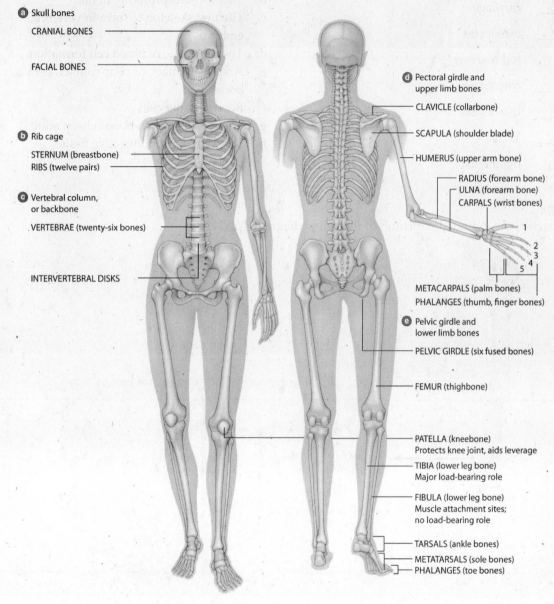

a Skull bones
  CRANIAL BONES
  FACIAL BONES

b Rib cage
  STERNUM (breastbone)
  RIBS (twelve pairs)

c Vertebral column, or backbone
  VERTEBRAE (twenty-six bones)
  INTERVERTEBRAL DISKS

d Pectoral girdle and upper limb bones
  CLAVICLE (collarbone)
  SCAPULA (shoulder blade)
  HUMERUS (upper arm bone)
  RADIUS (forearm bone)
  ULNA (forearm bone)
  CARPALS (wrist bones)
  1 2 3 4 5
  METACARPALS (palm bones)
  PHALANGES (thumb, finger bones)

e Pelvic girdle and lower limb bones
  PELVIC GIRDLE (six fused bones)
  FEMUR (thighbone)
  PATELLA (kneebone)
    Protects knee joint, aids leverage
  TIBIA (lower leg bone)
    Major load-bearing role
  FIBULA (lower leg bone)
    Muscle attachment sites;
    no load-bearing role
  TARSALS (ankle bones)
  METATARSALS (sole bones)
  PHALANGES (toe bones)

11. ____ Enclose and protect the brain.

12. ____ These bones are involved in locomotion.

13. ____ Support the skull and upper extremities.

14. ____ Contains bones with extensive muscle attachments

15. ____ Contains fibrous, cartilaginous structures that absorb movement-induced stresses.

16. ____ Supports the weight of the backbone.

17. ____ Protects the heart and lungs.

Matching

Match each of the following terms with its correct definition. [pp.524-525]

18. ____ yellow marrow
19. ____ cartilage
20. ____ osteocytes
21. ____ red marrow
22. ____ osteoblasts
23. ____ osteoclasts
24. ____ joints
25. ____ ligaments

a. Cells that break down bone.
b. This consists primarily of fat.
c. The first skeleton to form in vertebrate embryos
d. The major site of blood cell formation.
e. Areas of contact, or near-contact, between bones.
f. Bone forming cells
g. The most common bone cells in adults.
h. Straps of dense connective tissue.

Labeling

Label each of the indicated structures of the below diagram. [p.524]

26. _____

27. _____

28. _____

29. _____

30. _____

31. _____

Concept Map

Provide the missing terms for the numbered items in the concept map of bone formation and remodeling below. [pp.524-525]

32. _____

33. _____

34. _____

35. _____

36. _____

37. _____

38. _____

39. _____

```
                                    calcium ions
                                         |
                                         in
                                         |
                                        bone
                                         |
                                   regulated by
                                         |
                                       (32)
                                         |
                                      using
                                         |
                              (33) feedback loops
                        _____|_____
                       |                             |
                   if blood                       if blood
                 calcium is high               calcium is low
                       |                             |
                      (34)                          (37)
                       |                             |
                    secretes                       secrete
                       |                             |
                      (35)                          (38)
                       |                             |
                 which inhibits                which stimulates
                       |                             |
                      (36)                          (39)
```

Matching

Match each of the following ailments to its correct description. [p.525]

40. ____ osteoarthritis

41. ____ strain

42. ____ sprain

43. ____ osteoporosis

44. ____ rheumatoid arthritis

a. when cartilage in movable joints wear off.

b. tearing of the ligaments or tendons in a joint.

c. a decrease in bone density.

d. membranes in joints become inflamed due to an autoimmune response.

e. stretching or twisting a joint too far

32.4 SKELETAL-MUSCULAR SYSTEMS [pp.526-527]
32.5 HOW DOES SKELETAL MUSCLE CONTRACT [pp.528-529]

Selected Terms

skeletal muscle [p.526], smooth muscle [p.526], cardiac muscle [p.526], tendonitis [p.526], anti-inflammatory drugs [p.526], cytoskeletal elements [p.528], Z lines [p.528]

Boldfaced Terms

[p.526] tendon _____

_____

[p.528] skeletal muscle fibers _____

_____

[p.528] myofibrils _____

_____

[p.528] sarcomeres _____

_____

[p.528] actin _____

_____

[p.528] myosin _____

_____

[p.528] sliding filament model _____

_____

True-False

If the statement is true, place a 'T' in the space provided. If the statement is false, correct the underlined term so that the statement is correct. [p.526]

1. _____ The functional partner of bone is <u>smooth</u> muscle.

2. _____ Skeletal muscles are composed of bundles of <u>muscle fibers</u>.

3. _____ <u>Tendons</u> attach skeletal muscle to bone.

4. _____ In the human body, most skeletal muscles act as a <u>lever</u> system.

Labeling

Label the muscles in the following diagram, then match named muscle to its correct function. You may want to consult the Figure 32.10 for additional information. [p.527]

5. _____ ( )   a. extends and rotates the thigh outward when walking and running.

6. _____ ( )   b. bends the thigh at the hip, bends lower leg at the knee.

7. _____ ( )

8. _____ ( )   c. bends the forearm at the elbow
      d. raises the arm

9. _____ ( )   e. straightens the forearm at the elbow
      f. lifts the shoulder blade and draws the head back

10. _____ ( )

11. _____ ( )   g. draws thigh backward and bends the knee
      h. draws the arm forward and in toward the body

12. _____ ( )

13. _____ ( )   i. flexes the thigh at hips and extends the leg at the knee

14. _____ ( )   j. compresses the abdomen

15. _____ ( )   k. depresses the thoracic cavity and bends the backbone.

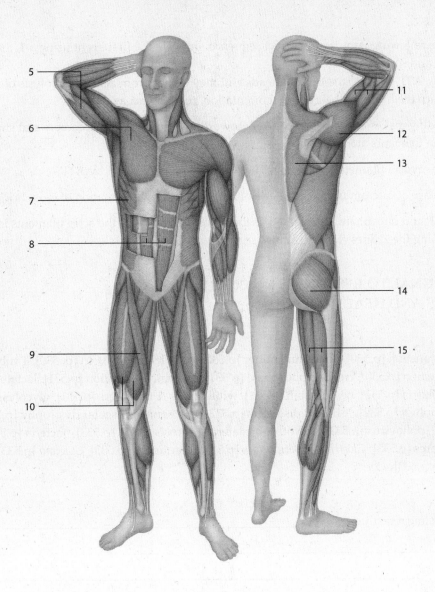

Matching

Match each of the following definitions to its correct term. [pp.528-529]

16. ____ The motor protein for muscle contraction.

17. ____ Threadlike, cross-banded structures within a muscle fiber.

18. ____ These run parallel to the muscle's long axis.

19. ____ Short, ATP driven movements of the myosin heads pull actin filaments towards the sarcomere center.

20. ____ The basic unit of muscle contraction.

a. myosin
b. sarcomeres
c. myofibrils
d. muscle fibers
e. sliding-filament model

Sequence

Place each of the following items in its correct sequence. Indicate the first event using a 1, the second event using a 2, and so on. [p.529]

21. _____ New ATP binds to the myosin heads and they detach from actin. Hydrolysis of the ATP will return the heads to their original orientation, ready to act again.

22. _____ Binding makes each myosin head tilt toward the center of the sarcomere and slide the bound actin filaments along with it.

23. _____ The myosin filaments are at rest. They were energized earlier by ATP.

24. _____ Release of calcium from a cellular storage system allows formation of cross bridges.

25. _____ ADP and phosphate are released as the myosin heads drag the actin filaments inward, pulling the Z lines closer together.

32.6 FROM SIGNAL TO RESPONSES [pp.530-531]
32.7 MUSCLES AND HEALTH [p.532]

Selected Terms

neuromuscular junction [p.530], neurotransmitter [p.530], acetylcholine (ACh) [p. 530], T tubules [p.530], sustained contraction [p.530], creatine phosphate [p.531], arerobic respiration [p.531], lactate fermentation [p.531], mitochondria [p.531], hemoglobin [p.531], white fibers & red fibers [p.531], glycolysis [p.532], muscular dystrophies [p.532], X-linked disorders [p.532], Duchenne muscular dystrophy [p.532], paralysis [p.533], poliovirus [p.533], amyotrophic lateral sclerosis (ALS) [p.533], bacteria [p.533], toxins [p.533], endospores [p.533], *Clostridium botulinium* [p.533], botulinum [p.533], *C. tetani* [p.533], tetanus [p.533], lockjaw [p.533]

Boldfaced Terms

[p.530] saccoplasmic reticulum _____

_____

[p.530] motor unit _____

_____

[p.530] muscle tension _____

_____

[p.530] muscle twitch _____

_____

Matching

Match each of the following terms to its correct definition. [pp.530-533]

1. ____ Muscle fatigue

2. ____ Muscle tension

3. ____ Acetylcholine

4. ____ Motor unit

5. ____ Botulinum

6. ____ Tetanus

7. ____ Muscular dystrophies

8. ____ Muscle twitch

a. A brief generation of contractile force.

b. The mechanical force exerted by a muscle on an object.

c. A neurotransmitter used in neuromuscular junctions.

d. The progressive weakening and degeneration of muscles.

e. A toxin that can prevent muscles from contracting.

f. A sustained series of muscle twitches.

g. A motor neuron and all of the muscle fibers that it controls.

h. A muscle that is in an ongoing state of muscle contraction.

Complete the Table

Indicate the disease that is caused by each of the organisms below. [p.533]

Clostridium tetani	9.
Clostridium botulinum	10.

SELF QUIZ

1. The basic unit of muscle contraction is the _____. [p.528]
   a. myofibril
   b. muscle fiber
   c. myosin
   d. sarcomere

2. Which of the following is not involved in the regulation of calcium levels in bone? [p.524]
   a. PTH
   b. vitamin D
   c. myosin
   d. calcitonin

3. In which of the following do internal body parts accept the applied force of contraction? [p.522]
   a. exoskeleton
   b. hydrostatic skeleton
   c. endoskeleton
   d. all of the above

4. Which of the following is the functional partner of bone? [p.526]
   a. cardiac muscle
   b. smooth muscle
   c. skeletal muscle
   d. all of the above

5. Which of the following components of bone is responsible for the formation of red blood cells? [p.524]
   a. yellow marrow
   b. osteoclasts
   c. osteoblasts
   d. red marrow
   e. osteocytes

6. The _____ attaches the bundles of muscle fiber to the bone. [p.549]
   a. myofibril
   b. tendon
   c. ligament
   d. joint

7. A brief generation of contractile force in a muscle is called _____. [p.526]
    a. tetanus
    b. muscle twitch
    c. muscle fatigue
    d. muscle tension

8. Which of the following correctly describes aerobic exercise? [p.531]
    a. long duration, intense activity
    b. long duration, not intense activity
    c. short duration, intense activity
    d. short duration, not intense activity

9. The sliding filament model describes which of the following? [p.529]
    a. the action of the sarcomere
    b. the function of the joints
    c. the skeletal model of an exoskeleton
    d. a form of exercise for muscle

10. Which of the following is not usually used as an energy source for muscle? [p.531]
    a. creatine phosphate
    b. glucose
    c. fatty acids
    d. amino acids

CHAPTER OBJECTIVES/ REVIEW QUESTIONS

1. Describe the general characteristics of the three primary types of skeletons. [p.522]

2. List some of the challenges of life on land that an endoskeleton helps to solve. [pp.522-523]

3. Identify the functions of the primary bone groups in humans. [p.523]

4. Understand the structure of bone and the role of red marrow, and yellow marrow. [pp.524-525]

5. Understand how bones connect with one another and the major diseases/ailments of these locations. [p.5525]

6. Understand the function of ligaments. [p.525]

7. Recognize the role and structure of a tendon. [p.526]

8. Know the three types of muscle. [p.526]

9. Understand the structure of a sacromere. [p.528]

10. Understand the process by which a muscle contracts. [pp.528-529]

11. Be able to point out major muscle groups of the human body. [p.527]

10. List the principles of the sliding filament model. [pp.528-529]

11. Define the terms muscle tension, muscle twitch, tetanus, and muscle fatigue as they relate to muscle contraction. [pp.530-531]

12. Understand the causes of the diseases tetanus and botulism. [p.533]

CHAPTER SUMMARY

    (1) _____ force exerted against some type of (2) _____ moves the animal body. Many invertebrates have a (3) _____ skeleton, which is a fluid-filled cavity. Others have an (4) _____ of hardened structures at the body surface. Vertebrates have an (5) _____, an internal skeleton of cartilage, bone, or both.

Bones are (6) _____ -rich organs that help the body move. They also protect and support soft organs, and store (7) _____. (8) _____ cells form in some bones. Cartilage or ligaments connect bones at (9) _____.

(10) _____ muscles are bundles of muscle fibers that interact with bones and with one another. Some cause movements by working as pairs or groups. Others oppose or (11) _____ the action of a partner muscle. (12) _____ attach skeletal muscles to bones.

A muscle fiber contains many (13) _____, each divided crosswise into (14) _____, the basic unit of contraction. Sarcomeres contain many parallel arrays of (15) _____ and myosin filaments. (16) _____ -driven interactions between the arrays shorten sarcomeres, which collectively accounts for (17) _____ of a whole muscle.

Muscle fibers in a muscle are organized in (18) _____ that contract in response to signals from one motor (19) _____. Cross-bridges form in all sarcomeres and collectively exert (20) _____. A muscle (21) _____ only when the tensile force exceeds other, opposing forces. (22) _____ enhances the properties of whole muscle, and aging and disease diminish them.

INTEGRATING AND APPLYING KEY CONCEPTS

1. Despite the risks of using anabolic steroids, many professional athletes continue to use steroids in an attempt to enhance their performance. From this chapter, speculate on a theoretical hormone that mimics one of the chemicals presented in the chapter. In addition, describe some problems that overuse of this chemical may present to the athlete over time.

2. While hormone-replacement therapy in women remains controversial, one of the benefits of an increase in estrogen is an increase in bone density. Why would this occur? What cells would it be stimulating?

33

CIRCULATION

INTRODUCTION

The human cardiovascular system is the result of millions of years of evolution and is one of the most important systems in the body. This system has evolved from an open circulatory system in arthropods and mollusks to a very sophisticated closed circulatory system in humans. This chapter discusses the evolution of the human cardiovascular systems, and then goes on to point out its critical function: the movement of substances in and out of cellular environments. Human blood is composed not only of red blood cells, which carry oxygen, but also of white blood cells for fighting pathogens and platelets for clotting blood. This chapter shows how these cells form and what their specific functions are. Then, the chapter goes on to discuss, in detail, the structure of the human heart and vascular system, as well as the lymphatic system. Finally, the importance of this system to human health is illustrated by discussions of diseases caused by abnormalities in the blood, or cardiovascular system. Some of these diseases include leukemia, anemia, hypertension and cardiovascular disease. An understanding of these pathologies enables us to better understand not only what causes disease, but also how to protect our own health by preventing these diseases.

FOCAL POINTS

- Figure 33.2 [p.538] compares open and closed circulatory systems.
- Figure 33.3 [p.539] compares fish, amphibian, reptiles, birds, and mammal hearts.
- Figure 33.4 [p.540] diagrams the major vessels of the human cardiovascular system.
- Figure 33.5 [p.541] animates the systemic and pulmonary circuits of the human cardiovascular system.
- Figure 33.6 [p.542] illustrates the human heart.
- Figure 33.9 [p.544] tabulates the typical components of human blood.
- Figure 33.11 [p.546] illustrates the structural components of human blood vessels.
- Figure 33.14 [p.548] shows fluid movement at the capillary bed.
- Figure 33.18 [p.550] shows a comparison between a normal artery and atherosclerotic plaque in an artery.
- Figure 33.21 [p.552] diagrams the components of the human lymphatic system.

INTERACTIVE EXERCISES

33.1 AND THEN MY HEART STOOD STILL [p.537]
33.2 INTERNAL TRANSPORT SYSTEMS [pp.538-539]

Selected Terms
heartbeat [p.537], pacemaker[p.537], cardiopulmonary resuscitation (CPR) [p.537], defibrillator [p.537], automated external defibrillator (AED) [p.537], diffusion [p.538], interstitial fluid [p.538], two, three, four chambered hearts [p.539], endothermic [p.538]

Boldfaced Terms

[p.538] circulatory system _____

_____

[p.538] heart _____

_____

[p.538] blood _____

_____

[p.538] open circulatory system _____

_____

[p.538] closed circulatory system _____

_____

[p.539] pulmonary circuit _____

_____

[p.539] systemic circuit _____

_____

Matching

Match each of the following statements to the correct term. You may want to consult Figure 33.3 in the textbook for additional information. [pp.538-539]

1. ____ carries oxygen poor, carbon dioxide rich blood to the lungs

2. ____ blood flow is slowest here as to allow the exchange of substances by diffusion

3. ____ may be either open or closed, depending on the type of organism.

4. ____ this is responsible for directly exchanging substances with cells.

5. ____ carries oxygen rich blood to the tissues.

a. pulmonary circuit
b. systemic circuit
c. capillaries
d. interstitial fluid
e. circulatory system

Choice

Choose which of the following circulatory systems each statement applies. [pp.538-539]

    a. fish
    b. amphibians
    c. birds and mammals

6. ____ Consists of two, separate circuits for the blood.

7. ____ Utilize a two-chambered heart.

8. ____ Utilize a three-chambered heart.

9. ____ Utilize a four-chambered heart.

10. ____ Blood pressure is lowest in this group.

11. ____ The development of a pulmonary circuit occurred in this group.

12. ____ Blood mixes in a common ventricle.

33.3 THE HUMAN CARDIOVASCULAR SYSTEM [pp.540-541]
33.4 THE HUMAN HEART [pp.542-543]

Selected Terms
"cardiovascular" [p.540], kardia [p.540], vasculum [p.540], thoracic cavity [p.542], pericardium [p.542], epithelium [p.542], atrioventricular valve [p.542], diaphragm [p.542], septum [p.542], cardiac muscle [p.542], aortic valve [p.542], contraction [p.543], "lub-dup" [p.543], heart murmur [p.543], signals [p.543], cardiac conduction system [p.543]

Boldfaced Terms
[p.540] arteries _____

_____

[p.540] arterioles _____

_____

[p.540] venule _____

_____

[p.540] veins _____

_____

[p.541] aorta _____

_____

[p.542] atrium _____

_____

[p.542] ventricle _____

_____

[p.542] superior vena cava _____

_____

[p.542] inferior vena cava _____

_____

[p.542] pulmonary artery _____

_____

[p.542] pulmonary vein _____

_____

[p.542] cardiac cycle _____

_____

[p.542] diastole _____

_____

[p.542] systole _____

_____

[p.543] sinoatrial node (SA)_____

_____

[p.543] atrioventricular node (AV)_____

_____

Labeling

First, name each of the components of the human circulatory system in the diagram below. Then choose the correct function for each component from the list provided. You may want to consult Figure 33.4 in the textbook for additional information. [p.540]

1. _____ ( )
2. _____ ( )
3. _____ ( )
4. _____ ( )
5. _____ ( )
6. _____ ( )
7. _____ ( )
8. _____ ( )
9. _____ ( )
10. _____ ( )
11. _____ ( )
12. _____ ( )
13. _____ ( )
14. _____ ( )
15. _____ ( )
16. _____ ( )
17. _____ ( )

a. delivers nutrient rich blood from small intestine to liver for processing
b. carries blood away from thigh and inner knee.
c. delivers blood to thigh and inner knee
d. delivers oxygenated blood from lungs to the heart
e. services the active cardiac muscles of heart.
f. carry blood away from pelvic organs and lower abdominal wall
g. delivers blood to kidneys
h. receives blood from brain and tissues of the head.
i. delivers blood to pelvic organs and lower abdominal wall.
j. receives blood from veins of upper body.
k. deliver oxygen-poor blood from heart to lungs
l. deliver blood to neck, head, brain.
m. carries oxygenated blood away from heart; the largest artery
n. delivers blood to upper extremities
o. carries processed blood away from kidneys

p. delivers blood to arteries leading to the digestive tract, kidneys, pelvic organs, lower extremities.

q. receives blood from all veins below diaphragm

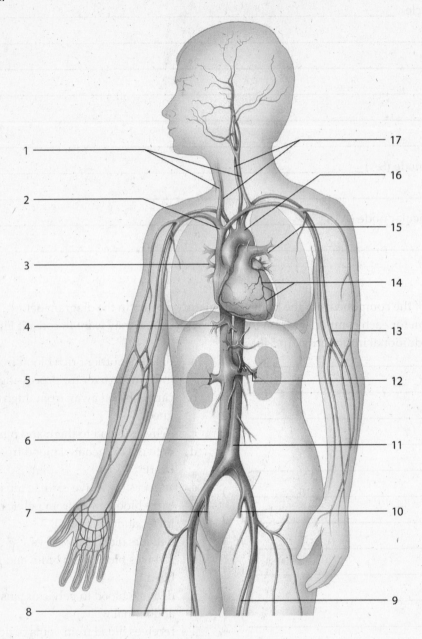

Labeling

Label each of the indicated components of the human heart. You might want to consult Figure 33.6 in the textbook fro additional information. [p.542]

18. _____
19. _____
20. _____
21. _____
22. _____
23. _____
24. _____
25. _____
26. _____
27. _____
28. _____
29. _____
30. _____

True-False

If the statement is correct, place a T in the space provided. If it is incorrect, correct the underlined word so that the statement is true. [pp.542-543]

31. _____ The contraction of the <u>right atrium</u> provides the driving force for blood circulation.

32. _____ The <u>AV</u> node of the heart functions as the cardiac pacemaker.

33. _____ The <u>AV</u> valves separate the atria from the ventricles.

34. _____ Semilunar and AV valves are <u>one-way</u> valves.

35. _____ The outermost connective tissue sac of the heart is called the <u>septum</u>.

33.5 CHARACTERISTICS AND FUNCTIONS OF BLOOD [pp.544-545]

Selected Terms

blood volume [p.544], heme group [p.544], hemoglobin [p.544], phagocytosis [p.544], blood clotting [p.544], menstruation [p.545], pathogen [p.545], malaria [p.545], sickle-cell anemia [p.545], neutrophils [p.545], eosinophil [p.545], basophils [p.545], leukemias [p.545], bone marrow [p.545], B & T lymphocyte [p.545], lymph node [p.545], lymphatic system [p.545], megakaryocytes [p.545], fibrinogen [p.545], cascade of enzyme reactions [p.545], vitamin K deficiency [p.545],

Boldfaced Terms

[p.544] plasma _____

_____

[p.544] red blood cells _____

_____

[p.545] cell count _____

_____

[p.545] white blood cells _____

_____

[p.545] platelets _____

_____

[p.545] hemostasis _____

_____

[p.545] fibrin _____

_____

[p.564] anemias _____

_____

Choice

Choose one of the following blood components for each statement. Answers may be used more than once. [pp.544-545]

    a. red blood cells
    b. plasma
    c. white blood cells
    d. platelets

1. ____ also called erythrocytes.

2. ____ also called leukocytes.

3. ____ the transport medium for blood cells and platelets.

4. ____ these initiate blood clotting.

5. ____ neutrophils, basophils and macrophages are examples.

6. ____ acts as a solvent for hundreds of plasma proteins.

7. ____ transport oxygen to aerobically transpiring cells.

8. ____ contains hemoglobin.

9. ____ identify "nonself" objects in the body.

10. ____ these have a 120 day lifespan.

Complete the Table

Complete the following table, which describes the components of blood. You might want to consult Figure 33.9 in the textbook for additional information. [p.544]

Components	Relative Amounts	Functions
Plasma Portion (50% - 60% of total volume)		
11.	91% - 92% of plasma volume	Solvent
12.	7% - 8%	Defense, clotting, lipid transport, roles in extracellular fluid volume, etc.
Ions, sugars, lipids, amino acids, hormones, vitamins, dissolved gases	13.	Roles in extracellular fluid volume, pH, etc.
Cellular Portion (40% - 50% of total volume)		
14	4,800,000–5,400,000 per microliter	O_2, CO_2 transport
15.	3,000–6,750	Phagocytosis
16.	1,000–2,700	Immunity
Monocytes (macrophages)	150–720	17.
Eosinophils	100–360	18.
19.	25–90	Roles in inflammatory response, anticlotting
20.	250,000–300,000	Roles in clotting

Choice

Indicate whether each of the following conditions is the result of a red blood cell disorder or a white blood cell disorder. [p.545]

a. red blood cell disorder

b. white blood cell disorder

21. ____ leukemia

22. ____ sickle cell anemia

33.6 BLOOD VESSEL STRUCTURE AND FUNCTION [p.546]
33.7 BLOOD PRESSURE [p.547]
33.8 CAPILLARY EXCHANGE [p.548]
33.9 VEIN FUNCTION [p.549]
33.10 CARDIOVASCULAR DISORDERS [pp. 550-551]
33.11 INTERACTIONS WITH THE LYMPHATIC SYSTEM [pp.576-577]

Selected Terms

flaplike valves [p.546], brachial artery [p.547], hypertension [p.547], stethoscope [p.547], 120/80 [p.547], vasodilation [p.547], cardiac output [p.547], hypertonic [p.548], edema [p.548], osmotic movement [p.548], dissolved proteins [p.548], exocytosis [p.548], backflow [p.549], varicose veins [p.549], hemorrhoids [p.549], embolus [p.549], electrocardiogram (ECG) [p.550], arrhythmias [p.550], bradycardia [p.550], tachycardia [p.550], atrial fibrillation [p.550], anticlotting medications [p.550], ventricular fibrillation [p.550], atherosclerosis [p.550], "hardening of the arteries" [p.550], plaque [p.550], low-density lipoproteins (LDL) [p.551], high-density lipoproteins (HDL) [p. 551], clot-dissolving drugs [p.551], coronary bypass surgery [p.551], balloon angioplasty [p.551], stent [p.551], cardiovascular disorders [p.551], tobacco [p.551], diabetes mellitus [p.551], obesity [p.551], plasma protein [p.552]

Boldfaced Terms

[p.546] pulse_____

_____

[p.546] vasodilation _____

_____

[p.546] vasocontriction _____

_____

[p.547] blood pressure _____

_____

[p.547] systolic pressure _____

_____

[p.547] diastolic pressure _____

_____

[p.552] lymph vascular system _____

_____

[p.552] lymph _____

_____

[p.553] lymph nodes _____

_____

[p.553] spleen_____

_____

Matching

Match each of the following Terms to the correct statement. [pp.546-547]

1. ____ vasoconstriction
2. ____ diastolic pressure
3. ____ cardiac output
4. ____ blood pressure
5. ____ systolic pressure
6. ____ vasodilation
7. ____ arteries
8. ____ venules
9. ____ veins
10. ____ arterioles
11. ____ capillaries

a. carry blood rapidly away from the heart's ventricles.
b. controls over blood flow operate here.
c. these form diffusion zones.
d. the small vessels between capillaries and veins
e. act as blood reservoirs
f. this is the pressure caused by ventricular contractions.
g. an increase in the diameter of the blood vessels
h. a decrease in the diameter of the blood vessels
i. the peak pressure against the walls of the arteries
j. the lowest arterial pressure.
k. how much blood the ventricles pump out

Labeling [p.548]

Label the following picture with the appropriate letters:

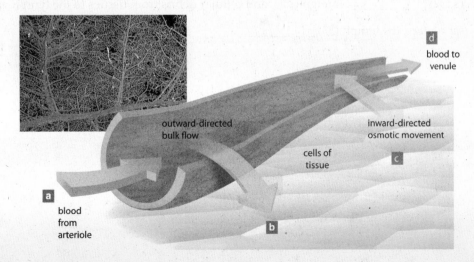

12. ____ ultrafiltration – fluid leaves capillaries for cells

13. ____ blood going to venule

14. ____ reabsorption – fluid entering capillaries from cells

15. ____ blood coming from arteriole

Matching

Match each of the following cardiovascular disorders to its correct statement. [pp.550-551]

16. ____ atherosclerosis

17. ____ diabetes and hypertension

18. ____ arrhythmias

19. ____ hemostasis

20. ____ hypertension

21. ____ arterial fibrillation

22. ____ ventricular fibrillation

a. chronically high blood pressure
b. the atria do not contract normally.
c. a serious condition in which the ventricle does not contract correctly
d. risk factors for coronary artery disease
e. abnormal rhythms in the heartbeat
f. the process that stops blood loss from damaged blood vessels.
g. high LDL levels increase the risk of this condition.

Fill-in-the-Blanks [pp. 552-553]

A portion of the lymphatic system, called the (23) _____ system, consists of many tubes that collect and deliver (24) _____ and solutes from (25) _____ fluid to ducts of the circulatory system. Its main components are (26) _____ capillaries and vessels. Tissue fluid that moves into these vessels is called (27) _____.

The lymph vascular system serves three functions. First, its vessels are drainage channels for water and (28) _____ proteins that have leaked out from the (29) _____ at capillary beds and must be delivered back to the blood circulation. Second, the system takes up (30) _____ that the body has absorbed from the (31) _____ intestine and delivers them to the general circulation. Third, it delivers (32) _____, foreign cells, and cellular debris from tissues to the lymph vascular system's disposal centers, the lymph (33) _____.

SELF QUIZ

1. The term varicose is associated with _____.
   [p.549]
   a. capillaries
   b. the lymphatic system
   c. the heart
   d. arteries
   e. veins

2. The _____ act as blood reservoirs in the body. [p.549]
   a. veins
   b. capillaries
   c. ventricles
   d. arterioles
   e. arteries

3. The pulmonary circuit first arose in which group of organisms? [p.539]
   a. fish
   b. birds and mammals
   c. reptiles
   d. amphibians

4. Which of the following serves as the cardiac pacemaker? [pp.542-543]
   a. the AV node
   b. the SA node
   c. the pericardium
   d. the left ventricle
   e. none of the above

5. The name of the major blood vessel that carries blood from the left ventricle of the heart is the _____ [p.542]
   a. femoral vein
   b. renal artery
   c. aorta
   d. jugular vein
   e. carotid arteries

6. Chronic elevated high blood pressure is called _____. [pp.550-551]
   a. hypertension
   b. arteriosclerosis
   c. angina pectoris
   d. edema
   e. atrial fibrillation

7. Which of the following makes up the largest percent of blood volume? [p.544]
   a. platelets
   b. red blood cells
   c. plasma
   d. white blood cells

8. The _____ returns excess interstitial fluid to the circulatory system. [pp.552-553]
   a. lymph vascular system
   b. pulmonary circuit
   c. systemic circuit
   d. renal vein
   e. vena cava

9. Neutrophils, basophils and lymphocytes are all _____ [p.544]
   a. red blood cells
   b. plasma proteins
   c. platelets
   d. white blood cells

10. Leukemia is a disease of the _____.
    [p.545]
    a. red blood cells
    b. platelets
    c. plasma proteins
    d. white blood cells

CHAPTER OBJECTIVES/ REVIEW QUESTIONS

1. Distinguish between open and closed circulatory systems. Provide examples of animals with one or the other. [p.538]

2. Describe how vertebrate circulatory systems have evolved from the fish model to the mammalian model. [pp.538-539]

3. Know the major arteries and veins of the human body. [p.540]

4. Know the major structures of the human heart. [p.542]

5. Explain the process of cardiac muscle contraction. [pp.542-543]

6. Describe the composition of human blood, using percentages of total volume. [p.544]

7. State where erythrocytes, leukocytes, and platelets are produced. [pp.544-545]

8. Recognize the major forms of red and white blood cell disorders. [p.545]

9. Explain hemostasis. [p.545]

10. Describe how the structures of arteries, capillaries, and veins differ. [p.546]

11. Explain what causes high pressure and low pressure in the human circulatory system. [p.547]

12. Define hypertension [p.547]

13. Describe the exchanges that occur in the capillary bed regions and the mechanisms that cause the exchanges. [p.548]

14. Define the major types of arrythmias. [p.550]

15. State the significance of high- and low-density lipoproteins to cardiovascular disorders. [pp.550-551]

16. Describe the composition and function of the lymphatic system. [pp.552-553]

CHAPTER SUMMARY

Many animals have either an open or a closed (1) _____ system that transports substances to and from all body (2) _____ .

Vertebrate blood is fluid (3) _____ tissue. It consists of red blood cells, white blood cells, (4) _____ , and diverse substances dissolved in (5) _____ , the transport medium. Red blood cells function in (6) _____ exchange, white blood cells and platelets help (7) _____ tissues.

The human heart has (8) _____ chambers. Blood flows into its two atria and then into two (9) _____ , which pump it into two separate circuits of blood vessels. One circuit extends through all body regions, the other through (10) _____ tissue only. Both circuits loop back to the (11) _____ .

The heart pumps blood (12) _____, on its own. Blood pressure is highest in the heart's (13) _____, drops as it flows through arteries, and is lowest in the (14) _____. Adjustments at (15) _____ regulate how much blood volume is distributed to tissues. Exchange of gases, wastes, and nutrients between blood and tissues takes place at (16) _____.

Ruptured or clogged blood (17) _____ or abnormal heart (18) _____ cause problems. Some problems have a genetic basis, most are related to age or (19) _____.

A (20) _____ vascular system delivers the excess fluid that collects in tissues to the blood. Lymphoid organs cleanse blood of (21) _____ agents and other threats to health.

INTEGRATING AND APPLYING KEY CONCEPTS

You observe that some people appear as though fluid had accumulated in their lower legs and feet. Their lower extremities resemble those of elephants. You inquire about what is wrong and are told that the condition is caused by the bite of a mosquito that is active at night. Construct a testable hypothesis that would explain (1) why the fluid was not being returned to the torso, as normal, and (2) what the mosquito did to its victims.

34

IMMUNITY

INTRODUCTION

The human body encounters millions of microbes, i.e. bacteria, viruses, fungi and protozoa, each day, yet we rarely become sick. Why is that? How can we constantly be exposed to disease causing organisms, yet remain relatively healthy? The answer is the immune system, an elaborate network of cells, tissue and organs dispersed throughout the vertebrate body that works to combat the potential invading organisms. This chapter discusses the organization, structure and function of the components of the mammalian immune system. From our exploration and study of the immune system, which really began in 1796 with the work of Edward Jenner, we have come to understand the exquisite design and impressive function of this system. Further, we also have come to understand autoimmune and allergic reactions; the consequences when the immune system does not respond appropriately to pathogens. Lastly, this knowledge has allowed us to manipulate the immune system in order to prevent disease. Like Jenner in the 1700s, we have been able to prevent many diseases by using vaccines, which are a way of "tricking" the immune system. A brief presentation is made about human blood typing. This and other ways of immune system manipulation offers much promise in the treatment and prevention of many diseases.

FOCAL POINTS

- Table 34.1 [p.558] shows a comparison of innate and adaptive immunity.
- Figure 34.3 [p.559] illustrates the various leukocytes (white blood cells).
- Table 34.2 [p.561] displays examples of surface barriers.
- Figure 34.8a [p.562] shows phagocytosis.
- Figure 34.9 [p.563] illustrates an innate immune response.
- Table 34.3 [p.564] displays structural classes of antibodies.
- Figure34.10 [p.564] illustrates antibody structure.
- Figure 34.11 [p.565] shows the process of antigen processing.
- Figure 34.12 [p.566] illustrates the primary and secondary immune response.
- Figure 34.13 [p.567] is an overview of the key interactions between antibody-mediated and cell-mediated responses.
- Figure 34.14 [p.567] gives backgrounds of the adaptive immune system.
- Figure 34.15 [p.568] animates an antibody mediated immune response.
- Figure 34.16 [p.569] illustrates clonal selection.
- Figure 34.17 [p.569] displays blood typing tests.
- Figure 34.18 [p.590] animates a cell mediated immune response.
- Table 34.4 [p.572] displays the recommended immunizations schedule for children.
- Table 34.5 [p573] shows examples of autoantibodies associated with autoimmune disorders.

INTERACTIVE EXERCISES

34.1 FRANKIE'S LAST WISH [p.557]
34.2 INTEGRATED RESPONSE TO THREATS [pp.588-559]
34.3 SURFACE BARRIERS [pp.560-561]

Selected Terms

cervix [p.557[, uterus [p.557], cancerous [p.557], gynecological testing [p.557], human papillomavirus (HPV) [p.557], genital HPV [p.557], Types 16 & 18 [p.557], radiation treatment [p.557], chemotherapy [p.557], pathogens [p.557], viruses, bacteria, fungi, & parasitic worms [p.558], infection [p.558], pathogen associated molecular patterns(PAMPs) [p.558], vertebrates [p.558], antigen [p.558], memory cells [p.558], secondary response [p.558], white blood cells [p.558], lymph nodes [p.558], spleen [p.558], immune system [p.559], interleukins, interferons & tumor necrosis factors [p.559], skin [p.560], normal flora [p.560], sebrum [p.560], *Propionbacterium acnes* [p.560], aerobic habitat [p.560], inflammation [p.560], acne [p.560], biofilm [p.560], glycoprotein [p.560], archea [p.560], lactic acid [p.560], cavities [p.560], fermentation [p.560], epidermis [p.561], *Porphyromonas gingivalis* [p.561], atherosclerosis [p.561], pneumonia [p.561], ulcers [p.561], colitis [p.561], whooping cough [p.561], meningitis [p.561], abscesses [p.561], *Clostridium tetani* [p.561], *Staphylococcus aureus* [p.561], MRSA (methicillin resistant) [p.561], *Lactobacillus* [p.561]

Boldfaced Terms

[p.558] immunity _____

_____

[p.558] complement _____

_____

[p.558] innate immunity _____

_____

[p.558] adaptive immunity _____

_____

[p.558] antigen _____

_____

[p.559] cytokines _____

_____

[p.559] neutrophils _____

_____

[p.559] macrophages _____

_____

[p.559] dendritic cells _____

_____

[p.559] eosinophils _____

_____

[p.559] basophils _____

_____

[p.559] mast cells _____

_____

[p.559] B cell_____

_____

[p.559] T cells _____

_____

[p.559] cytotoxic T cells _____

_____

[p.559] natural killer cells (NK) _____

_____

Matching

Match each of the following cell types to its correct function. [pp.558-559]

1. ____ macrophages
2. ____ neutrophils
3. ____ natural killer cells
4. ____ B lymphocytes and T lymphocytes
5. ____ eosinophils
6. ____ basophils and mast cells

a. destroy parasitic worms
b. "big eaters", can destroy up to 100 bacterial cells
c. the most common immune system cell
d. secrete chemicals to assist inflammation
e. manufacture proteins that chemically recognize pathogens
f. bind to tumor cells, virus-infected cells and bacterial cells

Complete the Table

For each of the following, list the mechanism by which the structure provides a first line of defense. [pp.560-561]

Structure	Defense Mechanism
Eyes	7.
Urinary tract	8.
Vertebrate skin	9.
Mouth	10.
Stomach	11.

34.4 INNATE IMMUNE RESPONSES [pp.562-563]
34.5 ANTIGEN RECEPTORS IN ADAPTIVE IMMUNITY [pp.564-565]
34.6 OVERVIEW OF ADAPTIVE IMMUNE RESPONSE [pp.566-567]

Selected Terms

Macrophages [p.562], complement [p.562], antigen [p.562], antibodies [p.562], basophils [p.562], neutrophils [p.562], prostaglandins [p.562], histamine [p.562], shivering or "chills" [p.563], Y-shaped proteins [p.564], B cells [p.564], IgG, IgA, IgF, IgF, & IgI [p.564], immunoglobulin [p.564], exocrine [p.565], "naïve" [p.565], MHC markers [p.565], specificity [p.566], diversity [p.566], memory [p.566], dendritic cells [p.567]

Boldfaced Terms

[p.562] inflammation _____

_____

[p.563] fever _____

_____

[p.564] T cell receptors _____

_____

[p.564] MHC markers _____

_____

[p.564] antibodies _____

_____

[p.565] B cell receptor _____

_____

[p.566] effector cells _____

_____

[p.566] memory cells _____

_____

[p.566] antibody-mediated immune response _____

_____

[p.567] cell-mediated immune response _____

_____

Choice

Choose the form of innate response that is associated with the statement. Some answers may be used more than once. [pp.562-563]

   a.   fever
   b.   antimicrobial proteins
   c.   inflammatory response

1. ____ capillaries become "leaky"

2. ____ proteins tag microbes for destruction

3. ____ the hypothalamic thermostat is reset

4. ____ endogenous pyrogens stimulate prostaglandin release

5. ____ mast cells in connective tissue release histamine.

6. ____ complement proteins attack microbes or interfere with reproduction

7. ____ edema is caused by a rush of fluid to the site of an infection.

Sequence

Place the following events of the inflammatory response in their correct sequence. The letter of the first event is placed in #8, the second event in #9, etc. [pp.562-563]

8. ____
9. ____
10. ____
11. ____
12. ____

a. bacteria invade cells and damage tissue
b. complement proteins attack bacteria
c. localized edema occurs
d. neutrophils and macrophages engulf debris and invades.
e. mast cells in tissue release histamine.

Fill-in-the-Blanks [p.566-567]

Four defining features characterize the (13) _____ immune system of all jawed vertebrates. These four characteristics are: (14) _____, (15) _____, (16) _____, and (17) _____.

Self vs. non self recognition starts with (18) _____ patterns that give each cell or virus its identity. The cells of the immune system can recognize these patterns as foreign. Specificity means that each B or T lymphocyte binds to only (19) _____ antigen.

Diversity refers to the billions of different antigen (20) _____. Memory refers to the ability of the immune system to (21) _____ an antigen.

Matching

Match each of the following to its correct statement. [pp.562-567]

22. ____ antigen-presenting cells
23. ____ B cells
24. ____ helper T cell
25. ____ MHC markers
26. ____ cytotoxic T cells
27. ____ antibodies

a. Induces T and B cells to start division and differentiation
b. Are involved in the cell-mediated responses.
c. Specific recognition proteins on the plasma membrane.
d. Bind pieces of antigen to MHC markers, forming emergency flags.
e. Are involved in the antibody-mediated responses.
f. A molecule that can bind to specific antigens.

Labeling

Label the following parts of the antibody molecule. You can consult Figure 34.10 in the textbook for more information. [p.564]

28. ____ antigen binding site

29. ____ constant region of the heavy chain

30. ____ variable region of the heavy chain

31. ____ variable region of the light chain

32. ____ constant region of the light chain

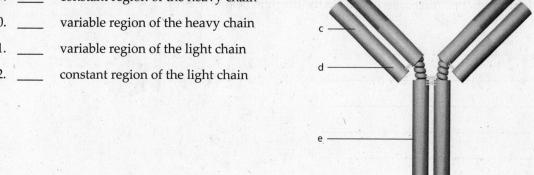

Matching

Match the class of antibody with its major function or characteristic. You might want to consult Table 34.3 in the textbook. [p.564]

33. ____ abundant in mucous and exocrine secretions

34. ____ signals the release of histamines and cytokines from the cell to which it is attached.

35. ____ main antibody in the blood; activates complete, neutralizes toxins etc.

36. ____ secreted as a pentamer, also functions as a B cell receptor.

37. ____ always membrane bound, functions as B cell receptor.

a. IgD
b. IgA
c. IgE
d. IgM
e. IgG

34.7 THE ANTIBODY-MEDIATED IMMUNE RESPONSE [pp.568-569]
34.8 BLOOD TYPING [p.569]
34.9 THE CELL MEDIATED RESPONSE [p.570-571]

Selected Terms

Staphyococcus aureus [p.568], polysaccharide[p.568], spleen [p.569], transfusion [p.569], hemoglobin [p.569], foreign [p.570], rejection of transplanted organ [p.570], apoptosis [p.570], antibody-mediated response [p.571], memory cells [p.571]

Boldfaced Terms

[p.569] agglutination _____

_____

Labeling

Label each of the indicated molecules or cells in the diagram below. [p.570]

1. _____

2. _____

3. _____

4. _____

5. _____

6. _____

7. _____

Fill in the Blank

If you are type A, your immune system will treat blood with type (8) _____ markers as foreign. In you are type B, your body will react against type (9) _____ markers. If your blood is type (10) _____, your body is familiar with both markers and won't react against either type; you are able to (11) _____ blood from anyone. If you are type O, both A and B markers are perceived as (12) _____. You can only receive blood from others who are also type (13) _____, but you can donate blood to anyone.

(14) _____ blood typing is based on the presence or absence of the Rh (15) _____. If you are type (16) _____, your body cells bear this marker. If you are type (17) they do not.

Usually people do not have (18) _____ against Rh markers, because their body has never been exposed to them. But an Rh- recipient of transfused (19) _____ blood will make antibodies, and they will remain in the blood.

Short Answer

20. Humans have four possible blood types: A, B, AB, and O. A court case has been filed by a mother with type O blood who has a son with type O blood. There are two possible fathers; Edward has type AB blood and Charles has type A blood. Which one of the men could be the father of the child?

_____

_____

34.10 ALLERGIES [p.571]
34.11 VACCINES [p.572]
34.12 ANTIBODIES AWRY [p.573]
34.13 AIDS [pp.574-575]

Selected *Terms*

mast cells [p.571], anaphylactic shock [p.571], allergic reaction [p.571], primary immune response [p.572], secondary immune response [p.572], smallpox epidemics [p.572], dairymaids [p.572], cowpox [p.572], Edward Jenner [p.572], "vaccination" [p.572], "bad" T-cell receptors [p.573], autoimmune disorders [p.573], Graves disease [p.573], neurological disorders (multiple sclerosis) [p.573], reactive T-cells [p.573], impaired immune function [p.573], severe combined immunodeficiencies [p.573], acquired immunodeficiency [p.573], pneumonia [p.574], *Pneumocystis jirovecii* [p.574], Kaposi's sarcoma [p.574], retrovirus [p.574], transmission [p.575], penis, rectum, mouth, & vagina [p.575], infected blood [p.575], blood [p.575], saliva [p.575], RNA nucleotide [p.575], AZT [p.575], reverse transcriptase [p.575], protease [p.575], viral vectors [p.575]

Boldfaced Terms

[p.571] allergen _____

_____

[p.571} allergy _____

_____

[p.572] immunization _____

_____

[p.572] vaccine _____

_____

[p.573] autoimmune response _____

_____

[p.574] AIDS _____

_____

Choice

List three vaccines that should be administered at 2 months of age. [p.572]

1. _____

2. _____

3. _____

Matching

Match each of the terms with its correct definition. [pp.571-575]

4. ____	passive immunization	a. a life-threatening response to an allergen
5. ____	anaphylactic shock	b. an autoimmune response of the thyroid gland
6. ____	asthma and hay fever	c. antibodies that fight infection but do not activate the immune system
7. ____	multiple sclerosis	d. a harmless substance that invokes inflammation
8. ____	AIDS	e. forms of allergies
9. ____	active immunization	f. a vaccine is used to immunize the individual
10. ____	rheumatoid arthritis	g. a secondary immune deficiency
11. ____	allergen	h. a autoimmune response in the nervous system
12. ____	Graves' disease	i. chronic inflammation of the joints and heart

SELF QUIZ

1. Which of the following is not an innate mechanism? [pp.559,562]
   a. macrophages
   b. B lymphocytes
   c. neutrophils
   d. eosinophils
   e. all of the above are non-specific

2. Which of the following is NOT associated with the immune system of vertebrates? [p. 566]
   a. diversity
   b. memory
   c. self vs. nonself recognition
   d. migration

3. Which of the following is not a physical defense mechanism? [p.560-561]
   a. lysozyme in saliva
   b. pH of urinary tract and stomach
   c. complement proteins
   d. keratin-packed epithelial cells

4. _____ are divided into two groups: T cells and B cells. [p.559]
   a. Macrophages
   b. Lymphocytes
   c. Platelets
   d. Complement cells
   e. Cancer cells

5. _____ immunoglobulins are secreted by B cells and comprise the main antibody in the blood. [p. 564]
   a. IgA
   b. IgE
   c. IgM
   d. IgG

6. The markers for every cell in the human body are referred to by the letters _____. [p.564]

   a. HIV
   b. MBC
   c. RNA
   d. DNA
   e. MHC

7. HIV primarily infects _____. [pp.574-575]
   a. neutrophils
   b. eosinophils
   c. macrophages, dendritic cells and T helper cells
   d. mast cells

8. Histamine is associated with which of the following? [p.585]
   a. antibody production
   b. activation of helper T cells
   c. inflammatory response
   d. immunological memory

9. Which of the following is not an autoimmune disease? [p.573]
   a. hay fever
   b. Graves' disease
   c. rheumatoid arthritis
   d. multiple sclerosis

10. A virus-infected cell of the body would be targeted for destruction by _____ cells. [p.559]
    a. cytotoxic T
    b. macrophages
    c. eosinophils
    d. B

CHAPTER OBJECTIVES/ REVIEW QUESTIONS

1. Distinguish between the major cells of the immune system. [pp.558-559]

2. Identify the major physical barriers to infection in the body. [pp.560-561]

3. Outline the process of the inflammatory response. [pp.562-563]

4. Understand the role of antimicrobial proteins and fever in the immune response. [p.563]

5. Recognize the cells associated with adaptive immunity. [p.564-565]

6. List the function of each class of immunoglobulins. [p.564]

7. Understand the role of lymphocytes in the immune response. [p.565]

8. Understand the concept of memory. [p.566]

9. Understand the adaptive immune response. [p.566-567]

10. Understand the antibody-mediated response. [p.568-569

11. Understand the major blood types and the Rh factor. [p569]

12. List the steps in a cell-mediated response. [pp. 570-571]

13. Understand allergies and how they affect humans. [p.571]

14. Understand vaccines, their history, and the timing of giving vaccines to children. [p.572]

15. Understand the major autoimmune disorders. [p.573]

13. Recognize the significance of AIDS. [pp.574-575]

CHAPTER SUMMARY

The vertebrate body has three lines of defense against invading microorganisms. These are (1) _____, _____, and _____. The (2) _____ immune response rids the body of most invaders before infection becomes established. The (3) _____ immune response targets specific pathogens and cancer cells. (4) _____, _____ and secretions at the body's surfaces function as barriers that exclude most microbes. (5) _____ immunity involves a set of general, immediate defenses. (6) _____ white blood cells, which engulf pathogens, are major components of the innate immune system. (7) B lymphocytes produce _____, which bind specifically to pathogens. (8) _____ lymphocytes directly destroy diseased body cells in a cell meditated response. The distinction between (9) _____ and non-self is critical in an appropriate immune response. (10) _____ and _____ can develop if the immune system responds inappropriately.

INTEGRATING AND APPLYING KEY CONCEPTS

1. Why would you not necessarily choose to treat a fever of 101.3 degrees F?

2. What reasons do some give for not having their children vaccinated?

35

RESPIRATION

INTRODUCTION

The respiratory system plays a number of important roles in the human body. Not only does it deliver oxygen to, and remove carbon dioxide from our tissues, it also plays an important role in maintaining internal homeostasis. This chapter explores the principles of a respiratory system and the major types of respiratory systems in both vertebrates and invertebrates. The unhealthy aspects of smoking are presented on the opening page. The important life saving Heimlich maneuver is illustrated and discussed. Respiratory diseases, disorders, carbon monoxide poisoning, and high altitude breathing problems are discussed in the closing pages of this chapter.

FOCAL POINTS

- Figure 35.3 [p.581] shows examples of invertebrate respiratory mechanisms.
- Figure 35.4 [p.582] illustrates the respiratory structures and functions of the gills of a bony fish and the oxygen/carbon dioxide % relating to the countercurrent flow.
- Figure 35.5 [p.583] shows how a frog uses its lungs.
- Figure 35.6 [p.583] shows a diagram of mammalian lungs.
- Figure 35.7 [p.583] shows the respiratory structures of a bird.
- Figure 35.8 [p.584] diagrams the major structures of the human respiratory system.
- Figure 35.10 [p.586] illustrates inhalation and exhalation of human lungs.
- Figure 35.11 [p.586] demonstrates the Heimlich maneuver for a choking victim.
- Figure 35.15 [p.588] diagrams a hemoglobin molecule.
- Figure 35.16 [p.589] illustrates partial pressures of oxygen and carbon dioxide in the atmosphere, blood, and tissues.

INTERACTIVE EXERCISES

35.1 UP IN SMOKE [pp.579]
35.1 THE PROCESS OF RESPIRATION [p.580]
35.3 INVERTEBRATE RESPIRATION [p.581]

Selected Terms
ciliated cells [p.579], pathogens [p.579], pollutants [p.579], asthma [p.579], bronchitis [p.579], nicotine [p.579], "bad" cholesterol [p.579], "good" cholesterol [p.579], carcinogens [p.579], respiratory systems [p.579], glucose + oxygen → carbon dioxide + water [p.580], diffusing [p.580], interstitial fluid [p.580], circulatory system [p.580], concentration gradient [p.580], inhalations & exhalations [p.580], hemoglobin [p.580], gastrovascular cavity [p.581], filamentous respiratory organs [p.581], feathery gills [p.581], external gills [p.581], chitin [p.581], insecticides [p.581], tracheal branches [p.581], spiracles [p.581]

Boldfaced Terms
[p.580] respiration_____

_____

[p.580] respiratory surface _____

_____

[p.580] respiratory proteins _____

_____

[p.581] gills _____

_____

[p.581] tracheal system _____

_____

Choice
For each of the following, choose the most appropriate factors affecting diffusion rates from the list below. [p.580]

      a.   respiratory proteins
      b.   respiration
      c.   respiratory surface

1. ____ this determines the overall size of the organism

2. ____ bind and release oxygen

3. ____ animals without respiratory organs are usually small and flat

4. ____ hemoglobin is an example

5. ____ the movement of air or water across a respiratory surface

Matching
Match each of the following terms to the correct statement. [pp.580-581]

6. ____ external body surface

7. ____ tracheal system

8. ____ hemoglobin

9. ____ gills

    a. the main respiratory pigment of vertebrates
    b. gas diffusion directly across the body surface
    c. a series of branching tubes inside invertebrates such as insects and spiders
    d. the primary respiratory surface of aquatic invertebrates

35.4 VERTEBRATE RESPIRATION [pp.582-583]
35.5 HUMAN RESPIRATORY SYSTEM [pp.584-585]

Selected Terms
oral cavity [p.582], bony gill [p.582], gill arch [p.582], filament [p.582], tetrapods [p.582], nostrils [p.582], amniotes [p.583], posterior air sac [p.583], anterior air sac [p.583], nasal passage [p.584], acid-base balance [p.584], alveolar duct [p.584], alveolar sac [p.584], pulmonary capillary [p.584], thoracic cavity [p.585], pleural membranes [p.585], "Adam's apple" [p.585], abdominal cavity [p.585], "bronchial tree" [p.585], intercostal muscles [p.585]

Boldfaced Terms

[p.582] countercurrent exchange _____

_____

[p.582] lung_____

_____

[p.583] alveoli _____

_____

[p.585] pharynx _____

_____

[p.585] larynx _____

_____

[p.585] glottis _____

_____

[p.585] epiglottis _____

_____

[p.585] trachea _____

_____

[p.585] bronchi _____

_____

[p.585] bronchioles _____

_____

[p.585] diaphragm _____

_____

Choice

For each of the following statements, choose the correct group of organisms to which the statement applies. Some questions may have more than one answer. [pp.582-583]

    a. fish
    b. birds
    c. amphibians
    d. mammals

1. ____ Larvae of this group have gills, but most adults have lungs.

2. ____ Use a countercurrent exchange system.

3. ____ Air in these organisms continuously flows over the respiratory surface

4. ____ These organisms "gulp" air into the lungs.

5. ____ Use paired lungs.

Respiration **363**

Labeling
First, identify each of the indicated structures in the human respiratory system. Then choose the correct function for each structure from the list provided. [pp.584-585].

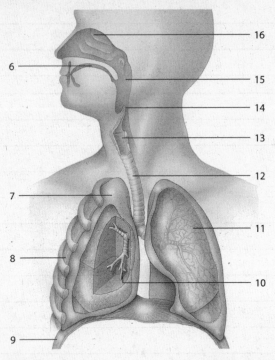

6. _____ ( )
7. _____ ( )
8. _____ ( )
9. _____ ( )
10. _____ ( )
11. _____ ( )
12. _____ ( )
13. _____ ( )
14. _____ ( )
15. _____ ( )
16. _____ ( )

a. a muscle located between the chest and abdominal cavities.
b. supplemental airway
c. airway connecting the nasal cavity and larynx
d. connects larynx with bronchi in lungs
e. warms, moistens and filters the incoming air.
f. skeletal muscles with roles in breathing.
g. branched structures of the lungs that end at the alveoli.
h. lobed organ that enhances gas exchange between internal and external environments.
i. closes off the larynx during swallowing.
j. double-layer membrane that separates the lungs from other organs
k. voice box

35.6 HOW YOU BREATH [pp.586-587]
35.7 GAS EXCHANGE AND TRANSPORT [pp.588-589]

Selected Terms

inhalation [p.586], exhalation [p.586], total lung volume [p.586], tidal volume [p.586], medulla oblongata [p.587], pacemaker [p.587], carbonic acid [p.587], blood pH [p.587], chemoreceptors [p.587], stimulus & response [p.587], residual volume [p.587], millimeters of mercury [p.588], carbaminohemoglobin [p.588], bicarbonate [p.588], alveolar epithelium [p. 588], capillary endothelium [p.588], heme group [p.588], carbon monoxide [p.589], acclimatization [p.589]

Boldfaced Terms

[p.586] respiratory cycle_____

_____

[p.586] Heimlich maneuver _____

_____

[p.586] vital capacity _____

_____

[p.588] respiratory membrane _____

_____

[p.588] partial pressure _____

_____

[p.588] oxyhemoglobin _____

_____

[p.589] carbonic anhydrase _____

_____

Choice

Choose whether each statement is associated with oxygen or carbon dioxide transport. [pp.588-589]

    a. carbon dioxide transport
    b. oxygen transport

1. ____ transport is enhanced by hemoglobin molecules.

2. ____ the majority is transported as bicarbonate

3. ____ the transported gas is released when the blood is warm and at a low pH.

4. ____ when bound, oxyhemoglobin is formed.

5. ____ carbaminohemoglobin is responsible for 30% of the transport.

6. ____ the enzyme carbonic anhydrase removes it from the red blood cell

Labeling

For each numbered item in the diagram below, provide the missing term or terms. You might want to consult Figure 35.13 in the textbook for additional information. [p.587]

7. _____

8. _____

9. _____

10. _____

11. _____

12. _____

13. _____

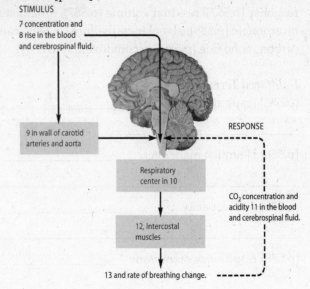

STIMULUS

7 concentration and 8 rise in the blood and cerebrospinal fluid.

9 in wall of carotid arteries and aorta

RESPONSE

Respiratory center in 10

CO_2 concentration and acidity 11 in the blood and cerebrospinal fluid.

12, Intercostal muscles

13 and rate of breathing change.

35.8 COMMON RESPIRATORY DISEASES AND DISORDERS [pp.590-591]

Selected Terms

interrupted breathing [p.590], sudden infant death syndrome (SIDS)[p.590], tuberculosis [p.590], *Mycobacterium tuberculosis* [p.590], pneumonia [p.590], "active TB" [p.590], antibiotics [p.590], bacteria, viruses, & fungi [p.590], bronchitis [p.590], inflammation [p.590], constriction [p.590], asthma [p.590], emphysema [p.590]

Matching

Match each of the following terms to its correct definition. [pp.590-591]

1. ____ interrupted breathing
2. ____ emphysema
3. ____ tuberculosis
4. ____ bronchitis
5. ____ smoking
6. ____ SIDS
7. ____ pneumonia

a. chronic infection of the epithelium lining of the bronchioles.
b. destruction of the alveoli
c. may be caused by defects in the respiratory control centers
d. a stoppage in the pattern of breathing
e. caused by a bacterial infection that causes problem breathing and bloody mucus
f. one of the leading causes of respiratory-related deaths
g. lung inflammation fills tissues with fluid

SELF QUIZ

1. The respiratory pigment of vertebrates is _____. [p.580;588]
   a. hemocyanin
   b. hemoglobin
   c. hemerythrin
   d. myoglobin

2. Which of the following is not involved in carbon dioxide transport? [pp.588-589]
   a. carbonic anhydrase
   b. carbaminohemoglobin
   c. bicarbonate
   d. oxyhemoglobin

3. If someone "forgets" to breathe, it is called _____. [p.590]
   a. emphysema
   b. hyperventilation
   c. bronchitis
   d. hypoxia
   e. apnea (interrupted breathing)

4. Which of the following is used by invertebrates as a respiratory system? [p.601]
   a. tracheal system
   b. gills
   c. spiracles
   d. body wall
   e. all of the above are used

5. Pneumonia can be caused by? [p.590]
   a. bacteria
   b. viruses
   c. fungi
   d. choices a and b, but not c
   e. choices a, b, and c

6. A countercurrent exchange system is found in the respiratory system of _____. [p.582]
   a. fish
   b. amphibians
   c. reptiles
   d. birds
   e. mammals

7. Under normal conditions, exhalation is a _____ process. [pp.586-587]
   a. active
   b. passive

8. Which structure serves as a passageway for both the respiratory and digestive systems? [pp.584-585]
   a. bronchial tubes
   b. esophagus
   c. nasal cavity
   d. pharynx
   e. larynx

9. This structure separates the abdominal and thoracic cavities. [pp.584-585]
   a. pharynx
   b. diaphragm
   c. intercostals muscles
   d. rib cage
   e. esophagus

10. The majority of gas exchange in humans occurs in the _____. [pp.585;588-589]
    a. tracheal system
    b. alveoli
    c. nasal cavity
    d. oral cavity

CHAPTER OBJECTIVES/ REVIEW QUESTIONS

1. Understand the factors that affect diffusion rates. [p.580]

2. Understand the role of respiratory pigments in gas exchange and know the major human respiratory pigment. [p.580]

3. Recognize the major respiratory systems of invertebrates. [p.581]

4. List the forms of respiratory structures found in vertebrates. [pp.582-583]

5. List all the principal parts of the human respiratory system and explain how each structure contributes to transporting oxygen from the external world to the bloodstream. [pp.584-585]

6. Be able to demonstrate the Heimlich maneuver. [p.586]

7. Explain the sequence of events in the process of inhalation and exhalation. [pp.586-587]

8. List the molecules and factors associated with oxygen and carbon dioxide transport. [pp.588-589]

9. Identify the causes of the major respiratory disorders. [pp.590-591]

10. Understand the effects of smoking. [p.591]

11. Understand the benefits of quitting smoking. [p.591]

CHAPTER SUMMARY

Aerobic respiration uses free (1) _____ and its (2) _____ wastes are removed before the internal environment's (3) _____ shifts dangerously. (4) _____ is the sum of the processes that move oxygen from the outside environment to all metabolically active (5) _____ and move carbon dioxide from those tissues to the outside.

Gas exchange occurs across the body surface or (6) _____ of aquatic invertebrates. In large invertebrates on land, it occurs across a moist, (7) _____ respiratory surface or at fluid-filled tips of branching (8) _____ that extend from the surface to internal tissues.

(9) _____, skin, and paired (10) _____ are gas exchange organs. Breathing ventilates (11) _____. (12) _____ is a transport medium. It picks up oxygen and give up carbon dioxide at the (13) _____ surface. It also exchanges gases with the (14) _____ fluid that bathes cells.

(15) _____ exchange is mot efficient when the rates of air flow and blood flow at the respiratory surface match. (16) _____ centers adjust the rate and (17) _____ of breathing.

Respiration can be disrupted by damage to respiratory centers in the (18) _____, physical obstructions, (19) _____, and inhalation of (20) _____, including (21) _____ smoke.

INTEGRATING AND APPLYING KEY CONCEPTS

1. Consider animals such are the amphibians that have aquatic larval forms (tadpoles) and terrestrial adults. Outline the respiratory changes that you think might occur as an aquatic tadpole metamorphoses into a land-going juvenile.

2. In the movie Waterworld, Kevin Costner's character develops gills to aid his underwater breathing. Explain what changes would be necessary in the human respiratory system in order to make this possible.

3. Blood doping, a process by which athletes attempt to increase their total red blood cell count, has become a major issue in professional athletics. Explain the benefit of an increased red blood cell count might be for an athlete.

4. How would you know if a person needed the Heimlich maneuver used on them? (Hint—can they talk?)

36

DIGESTION AND HUMAN NUTRITION

INTRODUCTION

How true is the saying "you are what you eat"? This chapter allows the student to explore this question by describing, in detail the structures and functions associated with the human digestive system. Following an in-depth discussion of the digestive processes, the chapter goes on to describe some diseases that are associated with various parts of the digestive system. Finally, the chapter discusses the human diet and the requirements of a healthy diet. Attention is given to the food groups, vitamins and minerals needed to maintain a healthy diet. Unfortunately, even with this knowledge readily available to all people, the majority of Americans are obese. The chapter ends with a discussion of obesity and the many lifestyle and genetic components that contribute to this disease. As always, this knowledge is presented as to afford the student the knowledge with which to better his of her own personal health.

FOCAL POINTS

- Figure 36.2 [p.596] Examples of animal digestive systems.
- Figure 35.5 [p.598] presents an overview of the human digestive tract.
- Figure 36.6 [p.599] illustrates the structure and function of teeth.
- Figure 36.7 [p.600] illustrates the location and structure of the stomach.
- Figure 36.8 [p.601] illustrates the structure of the lining or the intestinal mucosa and the villi at the surface.
- Figure 36.9 [p.602] animates the digestion and absorption that occurs in the small intestine.
- Figure 36.11 [p.604] shows the cecum and appendix of the large intestine.
- Figure 36.13 [p.606] summarizes the recommended USDA nutritional guidelines.
- Figure 36.14 [p.607] shows a nutritional facts label.
- Tables 36.2 & 36.3 [p.608-609] categorize the vitamins and minerals essential to a healthy diet.
- Figure 36.15 [p.610] shows how to estimate an "ideal" weight for adults.

INTERACTIVE EXERCISES

36.1 THE BATTLE AGAINST BULGE [p.595]
36.2 ANIMAL DIGESTIVE SYSTEM [pp.596-597]
36.3 THE HUMAN DIGESTIVE SYSTEM [pp.598-599]

Selected Terms
obesity [p.595], fat [p.595], obesity-related gene *fto* [P.595], homeostasis [p.596], liver [p.596], gall bladder [p.596], pancreas [p.596], enzymes [p.596], small intestine [p.596], beaks [p.596], keratin[p.596], carnivores [p.596], crown [p.596], molar [p.596], gizzard [p.597], regurgitated [p.597], salivary gland [p.598], chemoreceptors [p.599], acid [p.599]

Boldfaced Terms
 [p.596] incomplete digestive system _____

[p.596] complete digestive system _____

_____

[p.597] ruminants _____

_____

[p.599] esophagus _____

_____

[p.599] peristalsis _____

_____

[p.599] stomach _____

_____

[p.599] small intestine _____

_____

[p.599] colon (large intestine) _____

_____

[p.599] rectum _____

_____

[p.599] anus _____

_____

Fill-in-the-Blanks [p.595]

Recall, from Chapter 23, that some invertebrates have (1) _____ digestive systems. Their

digestive systems are (2) _____. This means that food enters and (3) _____ leave through a

single opening at the body's surface. In a flatworm, the saclike, branching (4) _____ cavity opens

at the start of a (5) _____, a muscular tube. Food enters the sac, is partially (6) _____ and

circulates to cells even as wastes are sent out. This two-way traffic does not favor regional (7)

_____.

Most species of animals have a (8) _____ digestive system. Basically, they have a tube

with an opening at one end for taking in (9) _____ and an opening at the other end for eliminating

unabsorbed residues and (10) _____. The tube is divided into (11) _____ regions for food

breakdown, (12) _____ and waste concentration.

Labeling

First, identify each numbered structure in the illustration. Then, match each structure with its function from the list provided. You can consult Figure 36.5 in your textbook for additional information.[p.598]

13. _____ ( )

14. _____ ( )

15. _____ ( )

16. _____ ( )

17. _____ ( )

18. _____ ( )

19. _____ ( )

20. _____ ( )

21. _____ ( )

22. _____ ( )

23. _____ ( )

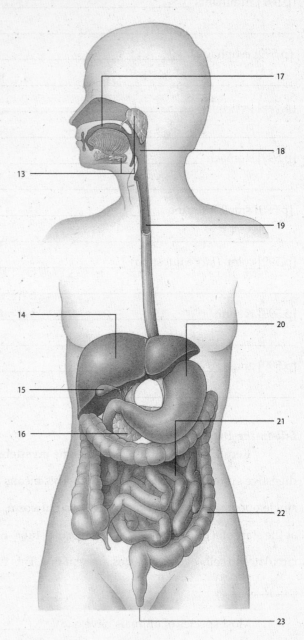

a. secretes digestive enzymes, buffers and insulin.

b. tube that moves food from pharynx to stomach

c. stores and concentrates bile.

d. secretes bile.

e. entrance to system, polysaccharide digestion begins.

f. major site of digestion and absorption

g. opening through which feces are expelled.

h. entrance to respiratory system and tubular section of digestive system.

i. muscular sac, protects against pathogens, protein digestion begins

j. secrete saliva, enzymes, buffers, and mucus.

k. concentrates and stores undigested matter.

Matching

Match each of the following statements with the correct term. [pp.598-599]

24. ____ a ring of smooth muscles that regulate movement of food.

25. ____ a mixture of water, salivary amylase and bicarbonate.

26. ____ site where protein digestion begins.

27. ____ the entrance to the esophagus and the lungs

28. ____ a muscular tube the moves food into the stomach

29. ____ site where carbohydrate digestion begins.

30. ____ site where the digestion of most nutrients is completed.

a. esophagus
b. small intestine
c. pharynx
d. mouth
e. stomach
f. sphincter
g. saliva

36.4 DIGESTION IN THE MOUTH [p.599]
36.5 FOOD STORAGE AND DIGESTION IN THE STOMACH [p.600]
36.6 STRUCTURE OF THE SMALL INTESTINE [pp.601-602]
36.7 DIGESTION AND ABSORPTION IN THE SMALL INTESTINE [pp.602-603]

Selected Terms

mechanical digestion [p.599], tooth [p.599], dentin-secreting cells [p.599], salivary amylase [p.599], bicarbonate [p.599], buffer [p.599], mucus [p.599], molars, incisors, premolars, & canines [p.599], crown, enamel, dentin, pulp cavity, root canal, periodonatal membrane root, & gingival [p.599], jaw [p.599], gastric juice [p.600], gastrin [p.600], pepsinogen [p.600], hydrochloric acid [p.600], denatures [p.600], gastroesophageal reflux [p.600], stomach ulcer [p.600], *Heliobacter pylori* [p.600], ibuprofen [p.600], aspirin [p.600], serosa [p.600], longitudinal, circular, & oblique muscles [p.600], submucosa [p.600], mucosa [p.600], pyloric sphincter [p.601], duodenum [p.601], absorption [p.601], monosaccharides [p.602], disaccharides [p.602], pancreatic amylase [p.602], fat digestion [p.602], amino acids [p.602], polypeptides [p.602], lactase [p.602], lactose [p.602], glucose [p.602], galactose [p.602], emulsification [p.602], bile salts [p.602], lipoproteins [p.602], triglycerides [p.602], trypsin [p.602], chymotrypsin [p.602], lactose intolerance [p.603], gallstones [p.603], pancreatitis [p.603], bilirubin [p.603], hemoglobin [p.603]

Boldfaced Terms

[p.599] salivary glands _____

_____

[p.600] sphincter _____

_____

[p.600] gastric fluid _____

_____

[p.600] chyme _____

_____

[p.601] villi _____

_____

[p.601] microvilli _____

_____

[p.601] brush border cells _____

_____

[p.603] bile _____

_____

[p.603] gallbladder _____

_____

[p.603] emulsification _____

_____

Complete the Table [pp.600-03]

Enzyme	Source	Where Active	Substrate
Salivary amylase	Salivary glands	Mouth stomach	(1).
Pancreatic amylase	Pancreas	(2)	Polysaccharides
(3)	Stomach lining	Stomach	Proteins
Trypsin and chymotrypsin	Pancreas	(4)	Proteins
Lipase	Pancreas	Small intestine	(5)

Matching

Match each of the following statements with the correct term. [pp.600-603]

6. ____ This chemical causes the stomach to release acid.

7. ____ The process by which bile salts break up fats for digestion.

8. ____ Secreted by the liver, this assists the process of fat digestion.

9. ____ A region of the small intestine

10. ____ A combination of hydrochloric acid, pepsinogen and other compounds.

11. ____ An accumulation of hard pellets within the gall bladder are called?

12. ____ The main components of bile.

a. gastric fluid
b. gastrin
c. duodenum
d. emulsification
e. bile
f. cholesterol and bilirubin
g. gall stones

Labeling [p.599]

Select the term from the list below and place its letter next to the number from the tooth diagram above.

13. ____

14. ____

15. ____

16. ____

a.   root
b.   pulp cavity
c.   crown
d.   gingiva

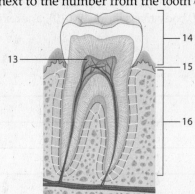

Labeling

Please match the following structures and functions with the appropriate labels from the figure. You might want to consult Figure 36.9 from your textbook for additional information. [p.602]

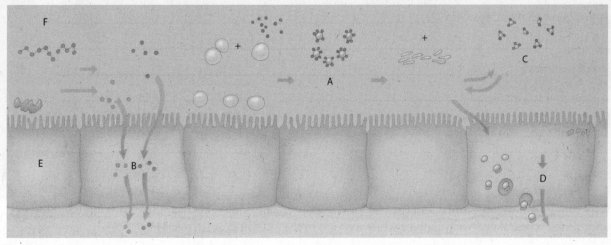

17. ____ interstitial fluid inside a villus

18. ____ monoglycerides & free fatty acids

19. ____ lumen of small intestine

20. ____ brush border cell

21. ____ lipoproteins

22. ____ emulsification droplets

36.8 THE LARGE INTESTINE [p.604]
36.9 THE FATE OF ABSORBED COMPOUNDS [p.605]
36.10 HUMAN NUTRITIONAL REQUIREMENTS [pp.606-607]
36.11 VITAMINS AND MINERALS [pp.608-609]
36.12 MAINTAINING A HEALTHY WEIGHT [pp.610-611]

Selected Terms

colon [p.04], cecum [p.604], ascending portion of large intestine [p.604], *Escherichia coli* [p.604], gut flora [p.604], Crohn's disease [p.604], chronic diarrhea [p.604], colon polyps [p.604], colonoscopy [p.604], cancer [p.604], organic metabolism [p.605]. adipose cells [p.605], metabolism [p.605], homeostasis [p.605], ammonia [p.605], amino acid breakdown [p.605], urea [p.605], refined grains [p.606], saturated fats [p.606], trans fatty acids (trans fats) [p.606], linoleic acid [p.606], alpha-linolenic acid [p.606], rheumatoid arthritis [p.607], oleic acid [p.607], iron [p.608], phytochemicals [p.609], phytonutrients [p.609], lutein & zeaxanthin [p.609], body mass index [p.610], anorexia nervosa [p.611], bulimia nervosa [p.611], vomiting [p.611]

Boldfaced Terms

[p.604] feces _____

_____

[p.626] appendix _____

_____

[p.604] organic metabolism _____

_____

[p.606] essential fatty acids _____

_____

[p.607] essential amino acids _____

_____

[p.608] vitamins_____

_____

[p.608] minerals_____

_____

True/False

If the statement is true, write a T in the blank. If the statement is false, correct it by changing the underlined word(s) and writing the correct word(s) in the answer blank. [p.604]

1. _____ The appendix has no known <u>digestive</u> functions.

2. _____ Water and <u>sodium ions</u> are absorbed into the bloodstream from the lumen of the large intestine.

3. _____ The risk of colon cancer increases with a <u>high</u> fiber diet.

4. _____ Bacteria in the large intestine are <u>beneficial</u>.

Fill in the Blank

List the recommended number of servings of each of the following foods. You might want to consult Figure 36.13 in the textbook for additional information. [p.606]

5. ____ dark green vegetables

6. ____ orange vegetables

7. ____ legumes

8. ____ starchy vegetables

9. ____ fruits

10. ____ milk products

11. ____ whole grains

12. ____ fish, poultry, lean meat

Complete the Table [pp.608-609]

Vitamin	Sources	Main Functions
Fat-soluble Vitamins		
A	13.	Synthesis of visual pigments, bone, teeth
D	14.	Promotes bone growth and mineralization; enhances calcium absorption
E	Whole grains, dark green vegetables, vegetable oils	15.
16.	Enterobacteria	Blood clotting, ATP formation
Water-soluble Vitamins		
Thiamin	Whole grains, legumes, lean meats	17.
Riboflavin	Whole grains, poultry, fish	18.
19.	Green leafy vegetables, potatoes, peanuts, poultry	Coenzyme action
B_6	20.	Coenzyme in amino acid metabolism.
Pantothenic acid	21.	Coenzyme in glucose metabolism
22.	Dark green vegetables, whole grains, lean meats	Coenzyme in nucleic acid and amino acid metabolism
B_{12}	23.	Coenzyme in nucleic acid metabolism
Biotin	Legumes, egg yolk	24.
C	25.	Collagen synthesis, structural role in bone, cartilage, teeth, used in carbohydrate metabolism

Matching

Match each of the following terms to the correct definition. [pp.610-611]

26. ____ body mass index
27. ____ obesity
28. ____ anorexia nervosa
29. ____ bulimia nervosa
30. ____ kilocalories (Calorie)
31. ____ apnea

a. a measurement that is designed to determine the health risk of weight.
b. a person who has access to food but places little effort on eating; can lead to death.
c. a halt in breathing
d. indicated by a BMI score >30.
e. a measurement of heat
f. a person who eats and then later vomits the food.

SELF QUIZ

1. The name of the enzyme(s) produced by the pancreas for protein digestion in the small intestine is _____ [p.600].
   a. salivary amylase
   b. lipase
   c. pancreatic nucleases
   d. trypsin and chymotrypsin
   e. aminopeptidase

2. The layer beneath the enamel on a tooth is ____. [p.599]
   a. dentin
   b. pulp cavity
   c. gingiva
   d. periodontal membrane

3. This organ produces bile in the body. [p.598]
   a. liver
   b. gall bladder
   c. pancreas
   d. small intestine
   e. stomach

4. The _____ contains pepsinogens and hydrochloric acid. [p.600]
   a. saliva
   b. mucus
   c. gastric fluid
   d. feces

5. Which of the following is not truly part of the digestive system? [p.598;604]
   a. esophagus
   b. colon
   c. appendix
   d. cecum
   e. oral cavity

6. Which of the following should be represented least in a healthy diet? [p.606]
   a. vegetables
   b. fruits
   c. grain products
   d. milk and milk products
   e. fats and refined sugars

7. Which of the following is not a fat-soluble vitamin? [p.608]
   a. D
   b. C
   c. E
   d. K
   e. A

8. A complete digestive system has both a mouth and an anus. [p.596]
   a. true
   b. false

9. The majority of digestion and absorption occurs in the _____. [pp.602-603]
   a. stomach
   b. liver
   c. small intestine
   d. large intestine
   e. pancreas

10. Which of the following is responsible for emulsifying fats? [p.602]
   a. amylase
   b. pepsin
   c. hydrochloric acid
   d. bile
   e. saliva

CHAPTER OBJECTIVES/ REVIEW QUESTIONS

1. List the five tasks that a digestive system needs to perform for an organism. [p.596]

2. Distinguish between incomplete and complete digestive systems, and tell which is characterized by (a) specialized regions, and (b) two-way traffic. [p.596]

3. List all parts (in order) of the human digestive system through which food actually passes. Then list the auxiliary organs that contribute one or more substances to the digestive process. [pp.598-599]

4. List the enzyme(s) that act in (a) the oral cavity, (b) the stomach, and (c) the small intestine. Then tell where each enzyme was originally made. [pp.598-603]

5. List the parts of a typical tooth. [p.599]

6. Understand the role of the villi and the brush border. [p.601]

7. Describe the cross-sectional structure of the small intestine, and explain how its structure is related to its function. [p.601]

8. Describe how the digestion and absorption of fats differ from the digestion and absorption of carbohydrates and proteins. [pp.602-603]

9. List the items that leave the digestive system and enter the circulatory system during the process of absorption. [pp.602-603]

10. Explain the disorders that can affect digestion in the small intestine. [p.603]

11. Describe events that occur in the human colon. Describe three disorders of the large intestine. [p.604]

12. List the fates of the various compounds after they leave the digestive system and enter the circulatory system. [p.605]

13. List the numerical range of servings permitted from each group, and also list some of the choices available. [p.628]

14. Define an essential fatty acid and an essential amino acid. [pp.606-607]

15. Be able to understand after reading a nutritional label exactly what it is telling you about this particular food product. [p.607]

16. Know the difference between the fat-soluble and water-soluble vitamins. [p.608]

17. Understand the role of key minerals in human metabolism. [p.609]

18. Understand what the BMI represents. [p.610]

19. Recognize new research trends into understanding the genetics of obesity. [pp.610-611]

CHAPTER SUMMARY

The digestive systems of some animals are (1) _____ -like, without a separate entrance and exit, while the digestive systems of most animals consist of a (2) _____ that extends throughout the body. In complex animals, the digestive system interacts with other (3) _____ to distribute (4) _____ and water, to dispose of (5) _____ and to maintain the internal environment.

In humans, digestion starts in the (6) _____, continues in the (7) _____ and is finished in the small intestine. Secretions from (8) _____ glands, the (9) _____ and the (10) _____ function in digestion. Most nutrients are absorbed in the (11) _____. The large intestine absorbs (12) _____ and concentrates and stores (13) _____. A healthy diet provides all (14) _____, (15) _____ and minerals necessary to support metabolism. Maintaining body weight requires balancing (16) _____ taken in with (17) _____ burned in metabolism and physical activity.

INTEGRATING AND APPLYING KEY CONCEPTS

1. Suppose that you were a chemist for a pharmaceutical company. You have an interest in developing a drug to fight a specific disease. This drug needs to be activated by the enzymes of the stomach, and then absorbed through the stomach lining. What nutrient classes would you pattern your new drug on? What if you wanted it absorbed in the small intestine?

2. For one week keep a food diary of EVERYTHING you eat. Calculate the amount of calories, fat, protein and fiber for each day. Also calculate the number of servings of vegetables (dark green, yellow, legumes, other) fruits, milk products, whole grains and meats. Compare your food intake with the suggested guidelines and evaluate your personal diet.

3. Visit a grocery store and read five Nutritional Facts label. Pay particular attention to serving size per container and the amounts of fat. It will change your life.

4. In a perfect world, how is the best way to lose weight in a healthy manner?

37

THE INTERNAL ENVIRONMENT

INTRODUCTION

The previous chapter discussed what happens to the food that enters the vertebrate body. The process of digestion was discussed and the student learned how the body uses the food it takes up to conduct various biological processes. However, as with any process, there is often waste produced. This chapter deals with how urine is produced in the kidneys and excreted from the body. This process in critical in the ability of the body to maintain a balanced environment in which cells can survive and function. Also critical to the balanced environment inside the vertebrate body is temperature control. Diseases of the kidney and treatments of these diseases are discussed. The mechanisms by which different organisms balance their body temperature are discussed, with particular detail regarding the warm blooded mammals.

FOCAL POINTS

- Figures 37.2 & 37.3 [p.616] illustrates the excretory systems of an earthworm and insect.
- Figure 37.5 [p.617] shows fluid-solute balance in bony fish.
- Figure 37.7 [p.618] diagrams the human urinary system and the human kidney.
- Figure 37.8 [p.619] illustrates a nephron and the associated blood capillaries.
- Figure 37.10 [p.621] illustrates how urine is formed.
- Figure 37.11 [p.622] diagrams two types of kidney dialysis.
- Table 37.1 [p.624] shows a comparison between heat and cold stress.
- Table 37.2 [p.625] summarizes the impact of cold stress.

INTERACTIVE EXERCISES

37.1 TRUTH IN A TEST TUBE [p.615]
37.2 MAINTAINING THE VOLUME AND COMPOSITION OF BODY FLUIDS [pp.616-617]
37.3 STRUCTURE OF THE URINARY SYSTEM [pp.618-619]

Selected Terms
Diabetes mellitus [p.615], "passing honey sweet water " [p.615], acidic & alkaline urine [p.615], pregnancy [p.615], luteinizing hormone [p.615], NCAA [p.615], "street drugs" [p.615], THC [p.615], interstitial fluid [p.616], solute composition [p.616], nephridium [p.616], Malpighian [p.616], marine invertebrate [p.616], tubular secretions [p.616], intercellular & extracellular fluid [p.616], waterproof skin [p.617], desert kangaroo rat [p.617], bottlenose dolphin [p.617], renal capsule [p.618], *renal* [p.618], renal pelvis [p.618], sphincter [p.618], renal cortex [p.618], microvilli [p.619], renal medulla [p.619], renal artery & vein [p.619]

Boldfaced Terms
[p.616] ammonia _____

_____

[p.617] uric acid _____

_____

[p.617] kidneys _____

_____

[p.617] urine _____

_____

[p.617] urea _____

_____

[p.618] ureter _____

_____

[p.618] urinary bladder _____

_____

[p.618] urethra _____

_____

[p.618] nephron _____

_____

[p.619] Bowman's capsule _____

_____

[p.619] proximal tubule _____

_____

[p.619] loop of Henle _____

_____

[p.619] distal tubule _____

_____

[p.619] collecting tubules _____

_____

[p.619] glomerulus _____

_____

[p.619] peritubular capillaries _____

_____

Labeling

Below are two figures of fishes. Label which one is a bony fish and whish one is a marine fish, based on the flow of water. You may want to consult Figure 37.5 in the textbook for more information. [p.617]

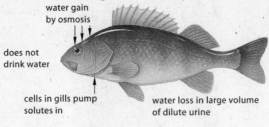

water gain by osmosis

does not drink water

cells in gills pump solutes in

water loss in large volume of dilute urine

A

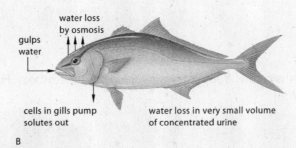

water loss by osmosis

gulps water

cells in gills pump solutes out

water loss in very small volume of concentrated urine

B

1. A is a _____ fish.
2. B is a _____ fish.

Matching [pp.616-617]

Match the following terms with the best definition.

3. ____ uric acid

4. ____ extracellular fluid

5. ____ urinary system

6. ____ urine

7. ____ kidneys

8. ____ ammonia

9. ____ urea

10. ____ intracellular fluid

a. the majority of the fluid in an animal; consists mainly of interstitial fluid
b. fluid inside the body cells
c. a pair of organs that filter blood
d. poison product of protein metabolism
e. most abundant metabolic waste product in human urine
f. ammonia is converted into this in land dwelling arthropods
g. the collection of excess water and solutes in kidneys
h. present in most vertebrates to filter water and solutes from the blood

Matching [p.616]

Use the letters (A, B, C, D) on the figure below that correspond to the correct component of the diagram. You may want to consult your textbook, Figure 37.2, for additional information. [p.616]

11. _____ human Body Fluids

12. _____ intracellular Fluid

13. _____ extracellular Fluid

14. _____ plasma

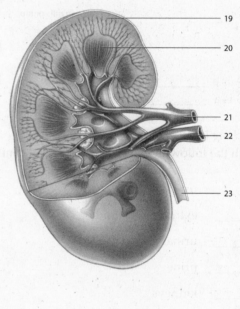

Labeling

Label each part of the diagrams below. You may want to consult the textbook Figure 37.7 a & b) for additional information. [pp.618]

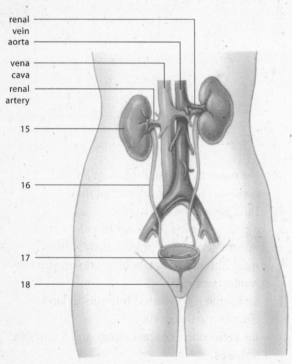

15. _____

16. _____

17. _____

18. _____

19. _____

20. _____

21. _____

22. _____

23. _____

37.4 URINE FORMATION [pp.620-621]
37.5 KIDNEY DISEASE [p.622]

Selected Terms

permeability [p.620], plasma proteins [p.620], blood cells [p.620], platelets [p.620], sodium ions, chloride ions, bicarbonate ions, & glucose [p.620], hormones [p.621], hypothalamus [p.621], adrenal glands [p.621], atrial natriuretic peptide [p.621], parathyroid hormone [p.621], "dialysis" [p.622], hemodialysis [p.622], peritoneal dialysis [p.622], kidney stones [p.621]

Boldfaced Terms

[p.620] glomerular filtration _____

_____

[p.620] tubular reabsorption _____

_____

[p.620] tubular secretion _____

_____

[p.621] antidiuretic hormone_____

_____

[p.621] aldosterone _____

_____

Labeling

Choose the location in the nephron where each statement occurs. Answers may be used more than once. You may want to consult Figures 37.8 and 37.9 in the textbook for additional information. [pp.619-620]

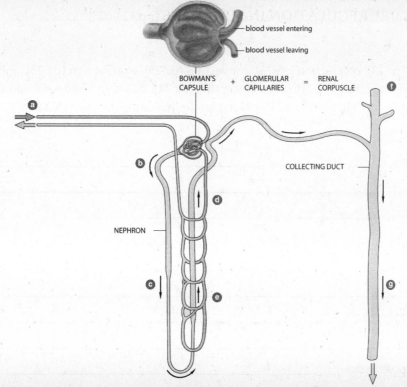

1. ____ Tubular reabsorption

2. ____ Tubular secretion

3. ____ ADH promotes water reabsorption.

4. ____ Filtration

5. ____ Aldosterone increases sodium reabsoprtion

6. ____ Cells pumps chloride ions out of the loop of Henle.

7. ____ Blood pumped from the heart travels to the renal artery, then into the kidneys.

8. ____ Urinary exretion

Matching
Match each of the following terms to the correct description.

9. ____ dialysis [p.622]

10. ____ kidney stones [p.622]

11. ____ diabetes mellitus [p.616]

12. ____ hemodialysis [p.622]

13. ____ peritoneal dialysis [p.622]

a. a external machine cleans the blood of toxins and wastes.

b. deposits of uric acid and calcium block the urethra or ureter

c. exchange of solutes across a semipermeable membrane between two solutions.

d. one of the major causes of kidney failure

e. a person's abdominal cavity is used to filter waste and toxins from the blood.

37.6 HEAT GAINS AND LOSSES [p.623]
37.7 TEMPERATURE REGULATION IN MAMMALS [pp.624-625]

Selected Terms
thermal radiation [p.623], conduction [p.623], convection [p.623], evaporation [p.623], insulation [p.623], panting [p. 624], sweat glands [p.624], thermoreceptors [p.624], negative feedback response [p.624], peripheral blood flow [p.624], fever [p.624], prostaglandin [p.624], pathogen [p.624]

Boldfaced Terms
[p.623] ectotherms_____

_____

[p.623] endotherms_____

_____

[p.623] heterotherms_____

_____

[p.625] shivering response _____

_____

[p.625] brown adipose tissue _____

[p.625] nonshivering heat production _____

_____

Matching
Match each of the following terms to the correct statement. [p.623-625]

1. ____ convection
2. ____ heterotherms
3. ____ thermal radiation
4. ____ evaporation
5. ____ evaporative heat loss
6. ____ fluffing hair or feathers
7. ____ ectotherms
8. ____ endotherms
9. ____ conduction
10. ____ shivering response
11. ____ nonshivering heat production

a. occurs as a result of sweating.
b. the transfer of heat between objects in direct contact.
c. creates a layer of still air.
d. rhythmic tremors of skeletal muscles to produce heat
e. organisms with high metabolic rates, such as mammals.
f. mitochondria in brown adipose tissue generate heat
g. emission of heat from an object in the form of radiant energy.
h. moving air or water moves heat.
i. organisms that balance between being endothermic and ectothermic.
j. the conversion of a liquid to a gas moves heat.
k. organisms that use the environment to gain heat.

SELF QUIZ

1. The functional unit of the kidney is the _____ [p.618]
   a. nephron
   b. nephridia
   c. renal artery
   d. kidney medulla

2. _____ is the hormone that promotes water reabsorption by the kidney. [p.621]
   a. insulin
   b. ADH
   c. aldosterone
   d. glycogen

3. Accumulations of uric acid and calcium in the kidneys may cause _____. [p.622]
   a. uremic toxicity
   b. metabolic acidosis
   c. kidney stones
   d. diabetes mellitus

4. The transfer of heat by the movement of air over a surface is called _____ . [p.623]
   a. evaporation
   b. transduction
   c. convection
   d. thermal radiation

The Internal Environment **387**

5. The ____ connect the kidneys and the urinary bladder. [p.618]
    a. urethra
    b. proximal tubule
    c. distal tubule
    d. ureter

6. A disease whose sign is "sweet" urine is called _____. [p.615]
    a. kidney stones
    b. kidney failure
    c. diabetes mellitus

7. _____ organisms use the external environment to gain body heat. [p.623]
    a. endothermic
    b. ectothermic
    c. heterothermic
    d. all

8. This procedure increases the dead air space next to the skin. [p.623]
    a. fluffing hair or feathers
    b. hyperthermia
    c. evaporative heat loss
    d. fever
    e. panting

9. Brown adipose tissue is involved in which of the following? [p.625]
    a. hypothermia
    b. nonshivering heat production
    c. shivering response
    d. fever production

10. A reduction in the core temperature of an organism is called _____. [p.625]
    a. hyperthermia
    b. hypothermia
    c. isothermia
    d. heterothermia

CHAPTER OBJECTIVES/ REVIEW QUESTIONS

1. Explain how the urinary system allows animals to adapt to life on land. [pp.616-617]

2. Give the location and function of the organs of the urinary system. [pp.618-619]

3. Know the structure of a human kidney. [pp.618-619]

4. Understand the process by which urine is formed. [pp.620-621]

5. Understand the role of ADH and aldosterone in maintaining the water balance. [p.621]

6. Understand the major diseases associated with the kidneys. [p.622]

7. Recognize the four methods of heat transfer. [p.623]

8. Distinguish between heterotherms, endotherms and ectotherms. [p.623]

9. Understand how mammals respond to heat stress. [pp.624-625]

10. Understand how mammals respond to cold stress. [pp.624-625]

CHAPTER SUMMARY

Animals continually produce metabolic (1) _____. They also continually gain and loose (2) _____ and solutes. Despite these continuous gains and losses, the overall composition and (3) _____ of (4) _____ fluid must be kept within a range that the cells can tolerate. This process is called (5) _____. In humans, the (6) _____ system interacts with other systems to accomplish these tasks. The human urinary system consists of two (7) _____, two ureters, a (8) _____ and a urethra. Inside each kidney, millions of (9) _____ filter water and solutes from the blood. Water and solutes not returned to the blood leave the body as (10) _____. Urine forms by three processes: (11) _____, (12) _____ and (13) _____. The hormones (14) _____ and (15) _____, as well as a thirst response influence whether urine is concentrated or dilute. Heat losses and heat gains determine an animal's (16) _____.

INTEGRATING AND APPLYING KEY CONCEPTS

1. The hemodialysis machine used in hospitals is expensive and time-consuming. So far, artificial kidneys capable of allowing people who have nonfunctional kidneys to purify their blood by themselves, without having to go to a hospital or clinic, have not been developed. Which aspects of the hemodialysis procedure do you think have presented the most problems in development of a method of home self-care? If you had an unlimited budget and were appointed head of a team to develop such a procedure and its instrumentation, what strategy would you pursue?

2. Do you think an artificial kidney could be developed? If you said yes, what would be your thoughts on how to design an artificial kidney? Certainly on your list would be size and filtration capacity. What else?

3. Given what you know, how could the hypothermia that killed many passengers on the Titanic have been prevented?

38

REPRODUCTION AND DEVELOPMENT

INTRODUCTION

Sexual reproduction is a defining characteristic of many species, including man. And while this process is costly from a biological standpoint, the rewards are great in that the products of sexual reproduction are genetically diverse and therefore able to respond to a changing environment. This chapter outlines the process involved in the animal life cycle, and then focuses, in detail, on the structure and function of the human reproductive system. The process of fertilization is outlined, as is the development of a human fetus. The student will also learn about the ability of humans to both prevent and promote pregnancy, using various physical and chemical methods. Also with a focus on human health, the chapter discusses many of the sexually transmitted diseases that may be spread through human sexual activities. A discussion of the birth process and the growth and maturation of the newborn conclude this chapter.

FOCAL POINTS

- Figure 38.4 [p.632] illustrates the components of the human male reproductive system and describes their functions.
- Figure 38.5 [p.633] shows sperm formation and the anatomy of a sperm.
- Figure 38.6 [p.634] diagrams the location of the human female reproductive system.
- Figure 38.7 [p.634] shows the human female reproductive system.
- Figure 38.8 [p.635] illustrates the cyclic events in a human ovary.
- Figure 38.9 [p.636] correlates the changes in the ovary and uterus in response to changing hormone levels.
- Table 38.1 [p.637] summarizes the events of the menstrual cycle lasting 28 days.
- Figure 38.10 [p.639] illustrates the process of fertilization in humans.
- Table 38.2 [p.640] shows various methods of contraceptives and their effectiveness.
- Table 38.3 [p.641] gives the frequency and causes of sexually transmitted diseases in the United States.
- Figure 38.12 [p.642] gives a flow chart of the development of vertebrates.
- Table 38.4 [p.642] shows the derivatives of vertebrate germ layers.
- Figure 38.13 [p.642] illustrates the development of a leopard frog.
- Figure 38.17 [p.646] shows how a body takes shape.
- Figure 38.20 [pp.648-649] illustrates the stages involved from fertilization to implantation for a human.
- Figure 38.22 [pp.650-651] shows the human embryo at various stages of development.
- Figure 38.23 [p.652] shows the life support system of a developing human.
- Figure 38.24 [p.653] illustrates a human birthing process.

INTERACTIVE EXERCISES

38.1 MIND-BOGGLING BIRTHS [p.629]
38.2 MODES OF ANIMAL REPRODUCTION [pp.630-631]

Selected Terms

infertility treatments [p.629], egg & sperm [p.629], miscarriage, stillbirth, premature death [p.629], octuplets [p.629], stable environment [p.630], budding [p.630], fragmentation [p.630], fragmentation [p.630], transverse fission [p.630], unfertilized egg [p.630], parthenogenesis [p.630], haploid [p.630], paternal & maternal genes [p.630], simultaneous hermaphrodites [p.631], amniotes [p.631], incubation [p.610], placental mammals [p.631]

Boldfaced Terms

[p.629] in vitro fertilization _____

_____

[p.630] asexual reproduction _____

_____

[p.630] sexual reproduction _____

_____

[p.630] hermaphrodites _____

_____

[p.631] external fertilization _____

_____

[p.631] internal fertilization _____

_____

[p.631] yolk _____

_____

[p.631] placenta _____

_____

Fill-in-the-Blanks [p.652]

In earlier chapters, we considered the genetic basis of sexual reproduction. Again, (1) _____ and the formation of gametes typically occur in two prospective parents. At (2) _____ a gamete from one parent fuses with a gamete from the other and form a (3) _____ the first cell of a new individual. We also looked at asexual reproduction, whereby a (4) _____ organism – one parent only – produces (5) _____ We now turn to examples of structural, behavioral, and ecological aspects of these two modes of animal reproduction.

Mutation aside, in cases of (6) _____, all offspring are (7) _____ the same as their individual parent. Phenotypically, they are also much the same. This can be advantageous if the (8) _____ does not vary over time. (9) _____ combinations that allowed the parent to (10)

_____ will be expected to do the (11) _____ for the offspring. Most (12) _____,

however, live where opportunities, (13) _____ and danger are variable. They reproduce (14)

_____ and offspring inherit different mixes of (15) _____ from female and male parents.

The resulting (16) _____ in traits can improve the likelihood that at least some of the offspring will

(17) _____ and (18) _____ [p.646] if prevailing conditions change.

Matching
Match the terms with either an example or its definition. [p.629-631]

19. ____ asexual reproduction

20. ____ hermaphrodite

21. ____ external fertilization

22. ____ sexual reproduction

23. ____ placenta

24. ____ in vitro fertilization

25. ____ internal fertilization

a. the combining of an human egg and sperm outside the body

b. reproduction that involves sperm and eggs

c. reproduction that does NOT involve sperm and eggs

d. an animal that is capable of producing sperm and eggs

e. a sexual reproduction that takes place outside of the body

f. a sexual reproduction that takes place inside the body

g. an organ that allows exchange of substances between the mother and offspring

38.3 REPRODUCTIVE FUNCTION OF HUMAN MALES [pp.632-633]

Selected Terms
sex hormones [p.632], scrotum [p.632], secondary sexual traits [p.632], cilia [p.632], ejaculatory duct [p.632], urethra [p.632], urination [p.632], intercourse [p.632], glans [p.632], foreskin[p.632], circumcision [p.632], spongy tissue [p.632], exocrine gland [p.633], fructose-rich fluid [p.633], seminal vesicles [p.633], prostate gland [p.633], rectum [p.633], cancer [p.633], spermatogonia [p.633], meiosis of primary spermatocytes [p.633], spermatids [p.633] haploid [p.633], flagellated [p.633], flagellum [p.633], "head" [p.633], fertilization [p.633], GnRH [p.633], midpiece [p.633], hypothalamus [p.633], anterior pituitary [p.633], luteinizing hormone [p.633], follicle stimulating hormone [p.633], negative feedback [p.633], secondary spermatocyte [p.633]

Boldfaced Terms
[p.632] gonads _____

_____

[p.632] testes _____

_____

[p.632] testosterone _____

_____

[p.632] puberty _____

_____

[p.632] epididymis _____

_____

[p.632] vas deferens _____

_____

[p.632] penis _____

_____

[p.632] semen _____

_____

[p.633] seminiferous tubules _____

_____

Matching

Match the letter of the structure on the diagram with its description below. You may consult Figure 38.4 in the textbook for more information. [p.632]

1. ____ bulbourethral gland

2. ____ scrotum

3. ____ prostate gland

4. ____ testis

5. ____ vas deferens

6. ____ seminal vesicle

7. ____ penis

8. ____ epididymis

9. ____ urinary bladder

10. ____ urethra

Dichotomous Choice

Circle one of two possible answers given between parentheses in each statement.

11. LH and FSH are secreted by the (anterior/posterior) lobe of the pituitary gland. [p.633]

12. The (testes/hypothalamus) governs sperm production by controlling secretion of testosterone, LH, and FSH. [p.633]

13. Feedback loops to the hypothalamus lead to (increased/decreased) testosterone secretion and sperm formation. [p.633]

38.4 REPRODUCTIVE FUNCTION OF HUMAN FEMALES [pp.634-635]

38.5 HORMONES AND THE MENSTRUAL CYCLE [pp.636-637]

Selected Terms

pelvic girdle [p.634], Fallopian tubes [p.634], pregnant [p.634], labia minora & majora [p.634], adipose tissue [p.634], erectile tissue [p.634], clitoris [p.634], myometrium [p.634], 28-day ovarian cycle [p.635], meiosis [p.635], polar body [p.635], "yellow body" [p.635], follicle cells [p.635], follicular phase [p.635], luteal phase [p.635] GnRH [p.636], FSH & LH [p.636], PMS [p. 637], aldosterone [p.637], sodium [p.637], prostaglandin [p.637], menstrual cramps [p.637], ibuprofen [p.637], endometriosis [p.637], uterine tumors (fibroids) [p.637], hot flashes [p.637], hormone replacement therapy [p.637]

Boldfaced Terms

[p.634] ovaries _____

_____

[p.634] oocytes _____

_____

[p.634] estrogens _____

_____

[p.634] progesterone _____

_____

[p.634] oviduct _____

_____

[p.634] uterus _____

_____

[p.634] endometrium _____

_____

[p.634] cervix _____

_____

[p.634] vagina _____

_____

[p.635] ovarian follicle _____

_____

[p.635] ovulation _____

_____

[p.635] corpus luteum _____

_____

[p.636] menstrual cycle _____

_____

[p.636] menstruation _____

_____

[p.637] menopause _____

_____

Labeling

Label each of the structures in the diagram below. You might want to consult Figure 38.7 in the textbook for additional information. [p.634]

1. _____

2. _____

3. _____

4. _____

5. _____

6. _____

7. _____

8. _____

9. _____

10. _____

11. _____

Labeling

Using the following diagram, label the following events that occur during a human female's 28-day cycle. You might want to consult Figure 38.9 from the text for more information.[p.636]

12. ____ ovulation

13. ____ mid-cycle peak of LH which triggers ovulation

14. ____ growth of follicle

15. ____ estrogen and progesterone levels high

16. ____ follicular phase

17. ____ luteal phase

18. ____ corpus luteum persists if implantation occur

19. ____ pituitary gland

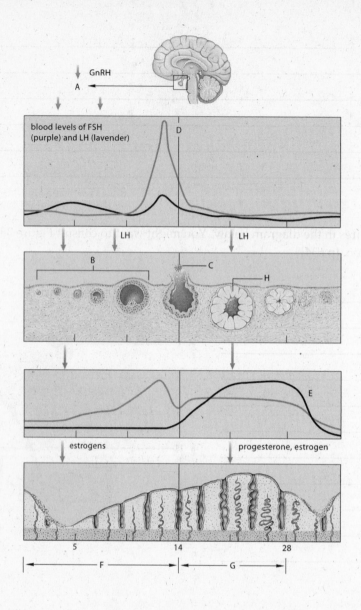

Ordering

Place the following events in order as they occur during the 28-day menstrual cycle.
Number 21 will be the first event, 22 the second and so on. [pp.636-637]

20. ____

21. ____

22. ____

23. ____

a. corpus luteum develops causing thickening of the endometrium.

b. menstrual flow occurs – the follicular phase

c. ovulation occurs – one oocyte released

d. rebuilding of the uterine lining; oocytes mature

38.6 WHEN EGG AND SPERM MEET [pp.638-639]
38.7 PREVENTING PREGNANCY [p.640]
38.8 SEXUALLY TRANSMITTED DISEASES [p.641]

Selected Terms

intercourse [p.638], erection [p.638], erectile dysfunction [p.638], Viagra [p.638], cone cells & retina [p.638], orgasm [p.638], oxytocin [p.638], endorphins [p.638], semen [p.638], ejaculation [p.638], jelly coat [p.638], infertile [p.638], fertilization [p.639], ovulation [p.639], oviduct [p.639], enzymes [p.639], follicle cell [p.639], cervical mucus [p.640], abstinence [p.640], rhythm method [p.640], withdrawal [p.640], vasectomy [p.640], tubal ligation [p.640], condom [p.640], spermicides [p.640], diaphragm [p.640], cervical cap [p.640], IUD [p. 671], birth control pill [p.640], stroke [p.640], oral contraceptives [p.640], birth control patch [p.640], progestin injection & implants [p.640], emergency contraception [p.640], induced abortion [p.667], morning after pills [p.640], Plan B [p.640], pathogens [p.641], sexual transmitted diseases [p.641], genital herpes [p.641], *genital warts* [p.641], *Trichomonas vaginalis* [p.641], chlamydia [p.641], *Chlamydia trachomatis* [p.641], *genital herpes* [p.641], *Herpes simplex* [p.672], *Neisseria gonorrhoeae* [p.641], *Treponema pallidum* [p.641], *syphilis* [p.641], *AIDS* [p.641], HIV [p.641], antiviral drugs [p.641], unprotected anal sex [p.641], vaginal sex [p.641], viruses, bacteria, & protozoan [p.641], oral sex [p.641], lubricant [p.641], cervical cancer [p.641], Pap smear [p.641], pelvic inflammatory disease (PID) [p.641], antibiotics [p.641], antiprotozoal drug [p.641]

Boldfaced Terms
[p.638] ovum _____

_____

Fill-in-the-Blanks [p. 638-639]

An average, an (1) _____ can put 300 million sperm in the (2) _____. Fertilization occurs if they arrive a few days before or after (3) _____ or any time in between. Less than thirty minutes after sperm arrive in the vagina, (4) _____ move them deep into the female's reproductive tract. A few (5) _____ actually reach the upper portion of the (6) _____, where eggs are usually fertilized.

Many sperm bind to the oocyte's (7) _____. Binding triggers the release of (8) _____ from the cap over each sperm's head. Collectively, these (9) _____ enzymes make a passage through the (10) _____. Usually only one sperm enters the secondary oocyte. Its entry triggers (11) _____ modifications that prevent other sperm from binding.

Upon sperm penetration, the secondary (12) _____ and the first (13) _____ complete meiosis II. Now there are three polar bodies and one mature egg, or (14) _____ The egg (15) _____ fuses with the sperm (16) _____. Together, chromosomes of both nuclei restore the (17) _____ number for a brand new zygote.

Matching

Choose the most appropriate answer for each term.[pp.670-671]

18. ____ abstinence

19. ____ rhythm method

20. ____ withdrawal

21. ____ douching

22. ____ vasectomy

23. ____ tubal ligation

24. ____ spermicides

25. ____ diaphragm

26. ____ condoms

27. ____ birth control pill

28. ____ projestin injections or implants

29. ____ morning after pill or Plan B

30. ____ IUDs

a. A flexible, dome-shaped device inserted into the vagina and positioned over the cervix before intercourse

b. Progestin injections or implants that inhibit ovulation

c. A woman's oviducts are cauterized or cut and tied off; extremely effective

d. Rinsing the vagina with a chemical right after intercourse; next to useless

e. Thin, tight-fitting sheaths worn over the penis during intercourse

f. Avoiding intercourse during a woman's fertile period

g. This pill interferes with implantation

h. Cutting and tying off each vas deferens of a man; extremely effective

i. An oral contraceptive made of synthetic estrogens and progestins that suppress oocyte maturation and ovulation

j. No sexual intercourse; foolproof

k. Removing the penis from the vagina before ejaculation; an ineffective method

l. Chemicals toxic to sperm are transferred from an applicator into the vagina just before intercourse

m. Coils inserted into the uterus by a physician; sometimes invite pelvic inflammatory disease

Matching

Match each of the following with *all* applicable diseases. [p.641]

31. ____ Can cause skin ulcers and damage the liver and bones

32. ____ Has no cure.

33. ____ Can cause severe cramps, fever, vomiting, and possibly sterility due to scarring and blocking of the oviducts.

34. ____ Caused by the spirochete, Treponema pallidum.

35. ____ Infected women typically have miscarriages, stillbirths, or syphilitic newborn

36. ____ Caused by Neisseria gonorrhoeae.

37. ____ Infection requires direct contact with the viral agent; about 45 million people in the United States are infected by it

38. ____ This is a spiral shaped bacteria

39. ____ The virus can be reactivated, producing painful sores, causing tingling or itching

40. ____ This bacteria can eventually damage the brain

41. ____ Antiviral drugs can promote healing of sores but cannot cure

42. ____ This is a secondary outcome of bacterial STDs

43. ____ Contrary to common belief, it can be contracted over and over again.

44. ____ A symptom is a yellow pus oozing from the penis or vagina

45. ____ Generally preventable by correct condom usage.

a. AIDS
b. Chlamydial infection
c. Genital herpes
d. Gonorrhea
e. Pelvic inflammatory disease
f. Syphilis

38.9 OVERVIEW OF ANIMAL DEVELOPMENT [pp.642-643]

Selected Terms

fertilization [p.642], blastocoel [p.642], tight junction [p.642], tadpole [p.643], chordate embryos [p.642], gray crescent [p.643], blastula [p.643], dorsal lip [p.643], yolk plug [p.643], notochord [p.643], neural tube [p.643], metamorphosis [p.643]

Boldface Terms

[p.642] cleavage _____

_____

[p.642] blastula _____

_____

[p.642] gastrulation _____

_____

[p.642] gastrula _____

_____

[p.642] germ layers _____

_____

[p.642] ectoderm _____

_____

[p.642] endoderm _____

_____

[p.642] mesoderm _____

_____

Sequence

Arrange the following events in correct chronological sequence. Write the letter of the first event next to 1, the letter of the second event next to 2, and so on. You may want to consult Figure 38.12 from the textbook for additional information. [p.642]

1. _____

2. _____

3. _____

4. _____

5. _____

6. _____

a. gastulation
b. fertilization
c. cleavage
d. growth, tissue specialization
e. organ formation
f. gamete formation

Complete the Table

7. Complete the table below by entering the correct germ layer (ectoderm, mesoderm, or endoderm) that forms the tissues and organs listed. You may want to consult Table 38.4 for additional information. [p.642]

Tissues/Organs	Germ Layer
Muscle, circulatory organs	a.
Nervous system tissues	b.
Inner lining of the gut	c.
Circulatory organs (blood vessels, heart)	d.
Outer layer of the integument	e.
Reproductive and excretory organs	f.
Organs derived from the gut	g.
Most of the skeleton	h.
Connective tissues of the gut and integument	i.

38.10 EARLY MARCHING ORDERS [pp.644-645]

Selected Terms

mRNA transcripts [p.644], mitotic spindle [p.644], cell cortex [p.644], plasma membrane [p.644], animal pole [p.644], vegetal pole [p.644], cleavage furrow [p.644], blastomeres [p.644], gray crescent [p.644], egg's polar axis [p.645],

Boldfaced Terms

[p.644] cytoplasmic localization _____

_____

Matching

Choose the most appropriate answer for each term. [pp.644-645]

1. ____ oocyte
2. ____ sperm
3. ____ gray crescent
4. ____ cytoplasmic localization
5. ____ cleavage
6. ____ blastula
7. ____ vegetal pole
8. ____ animal pole
9. ____ gastrulation
10. ____ germ layers
11. ____ rotation

a. process of organization that gives rise to 3 germ layers; initiated by signals from the dorsal lip

b. ectoderm, mesoderm, and endoderm

c. embryonic stage characterized by blastomeres and a fluid-filled cavity, the blastocoel

d. consists of paternal DNA and cellular equipment enabling this cell to reach and penetrate an egg

e. an area of intermediate pigmentation and maternal messages in some animal eggs

f. contains little yolk, pigment most concentrated here

g. by virtue of where they form, blastomeres end up with different maternal messages;

helps seal the developmental fate of each cell's descendants

h. process that puts different parts of the egg cytoplasm into different blastomeres – not random.

i. term for an immature egg

j. in a yolk rich egg, this contains most of the yolk

k. occurs following fertilization of the egg with sperm; the cortex of the egg moves to reveal the gray crescent

38.11 SPECIALIZED CELLS, TISSUES, AND ORGANS [pp.646-647]

Selected Terms

crystalline [p.646], entire genome [p.646], intercellular signals [p.646], master genes [p.646], target cells [p.646], cell migrations [p.646], neural groove [p.646], neral tube [p.646], top-to- bottom axis [p.647], development [p.647], mutations [p.647], apical ectodermal ridge (AER) [p.647]

Boldfaced Terms

[p.646] morphogens _____

_____

[p.646] embryonic induction _____

_____

[p.646] apoptosis _____

_____

[p.647] pattern formation _____

_____

[p.647] homeotic genes _____

_____

Matching [p. 646-647]

1. ____ selective gene expression
2. ____ morphogens
3. ____ embyronic induction
4. ____ homeotic genes
5. ____ apoptosis
6. ____ AER
7. ____ pattern formation
8. ____ cell differentiation
9. ____ morphogenesis

a. different cell lineages express different subsets of genes
b. the process by which cell lineages become specialized
c. signaling molecules that are the products of master genes
d. genes that regulate development of specific body parts
e. apical ectodermal ridge – induces mesoderm under it to from limb
f. embryonic cells produce signals that alter the behavior of neighboring cells
g. the process by which tissue and organs form
h. the process by which certain body parts form in certain places
i. programmed cell death

38.12 EARLY HUMAN DEVELOPMENT [pp.648-649]
38.13 EMERGENCE OF DISTINCTLY HUMAN FEATURES [pp.650-651]
38.14 FUNCTION OF THE PLACENTA [p.652]
38.15 BIRTH AND LACTATION [p.653]

Selected Terms

cleavage [p.648], identical twins [p.648], endometrium [p.648], menstruation [p.648], HCG [p.648], corpus luteum [p.648], embryonic disk [p. 648], "dipstick" [p.648], implantation [p.648], jelly layer [p.648], yolk sac [p.648], chorionic villi [p.648], neural folds [p.649], amniotic cavity [p.649], chorionic cavity [p.649], primitive streak [p.649], somites [p.649], amniotes [p.649], umbilical cord [p.649], notochord [p.649], pharyngeal arches [p.649], stethoscope [p.650], upper limb bud [p.650], embryo [p.650], foot plate, trimester [p.650], [p.650], thirty-eights [p.651], premature births [p.651], embryonic period [p.651], fetal period, [p.651], maternal diet [p.652], folate [p.652], iodine [p.652], cretinism [p.652], rubella (German measles) [p.652], HIV [p.652], alcohol [p.652], caffeine [p.652], tobacco smoke [p.652], Accutane [p.652], Paxil [p.652], amniotic fluid, [p.653], oxytocin [p.653], "afterbirth" [p.653], navel [p.653], prolactin [p.653]

Boldfaced Terms

[p.648] blastocyst _____

_____

[p.648] implantation _____

_____

[p.649] amnion _____

_____

[p.649] chorion _____

_____

[p.649] allantois _____

_____

[p.650] fetus _____

_____

[p.653] labor _____

_____

[p.653] lactation _____

_____

Matching

Choose the most appropriate answer for each term. [pp.648-651]

1. ____ embryonic period
2. ____ fetus
3. ____ fetal period
4. ____ implantation
5. ____ allantois
6. ____ amnion
7. ____ chorion
8. ____ yolk sac

a. Outermost membrane; will become part of the spongy blood-engorged placenta
b. Directly encloses the embryo and cradles it in a buoyant protective fluid
c. In humans, some of this membrane becomes a site for blood cell formation, some will give rise to germ cells, the forerunners of gametes
d. In humans, the urinary bladder as well as blood vessels form from it
e. The blastocyst adheres to the uterine lining and sends out projections that invade the mother's tissues
f. The time span from the third to the end of the eighth week of pregnancy
g. Term for the new, distinctly human individual at the end of the embryonic period
h. From the start of the ninth week until birth, organs enlarge and become specialized

Sequence

Arrange the following human developmental stages of early human embryo formation in the proper chronological sequence (day 1 to day 14). Write the letter of the first stage next to 9. The letter of the final process is written next to 16. [pp.648-651]

9. ____
10. ____
11. ____
12. ____
13. ____
14. ____
15. ____
16. ____

a. the first cleavage furrow divides the cell
b. the chorionic cavity starts to form
c. the morula forms
d. implantation begins
e. the yolk sac, embryonic disk and amniotic cavity have started to form
f. the structures of the placenta start to form
g. the cells huddle into a compact ball, with junctions allowing communication
h. the blastocoel forms

Sequence

Arrange the following human developmental stages in the emerging vertebrate body plan in the proper chronological sequence (day 15 to days 24-25). Write the letter of the first stage next to 17, and so on. [p.649]

17. ____

18. ____

19. ____

a. Morphogenesis occurs; the neural tube and somites form; mesoderm somites give rise to the axial skeleton, skeletal muscles, and much of the dermis.

b. Pharyngeal arches form to contribute to formation of the face, neck, mouth, nasal cavities, larynx, and pharynx.

c. The primitive streak forms to mark the onset of gastrulation in vertebrate embryos.

Labeling

Identify each numbered placental component in the illustration below. You may want to consult Figure 38.23 for more information. [p.652]

20. _____

21. _____

22. _____

23. _____

24. _____

25. _____

26. _____

27. List two pathogens that could infect an embryo or fetus if the mother is exposed?

_____

_____

28. List four toxins that can cross the placenta.

_____

_____

_____

_____

29. List two prescription drugs that can cross the placenta.

_____

_____

Sequence

Arrange the following human developmental stages demonstrating the emergence of distinctly human features in the proper chronological sequence (four weeks to full term or completion of the fetal period). Write the letter of the first stage next to 30. The letter of the final process is written next to 54. [pp.650-651]

30. ____

31. ____

32. ____

33. ____

34. ____

a. Movements begin as nerves make functional connections with developing muscles; legs kick, arms wave, fingers grasp, and the mouth puckers.

b. The length of the fetus increases from 16 centimeters to 50 centimeters, and weight increases from about 7 ounces to 7.5 pounds.

c. The human embryo has a tail and pharyngeal arches, and limbs; fingers and toes are sculpted from embryonic paddles; the circulatory system becomes more complex, and the head develops.

d. The human embryo is distinctly human as compared to other vertebrate embryos; upper and lower limbs are well formed; fingers and then toes have separated; primordial tissues of all internal and external structures have developed; the tail has become stubby.

e. Human features are visible at the boundary of the embryonic and fetal periods; the embryo floats in amniotic fluid and the chorion covers the amnion.

Choice [p.652]

For the following questions, choose from the following threats to human development:

    a. nutrition
    b. infections
    c. prescription drugs
    d. alcohol
    e. antidepressants
    f. cigarette smoke
    g. anti-acne drugs
    h. all of the preceding threats

35. ____ About 60-70 percent of newborns of these women have FAS.

36. ____ In the case of rubella, there is a 50 percent chance some organs won't form properly.

37. ____ In cases of exposure to the second-hand type, children were smaller, died of more postdelivery complications, and had twice as many heart abnormalities.

38. ____ Increasing B-complex intake of the mother before conception and during early pregnancy reduces the risk that an embryo will develop severe neural tube defects.

39. ____ Taking these early in pregnancy increases the likelihood of heart malformations.

40. ____ Women who used the tranquilizer thalidomide during the first trimester gave birth to infants with missing or severely deformed arms and legs.

41. ____ A pregnant woman must eat enough so that her body weight increases by 20-25 pounds, on average.

42. ____ Reduced brain and head size, mental retardation, facial deformities, poor growth and poor coordination, and often heart defects

Matching
Match each of the following terms with its correct definition. [p.653]

43. ____ oxytocin

44. ____ lactation

45. ____ labor

46. ____ "afterbirth"

a. milk production
b. placenta expelled from the uterus
c. binds to smooth muscle of the uterus, causing contractions.
d. the birth process

SELF QUIZ

1. The process of cleavage most commonly produces a(n) _____. [p.642]
   a. zygote
   b. blastula
   c. gastrula
   d. third germ layer
   e. organ

2. The formation of three germ (embryonic) tissue layers occurs during_____. [p.642]
   a. gastrulation
   b. cleavage
   c. pattern formation
   d. morphogenesis
   e. neural plate formation

3. Muscles differentiate from _____ tissue. [p.642]
   a. ectoderm
   b. mesoderm
   c. endoderm
   d. parthenogenetic
   e. yolky

4. The gray crescent is _____. [p.644]
   a. formed where the sperm penetrates the egg
   b. part of only one blastomere after the first cleavage
   c. the yolky region of the egg
   d. where the first mitotic division begins
   e. formed opposite from where the sperm enters the egg

5. The type of cleavage observed in mammal eggs is _____. [p.644]
   a. incomplete
   b. animal-vegetal axis
   c. radial
   d. rotational
   e. cytoplasmic localization

6. Of the following, _____ is *not* an aspect of morphogenesis. [p.646-647]
   a. active cell migration
   b. expansion and folding of whole sheets of cells
   c. apoptosis
   d. shaping the proportion of body parts
   e. initiating production of a protein by some cells and not others

7. The differentiation of a body part in response to signals from an adjacent body part is _____. [p.646]
   a. contact inhibition
   b. ooplasmic localization
   c. embryonic induction
   d. pattern formation
   e. none of the above

8. Morphogens are _____. [p.646]
   a. master genes
   b. agents of apoptosis
   c. agents of cell differentiation
   d. concentrations of gap gene products
   e. degradable molecules that diffuse as long-range signals

For questions 9-12, choose from the following answers: [all from pp.632]
   a. seminal vesicles
   b. seminiferous tubules
   c. vas deferens
   d. epididymis
   e. prostate

9. The _____ connects a structure on the surface of the testis with the ejaculatory duct.

10. Contribute fructose-rich fluid to semen _____.

11. Sperm formation begins in the _____.

12. Sperm mature in the _____.

For questions 13-15, choose from the following answers:
   a. blastocyst
   b. amnion
   c. yolk sac
   d. oviduct
   e. cervix

13. The _____ lies between the uterus and the vagina. [p.634]

14. The _____ is a pathway from the ovary to the uterus. [p.634]

15. The _____ results from the process known as cleavage. [p.642]

For questions 16-20, choose from the following answers: [p.641]
   a. AIDS
   b. Chlamydial infection
   c. Genital herpes
   d. Gonorrhea
   e. Syphilis

16. _____ is a disease caused by a species of *Neisseria*; it is curable by prompt diagnosis and treatment.

17. _____ is a disease caused by *Treponema* that produces a localized ulcer (a chancre)

18. _____ is an incurable disease caused by the human immunodeficiency virus.

19. _____ is a disease caused by an intracellular parasite that lives in the genital and urinary tracts; also causes NGU.

20. _____ is an extremely contagious viral infection that causes sores on the mouth area and genitals; acyclovir may be effective in decreasing healing time and decreasing pain.

CHAPTER OBJECTIVES/REVIEW QUESTIONS

1. Understand how asexual reproduction differs from sexual reproduction. Know the advantages and problems associated with having separate sexes. [p.630].

2. Explain why evolutionary trends in many groups of organisms tend toward developing more complex, sexual strategies rather than retaining simpler, asexual strategies. [pp.630-631]

3. Name the primary human male reproductive organs; list secondary sexual traits determined by these organs. [pp.632-633]

4. Follow the path of a mature sperm from the seminiferous tubules to the urethral exit. List every structure encountered along the path, and state the contribution to the nurturing of the sperm. [pp.632-633]

5. Name the four hormones that directly or indirectly control male reproductive function.

6. Diagram the negative feedback mechanisms that link the hypothalamus, anterior pituitary, and testes in controlling gonadal function. [pp.632-633]

7. Diagram the structure of a sperm, label its components, and state the function of each. [p.633]

8. Name the major structures of the female reproductive system and give the reproductive function of each. [pp.634-635]

9. Distinguish the follicular phase of the menstrual cycle from the luteal phase, and explain how the two cycles are synchronized by hormones from the anterior pituitary, hypothalamus, and ovaries. [pp.636-637]

10. State which hormonal event brings about ovulation and which other hormonal events bring about the onset and finish of menstruation. [pp.636-637]

11. List the physiological factors that bring about erection of the penis during sexual stimulation and the factors that bring about ejaculation. [p.638]

12. List the similar events that occur in both male and female orgasm. [p.638]

13. Describe two different types of sterilization. [p.640]

14. Identify the three most effective birth control methods used in the United States and the four least effective birth control methods. [p.640]

15. For each STD described in the text, know the causative organism and the symptoms of the disease. [p.641]

16. Name each of the three embryonic tissue layers and the organs formed from each. [p.642]

17. Describe early embryonic development and distinguish among the following: gamete formation, fertilization, cleavage, gastrulation, organ formation, and growth and tissue specialization. [pp.642-643]

18. Define cytoplasmic localization [p.644]

19. Define *differentiation, and* give one example of cells in a multicellular organism that have undergone differentiation. [p.646]

20. Define morphogenesis [p.646]

21. Explain the processes of embryonic induction and pattern formation. [pp.646-647]

22. Explain the role of master genes, homeotic genes and the constraits that are present in the development of an embryo. [pp.646-647]

23. Describe the process of implantation. [p.648]

24. Describe the events that occur during the first month of human development; include the developed structures of the blastocyst as seen at about 15 days. [pp.648-649]

25. List and describe the four extraembryonic membranes and their functions.[pp.648-649]

26. Characterize the human developmental events occurring from day 15 to day 25. [p.649]

27. Distinguish between the embryonic period and the fetal period. [pp.650-651]

28. List the roles of the placenta in pregnancy. [p.652]

29. Explain why the mother must be particularly careful of her diet, health habits, and life-style during the first trimester after fertilization (especially during the first six weeks). [p.653]

30. Generally describe the process of labor or delivery. [p.653]

31. Explain the role of the hormones oxytocin. [p.653]

CHAPTER SUMMARY

Biologically speaking, (1) _____ reproduction is much more "expensive" than (2) _____ reproduction. The rewards are great, however, in that sexual reproduction produces genetically (3) _____ offspring. The six stages of animal life cycles are, in order; (4) _____, (5) _____, (6) _____, (7) _____, (8) _____ and (9) _____. The primary human reproductive organs are the (10) _____ producing testes in the male and the oocyte-producing (11) _____ in the female. The (12) _____ and the pituitary gland control the production and release of sex hormones in both males and females. From puberty onward, females are fertile on a (13) _____ basis. (14) _____ leads to pregnancy. Human sexual behavior can transmit (15) _____ which can cause disease. A pregnancy begins with (16) _____ and the implantation of a (17) _____ in the uterine lining. A (18) _____ connects the embryo with its mother. Aging and death are the result of (19) _____ factors and the ongoing environmental assaults on (20) _____ and tissues.

INTEGRATING AND APPLYING KEY CONCEPTS

1. If embryonic induction did not occur in a human embryo, how would the eye region appear? What would happen to the forebrain and epidermis? If controlled cell death did not happen in a human embryo, how would its hands appear? its face?

2. What rewards do you think a society should give a woman who has at most two children during her lifetime? In the absence of rewards or punishments, how can a society encourage women not to have abortions and yet ensure that the human birth rate does not continue to increase?

3. What are the advantages of asexual reproduction? The disadvantages?

4. What are the advantages of sexual reproduction? The disadvantages?

39

ANIMAL BEHAVIOR

INTRODUCTION

The chapter explores the differences between instinctive and learned behavior and gives examples of many different types of behaviors in the animal kingdom. It begins with an introduction to aggressive behavior that is displayed by the Africanized honey bees as they guard their hive and respond to injuries caused to the hive's members. Next a question introduces the genetic basis for behavior. Garter snakes and slugs are offered as an example to illustrate the concept of genetics and its relations with behavior. Instinct and learning with various supporting examples are presented next. A historical photograph of Konrad Lorenz and his famous "imprinted" ducking is certainly worth remembering. This chapter covers communication signals and presents several good examples of how these work, from sounds to postures. Mating and reproductive success and living in groups are integral components of behavior and are covered toward the end of this chapter. Anyone with an interest in animals will certainly enjoy the numerous examples and photographs offered in this chapter.

FOCAL POINTS

- Figure 39.6 [p.660] shows a famous photograph of Konrad Lorenz with the imprinted ducklings.
- Figure 39.11 [p.663] illustrates the waggle dance of honeybees.
- Numerous color photographs of various animals displaying their behaviors.

INTERACTIVE EXERCISES

39.1 AN AGGRESSIVE DEFENSE [p.657]
39.2 BEHAVIOR'S GENETIC BASIS [pp.658-659]
39.3 INSTINCT AND LEARNING [pp.660-661]

Selected Terms
"aggressive" [p.657], Africanized honeybees [p.657], killer bees [p.657], hybrid [p.657], European honeybees [p.657], pollinator [p.657], proximate cause [p.657], sensory receptor [p.658], genes [p.657], "rovers" [p.658], "sitters" [p.658], dominate allele [p.658], *foraging* [p.658], social bond [p.658], ultimate cause [p.658], genetic polymorphism [p.658], promiscuous [p.659], oxytocin [p.659], *foraging gene* [p.659], monogamous [p.659], arginine vasopressin [p.659], polygenic basis [p.659], environment [p.659], trill-seeking [p.659], autism [p.659], brood parasitism [p.660], competition [p.60], instinctive behavior [p.660], male tutor [p.660], involuntary response [p.661]

Boldfaced Terms
[p.657] pheromone _____

_____

[p.657] ecology _____

_____

[p.658] stimulus _____

_____

[p.660] instinctive behavior _____

_____

[p.660] fixed action pattern _____

_____

[p.660] learned behavior _____

_____

[p.660] imprinting _____

_____

[p.661] habituation _____

_____

Matching
Choose the most appropriate answer for each term. [pp.660-661]

1. _____ involuntary response
2. _____ instinctive behavior
3. _____ stimulus
4. _____ fixed action pattern
5. _____ learned behavior
6. _____ imprinting

a. Animals process and integrate information gained from experiences, then use it to vary or change responses to stimuli.
b. Information from the environment that a receptor has detected.
c. When a stimulus becomes associated with a stimulus that is presented at the same time.
d. Time-dependent form of learning; triggered by exposure to sign stimuli and usually occurring during sensitive periods of young animals.
e. A behavior performed without having been learned by actual environmental experience.
f. A program of coordinated muscle activity that runs to completion independently of feedback from the environment.

Dichotomous Choice
Circle one of two possible answers given between parentheses in each statement. [pp.658-659]

7. For garter snake populations living along the California coast, the food of choice is (the banana slug/tadpoles and small fishes).

8. In Stevan Arnold's experiments, newborn garter snakes that were offspring of coastal parents usually (ate/ignored) a chunk of slug as the first meal.

9. Newborn garter snake offspring of (coastal/inland) parents ignored cotton swabs drenched in essence of slug and only rarely ate the slug meat.

10. The differences in the behavioral eating responses of coastal and inland snakes (were/were not) learned.

11. Hybrid garter snakes with coastal and inland parents exhibited a feeding response that indicated a(n) (environmental/genetic) basis for this behavior.

Complete the Table [pp.660-661]
12. Complete the following table to consider examples of instinctive and learned behavior.

Category	Examples
[p.660] a. Instinctive behavior	
[p.660] b. Learned behavior	

39.4 ADAPTIVE BEHAVIOR [p.662]
39.5 COMMUNICATION SIGNALS [pp.662-663]
39.6 MATES, OFFSPRING, AND REPRODUCTIVE SUCCESS [pp.664-665]
Selected Terms:
adaptive trait [p.662], bloodsucking mites [p.662], parasite [p.662], chemical, acoustical, visual signals, & tactile cues [p.662], alarm calls [p.662], waggle dance [p.663], courtship-related signals [p.663], microevoluntionary [p.664], sex pheromones [p.664], "nuptial gift" [p.664], displaying males & males defending territory [p.664], parental behavior [p.665]

Boldfaced Terms
[p.662] communication signals _____

_____

[p.664] lek_____

_____

[p.664] territory _____

_____

[p.665] sexual dimorphism _____

_____

Choice
Choose the appropriate type of communication cue for each statement. [pp.662-663]
   a. visual
   b. chemical
   c. tactile
   d. acoustical
1. ____ baboon threat display

2. ____ touch

3. ____ pheromones

4. ____ bird songs and prairie dog barks

5. ____ courtship displays in birds

Complete the Table

Complete the following table to supply the common names of the animals that fit the text examples of sexual selection. [pp.664-665]

Animals	Descriptions of Sexual Selection
6.	Females select the males that offer them superior material goods; females permit mating only after they have eaten the "nuptial gift" for about five minutes.
7.	Males congregate in a lek or communal display ground; each male stakes out a few square meters as his territory; females are attracted to the lek to observe male displays and usually select and mate with only one male.
8.	Females of a species cluster in defendable groups at a time they are sexually receptive; males compete for access to the clusters; combative males are favored.

39.7 LIVING IN GROUPS [pp.666-667]
39.8 WHY SACRIFICE YOURSELF? [pp.668-669]

Selected Terms:
cooperative responses [p.666], counterattack [p.666], "fishing sticks" [p.666], subordinate [p.667], dominance [p.667], cannibalize [p.667], pack [p.667], sterility [p.668], nonbreeding members [p.668], "queen" [p.668], "king" [p.668], sibling [p.668], evolution [p.668], *Heterocephalus glaber* [p.668], "self-sacrifice" [p.669], inbreeding [p.669], infanticide [p.669]

Boldfaced Terms
[p.666] selfish-herd _____

_____

[p.667] dominance hierarchy _____

_____

[p.668] altruistic behavior _____

_____

[p.668] theory of inclusive fitness _____

_____

Matching

Match each of the following terms to its correct definition. [pp.666-669]

1. ____ dominance hierarchy
2. ____ selfish herd
3. ____ theory of inclusive fitness
4. ____ altruistic behavior

a. enhances another's reproductive success at the expense of the individual

b. individuals hiding behind one another

c. genes associated with altruistic behavior are selective for because they promote the success of close relatives

d. animals that live in groups sometimes form these

SELF QUIZ

1. The observable, coordinated responses that animals make to stimuli are what we call _____. [p.658]
   a. imprinting
   b. instinct
   c. behavior
   d. learning

2. In _____, a particular behavior is performed without having been learned by actual experience in the environment. [p.660]
   a. natural selection
   b. altruistic behavior
   c. sexual selection
   d. instinctive behavior

3. Newly hatched goslings follow any large moving objects to which they are exposed shortly after hatching; this is an example of _____. [p.660]
   a. homing behavior
   b. imprinting
   c. piloting
   d. migration

4. A young toad flips its sticky-tipped tongue and captures a bumblebee that stings its tongue; in the future, the toad leaves bumblebees alone. This is _____. [p.660]
   a. instinctive behavior
   b. a fixed reaction pattern
   c. altruistic
   d. learned behavior

5. The claiming of the more protected central locations of the bluegill colony by the largest, most powerful males suggests _____. [p.666]
   a. cooperative predator avoidance
   b. the selfish herd
   c. a huge parent cost
   d. self-sacrificing behavior

6. A chemical odor in the urine of male mice triggers and enhances estrus in female mice. The source of stimulus for this response is a _____. [p.657;662]
   a. tactical signal
   b. pheromone
   c. acoustical signal
   d. visual signal

7. Female insects often attract mates by releasing sex pheromones. This is an example of a(n) _____ signal. [p.662]
   a. chemical
   b. visual
   c. acoustical
   d. tactile

8. The branch of biology that deals with behavior of animals as it relates to their environment is _____. [p.657]
   a. ecology
   b. behavioral ecology
   c. biogeography
   d. sociobiology

9. The scientists who discovered imprinting with ducklings was _____. [p.660]
   a. Ludwig Huber
   b. Bernhard Voelkl
   c. Konrad Lorenz
   d. Stevan Arnold

10. An example of instinctive behavior is _____. [p.660]
    a. shown by ducklings following Konrad Lorenz
    b. shown by two male lobsters battling at first meeting
    c. shown by a marmoset opening a container by using its teeth
    d. shown by a male baboon showing his teeth
    e. shown by a young cuckoo bird tossing out an egg from a nest

CHAPTER OBJECTIVES/REVIEW QUESTIONS

1. Define *stimuli*. [p.658]

2. Distinguish between instinctive and learned behaviors [p.660]

3. Explain the process of imprinting. [p.660]

4. Define adaptive behavior. [p.662]

5. List the forms of communication signals and give an example of each. [pp.662-663]

6. Explain the purpose of a pheromone. [p.662]

7. Explain the similarities and differences between sexual selection and natural selection. [p.664-665]

8. Why is parental care important? [p.665]

9. Define *selfish herd*; cite an example. [p.666]

10. Explain why a dominance hierarchy may be beneficial. [p.667]

11. What are the basic premises of inclusive fitness? [p.668]

12. Explain the benefits of altruistic behavior. [p.668]

13. Explain "self-sacrifice"? [p.669]

CHAPTER SUMMARY

An individual's (1) _____ starts with interactions among gene products, such as hormones and (2) _____. Most forms of behavior have innate components, but they can be modified by (3) _____ factors.

When behavioral traits have a (4) _____ basis, they may evolve by way of (5) _____.

(6) _____ behavior depends on evolved modes of (7) _____. Communication signals hold clear meaning for both the sender and the receiver of signals.

Life in social groups has reproductive (8) _____ and costs. Not all environments favor the (9) _____ of such groups. Self-sacrificing behavior has evolved among a few kinds of animals that live in large family groups.

(10) _____ behavior was shaped by the same evolutionary forces shape the behavior of other (11) _____. Only humans consistently make (12) _____ choices about their behavior.

INTEGRATING AND APPLYING KEY CONCEPTS

1. Think about communication signals that humans use and list them. Do you believe a dominance hierarchy exists in human society? Think of examples.

2. How do animals indicate their emotions? Give some examples.

3. How do humans indicate their emotions? Give some examples.

40

POPULATION ECOLOGY

INTRODUCTION

Population biologists and population geneticists are very interested in how groups of a species interact with their environment, and how the environment plays a role population size. Demographics and how to conduct a plot sample of a population are presented. The various characteristics that define a population are discussed. Population growth is explained and exponential growth is illustrated, along with a graph that clearly shows that this type of growth is NOT sustainable. This chapter explores some of the principles of population ecology. The chapter concludes with a look at human population growth, where it was and where it is heading.

FOCAL POINTS

- Figure 40.3 [p.675] shows population distribution patterns.
- Figure 40.4 [p.676] illustrates exponential growth.
- Figure 40.7 [p.678] illustrates logistic growth and carry capacity.
- Table 40.1 [p.680] shows data for a life table for an annual plant cohort.
- Table 40.2 [p.680] shows a life table for humans in the United States.
- Figure 40.9 [p.681] illustrates and gives examples of the three generalized survivorship curves.
- Figure 40.13 [p.684] shows the growth curve for the world human population.
- Figure 40.16 [p.685] diagrams the age structure of three populous countries in the world.

INTERACTIVE EXERCISES

40.1 A HONKING MESS [p.673]
40.2 POPULATION DEMOGRAPHICS [pp.674-675]

Selected Terms
Canada geese [p.673], *Branta canadensis* [p.673], nonmigratory [p.673], U. S. Fish and Wildlife Service [p.673], sampling techniques [p.674], euglena [p.674], organism [p.674], clumped distribution [p.674], near uniform distribution pattern (equally spaced apart) [p.674], random distribution [p.674], asexual reproduction [p.674], coral [p.674], aspen tree clones [p.674], flocks & schools [p.675], deserts [p.675], American Southwest [p.675], pre-reproductive, reproductive, & post reproductive [p.675]

Boldfaced Terms
[p.673] population _____

_____

[p.674] demographics _____

_____

[p.674] population size _____

_____

[p.674] plot sampling _____

_____

[p.674] mark-recapture method _____

_____

[p.674] population density _____

_____

[p.674] population distribution _____

_____

[p.675] age structure _____

_____

[p.675] reproductive base _____

_____

Short Answer

1. Briefly, state why Canada geese have become a problem. [p.673]]

_____

_____

_____

Matching

Match each term with its description. [pp.674-675]

2. ____ clumped distribution

3. ____ demographics

4. ____ population size

5. ____ age structure

6. ____ pre-reproductive, reproductive, and post-reproductive ages

7. ____ reproductive base

8. ____ population density

9. ____ population distribution

10. ____ population

11. ____ nearly uniform distribution

12. ____ corals and aspen trees

13. ____ capture-recapture method

14. ____ random distribution

a. Includes pre-reproductive and reproductive age categories

b. A measured number of individuals in some specified area

c. Examples of uniform distribution pattern

d. The number of individuals in some specified area or volume of a habitat

e. The number of individuals in each of several age categories

f. The vital statistics of the population

g. An example of age categories in a population

h. A way of sampling population density; capturing mobile individuals and marking them in some fashion and recapturing them after some period of time

i. A type of distribution that is usually the result of limited conditions or resources

j. A group of organisms of the same species that live in the same area and interbreed

k. The general pattern in which individuals are dispersed in a specified area
l. Individuals are more evenly spaced that expected by chance alone

m. Caused by uniform conditions and resources are plentiful

Population Problem

15. To determine the size of a fruit fly population, a researcher captures 400 flies and marks them with a fluorescent tag. Five days later the researcher repeats the capture. Out of 500 flies, 80 have tags. Based on this information, what is the estimated population size? [p.674]

_____

_____

40.3 POPULATION SIZE AND EXPONENTIAL GROWTH [pp.676-677]
40.4 LIMITS ON POPULATION GROWTH [pp.678-679]

Selected Terms
resources [p.676], per capita [p.676], r [p.676], J-shaped curve [p.677], limiting factors [p.677], S-shaped curve [p.678], competition [p.678], tsunami [p.679], reindeer [p.679], St. Matthew Island [p.679]

Boldfaced Terms
[p.676] immigration _____

_____

[p.676] emigration _____

_____

[p.676] zero population growth _____

_____

[p.676] per capita growth rate _____

_____

[p.676] exponential growth _____

_____

[p.676] biotic potential _____

_____

[p.678] limiting factor _____

_____

[p.678] carrying capacity _____

_____

[p.678] logistic growth _____

_____

[p.678] density-dependent factors _____

_____

[p.679] density-independent factors _____

_____

Growth Rate Problem

Consider the equation $G = rN$, where G = the population growth rate per unit time, r = the net population growth rate per individual per unit time, and N = the number of individuals in the population. [p.676]

1. Assume that 5,000 rats live in the streets of a city. The rats give birth to 2,000 rats every month.
    Every month 500 of the 5,000 rats die. Calculate the following:
    a.   birth rate = _____
    b.   death rate = _____
    c.   the value of r = _____

Matching

Match each of the following terms to the correct statement. [p.676-679]

2. ____   immigration

3. ____   zero population growth

4. ____   biotic potential

5. ____   emigration

6. ____   logistic growth

7. ____   limiting factor

8. ____   per capita growth rate

9. ____   doubling time

10. ____   exponential growth

11. ____   carrying capacity

a.   an essential resource that is in short supply
b.   an S-shaped population growth curve
c.   the time that it takes for a population to double in size
d.   the maximum number of individuals that a given environment can sustain indefinitely
e.   the arrival of new individuals from other populations
f.   individuals leaving a population to live elsewhere
g.   a J-shaped population curve
h.   the difference between the per capita birth rate and per capita death rate
i.   the maximum per capita rate of increase for a species
j.   when the number of births equals the number of deaths

Choice

Choose from the following [pp.678-679]
    a.   density-dependent factors
    b.   density-independent factors

12. ____   A sudden freeze

13. ____   Overcrowding

14. ____   Bubonic plague

15. ____   Application of pesticides in your backyard

16. ____   Predators, parasites, and pathogens

17. ____   Droughts, floods, and earthquakes

40.5 LIFE HISTORY PATTERNS [pp.680-681]
40.6 EVIDENCE OF EVOLVING LIFE HISTORY PATTERNS [pp.682-683]

Selected Terms

type I curve [p.680], type II curve [p.680], type III curve [p.680], survivorship [p.680], death rate [p.680], "birth" rate [p.680], natural selection [p.681], guppies [p.682], *Poecilla reticulate* [p.682], cichlids [p.683], killifish [p.683], Atlantic codfish [p.683], *Gadus morhua* [p.683]

Boldfaced Terms

[p.680] life history pattern _____

_____

[p.680] cohort _____

_____

[p.680] survivorship curve _____

_____

Matching

Match each term with its description. [pp.680-681]

1. ____ life history pattern
2. ____ cohort
3. ____ survivorship curve
4. ____ type I curves
5. ____ type II curves
6. ____ type III curves

a. Shows, for example, the number of individuals actually reaching specified ages
b. Reflect high survivorship until fairly late in life, then a large increase in deaths
c. A group of individuals, recorded from the time of birth until the last one dies; data is gathered such as the number of offspring born to individuals in each age interval
d. Signify a death rate that is highest early on; typical of species that produce many small offspring and do little, if any, parenting
e. For each species, a set of adaptations that influence survival, fertility, and age at first reproduction
f. Reflect a fairly constant death rate at all ages; typical of organisms just as likely to be killed or die of disease at any age, such as lizards, small mammals, and large birds

True-False

If the statement is true, write a T in the blank. If the statement is false, correct it by writing the correct word(s) for the underlined word(s) in the answer blank. [pp.682-683]

7. _____ Reznick and Endler showed that life history traits, like other characteristics, <u>cannot</u> be inherited.

8. _____ Guppy populations deal with predators, especially pike-cichlids and <u>killifish</u>.

9. _____ Smaller killifish prey on smaller, immature guppies but not on the larger adults; pike-cichlids live in other streams and prey on <u>larger</u> and sexually mature guppies.

10. _____ Reznick and Endler hypothesized that predation is a selective agent acting on <u>killifish</u> life history patterns.

11. _____ In pike-cichlid streams, guppies grow <u>slower</u> and are smaller at maturity, compared to guppies in killifish streams.

12. _____ Guppies hunted by pike-cichlids reproduce earlier in life, reproduce more often, and have <u>more</u> young per brood

13. _____ Following laboratory experiments in the United States, the researchers concluded the differences between guppies preyed upon by different predators have a(n) <u>environmental</u> foundation.

14. _____ Guppies introduced to the upstream experimental site were taken from a population that had evolved downstream from the waterfall, with the larger <u>killifishes</u>.

15. _____ Eleven years (thirty to sixty guppy generations) later, researchers revisited the stream. They found that the experimental population <u>had</u> <u>not</u> evolved.

16. _____ Later laboratory experiments involving two generations of guppies confirmed that the differences studied in guppies has a <u>genetic</u> basis

40.7 HUMAN POPULATION GROWTH [pp.684-685]
40.8 POPULATION GROWTH AND ECONOMIC EFFECTS [pp.686-687]

Selected Terms
technologies [p.684], domestication of wild grasses [p.684], infant mortality [p.686], life expectancy [p.686], preindustrial stage [p.686], transitional stage [p.686], industrial stage [p.686], postindustrial stage [p.686], carrying capacity [p.687]

Boldfaced Terms
[p.685] total fertility rate _____

_____

[p.685] replacement fertility rate _____

_____

[p.686] demographic transition model _____

_____

[p.687] ecological footprint _____

_____

Short Answer

1. List the three possible reasons humans began sidestepping controls. [p.684]

_____

_____

_____

2. Explain why continued human population growth cannot continue indefinitely. [p.685]

_____

_____

_____

3. What is meant by total fertility rate? [p.685]

_____

_____

_____

Matching

Match each age structure diagram with its description below. You may want to consult Figure 40.14 in the textbook. [p.685]

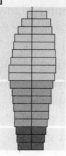 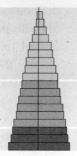 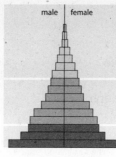 

a   b   c   d

4. _____ zero growth

5. _____ rapid growth (more than a third of the world population)

6. _____ negative growth

7. _____ slow growth

Choice

Refer to Figure 40.16 [p.701]. Using the age structure diagrams shown above, choose the diagram that best fits the countries listed below.

8. _____ China

9. _____ United States

10. _____ India

Sequence

Arrange the following stages of the demographic transition model in correct chronological sequence. Write the letter of the first step next to 11, the letter of the second step next to 12, and so on. [p.700]

11. ____

12. ____

13. ____

14. ____

a. Industrial stage: population growth slows dramatically and industrialization is in full swing; a slowdown starts mainly because people want to move from the country to cities, and urban people want smaller families

b. Preindustrial stage: living conditions are the harshest before technological and medical advances become widespread; birth and death rates are high, so the growth rate is low

c. Postindustrial stage: population growth rates become negative; the birth rate falls below the death rate, and population size slowly decreases

d. Transitional stage: industrialization begins, food and health care improve; death rates drop, birth rates stay high in agricultural societies

SELF QUIZ

1. The number of individuals that contribute to a population's gene pool is _____. [pp.673-675]
   a. the population density
   b. the population growth
   c. the population birth rate
   d. the population size
   e. the population distribution

2. How are the individuals in a population *most often* dispersed? [pp.674-675]
   a. clumped
   b. very uniform
   c. nearly uniform
   d. random
   e. very random

3. Assuming immigration is balancing emigration over time, _____ may be defined as an interval in which the number of births is balanced by the number of deaths. [p.690]
   a. the lack of a limiting factor
   b. exponential growth
   c. saturation
   d. zero population growth
   e. logistic growth

4. A population that is growing exponentially in the absence of limiting factors can be illustrated accurately by a(n) _____. [pp.676-677]
   a. S-shaped curve
   b. J-shaped curve
   c. curve that terminates in a plateau phase
   d. tolerance curve
   e. resource curve

5. The population growth rate (G) is equal to the _____ net population growth rate per individual (r) and number of individuals (N). [p.676]
   a. sum of
   b. product of
   c. doubling of
   d. difference between
   e. tripling of

6. Assuming the birth and death rate remain constant, both can be combined into a single variable, r, or _____. [p.676]
   a. the per capita rate
   b. the minus migration factor
   c. the number of individuals
   d. the net reproduction per individual per unit
   e. population growth

7. Which of the following is *not* characteristic of logistic growth? [p.677]
   a. S-shaped curve
   b. leveling off growth as carrying capacity is reached
   c. unrestricted growth
   d. slow growth of a low-density population followed by rapid growth
   e. limiting factors

8. The maximum number of individuals of a population (or species) that a given environment can sustain indefinitely defines _____. [p.678]
   a. the carrying capacity of the environment
   b. exponential growth
   c. the doubling time of a population
   d. density-independent factors
   e. logistic growth

9. The survivorship curve typical of industrialized human populations is type _____. [p.681]
   a. I
   b. II
   c. III
   d. both I and II
   e. both I and III

10. "Industrialization begins, food production and health care improve anddeath rates slow," describes the _____ stage of the demographic transition model. [pp.686-687]
   a. preindustrial
   b. transitional
   c. industrial
   d. postindustrial
   e. includes both preindustrial and transitional

CHAPTER OBJECTIVES/REVIEW QUESTIONS

1. Explain what lessons can be learned from the crashing of the plane because of Canada geese. [p.673]

2. Be able to define: *demographics, population size, age structure, population size, age structure, reproductive base, population density, population distribution, crude density, quadrats,* and *capture-recapture methods.* [pp.674-675]

3. List and describe the three patterns of dispersion illustrated by populations in a habitat. [p.674]

4. Be able to calculate population size using the mark-recapture method. [p.674]

4. Distinguish immigration from emigration and define zero population growth. [p.676]

5. Understand how to calculate the per capita growth rate (r). [p.676]

6. Understand the characteristics of a population in exponential growth. [pp.676-677]

7. Understand the concept of biotic potential. [p.677]

8. List several examples of limiting factors and explain how they influence population curves. [pp.678-679]

9. Define carrying capacity [p.678]

10 Tell how the pattern of logistic growth shows how carrying capacity can affect population size. [pp.678-679]

11. Explain how density-dependent factors and density-independent factors relate to population size; cite two examples of each. [pp.678-679]

12. Understand the concept of life history patterns and survivorships curves. [pp.680-681]

13. After consideration of the research results obtained on guppies and their predators by Reznick and Endler, provide an explanation for these differences. [pp.682-683]

14. Be able to list three possible reasons why growth of the human population is out of control. [pp.684-685]

15. List reasons why continued growth of the human population cannot be sustained indefinitely. [pp.684-685]

16. Define total fertility rate. [p.685]

17. Understand the age structure diagrams for populations growing at different rates. [p.685]

18. List and describe the four stages of the demographic transition model. [pp.686-687]

19. Define and give an example of ecological footprint. [p.687]

CHAPTER SUMMARY

Ecologists explain population (1) _____ in terms of population (2) _____, density, (3) _____, and number of individuals in different (4) _____ categories. They have methods of estimating population size and (5) _____ in the field.

A population's size and (6) _____ base influence its rate of growth. When the population is increasing at a rate proportional to its size, it is undergoing (7) _____ growth.

Over time, exponential growth typically overshoots the (8) _____ - the maximum number of individuals of a population that (9) _____ resources can sustain indefinitely. Some populations (10) _____ after a crash. Others never recover.

Resource availability, (11) _____, and predation are major factors that can (12) _____ population growth. These (13) _____ factors differ among species and shape their life (14) _____ patterns.

Human populations have (15) _____ limits to growth by way of (16) _____ expansion into new habitats, (17) _____ interventions, and (18) _____ innovations. No population can expand indefinitely when (19) _____ dwindle.

INTEGRATING AND APPLYING KEY CONCEPTS

1. Following your study of this chapter and considering what you have learned about the problems facing planet Earth, prepare a list of positive changes you could make in your own living habits. These changes would help solve the serious problems we face as a human population. Share your knowledge and your list with family and friends and request that they each prepare a similar list. Above all, encourage everyone to *act* on the personal changes they propose..

2. Some researchers have proposed that we are rapidly approaching "peak oil" – a point where the discovery of new reserves can no longer meet demand. Explain the effect that peak oil may have on the demographic transition model.

3. How would you propose limiting the human population growth? What would be the upside and downside of your idea?

41
COMMUNITY ECOLOGY

INTRODUCTION

This chapter is begins with an example of an introduced (exotic) species – fire ants and the problems that have followed in their aftermath. We have all heard stories about "radical" environmentalists that take drastic measures to save a species, the spotted owl, for example, or even a single tree. What drives these people to risk so much for seemingly so little? This chapter describes the relationships that exist in a community of species. This includes the shaping of communities, as well as the many interactions that go on within a single community. The chapter then goes on to describe the effects of introduced organisms, such as kudzu, on a community and population. Many introduced species have had devastating results, and these examples certainly serve as a warning for future activities. Finally, the text discusses biodiversity and introduces the field of conservation biology, which attempts to promote and sustain environmental diversity while also accommodating the survival needs of man. The focus is on the fact that all members of a community are critical to that community maintaining balance is paramount to species survival.

FOCAL POINTS

- Table 41.1 [p.692] compares the different types of direct two-species interactions.
- Figure 41.6 [p.695] illustrates competitive exclusion as shown by Gause and his *Paramecium spp*.
- Figure 41.8 [p.696] shows a graph of the number of Canadian lynx and snowshoe hares.
- Figure 41.17 [p.702] illustrates the effect of competition and predation in a specific environment.
- Figure 41.20 [p.704] shows ant species richness by latitude.
- Figure 41.22 [p.705] illustrates island biogeography patterns.

INTERACTIVE EXERCISES

41.1 FIGHTING FOREIGN FIRE ANTS [p.691]
41.2 COMMUNITY STRUCTURE [p.692]
41.3 MUTUALISM [p.693]
41.4 COMPETITIVE INTERACTIONS [pp.694-695]
41.5 PREDATION AND HERBIVORY [pp.696-697]

Selected Terms
Solenopsis invicta [p.691], fire ants [p.691], stinging [p.691], venom [p.691], bobwhites & vireos [p.691], pesticides [p.691], biological controls [p.691], phorid flies [p.691], parasitoids [p.691], ecologists [p.691], abundant & rare species [p.692], physical factors [p.692], parasitism [p.692], parasitoidism [p.692], pollinators [p.693], lichens [p.693], natural selection [p.693], sea anemone [p.693], nematocysts [p.693], cyanobacteria [p.693], *Heteractis magnifica* [p.693], *Amphiprion perideraion* [p.693], yucca moth [p.693], Charles Darwin [p.694], animal's niche [p.694], exploitative competition [p.694], G. Gause [p.694], protozoan [p.694], *Paramecium* [p.694], coexist [p.695], directional selection [p.695], ground finch [p.695], *Geospiza fortis* [p.695], *G. magnirostris* [p.695], Galapagos Islands [p.695], predator & prey [p.696], defensive adaptations [p.696], camouflage [p.697], ricin [p.697] toxin [p.697], caffeine[p.697], nicotine [p.697], koalas [p.697]

Boldfaced Terms

[p.691] community _____

_____

[p.692] habitat _____

_____

[p.692] commensalism _____

_____

[p.692] symbiosis _____

_____

[p.693] mutualism _____

_____

[p.694] interspecific competition _____

_____

[p.694] ecological niche _____

_____

[p.694] competitive exclusion _____

_____

[p.695] resource partitioning _____

_____

[p.695] character displacement _____

_____

[p.696] predation _____

_____

[p.696] mimicry _____

_____

[p.697] herbivory _____

_____

Short Answer

1. Describe the biological controls being enlisted by ecologists in an attempt to control *Solenopsis*, the genus name of the dreadful fire ants. [p.691]

    _____

    _____

    _____

Fill in the Blanks [p.692]

Every community has a characteristic structure, which arises as an outcome of five factors. First, the (2) _____ and topography interact to dictate temperatures, rainfall, type of soil, and other conditions. Second, the kinds and quantities of (3) _____ and other resources that become available through the year affect which species live there. Third, (4) _____ traits of each species help it (5) _____ and exploit specific resources on the habitat. Fourth, species interact in ways that cause shift in (6) _____ and abundances. Fifth, the timing and history of disturbances, both natural and human-induced, affect (7) _____ structure.

Matching
Match each term with its description. [pp.692-695]

8. _____ habitat

9. _____ niche

10. _____ prey

11. _____ natural selection

12. _____ indirect interactions

13. _____ commensalism

14. _____ mutualism

15. _____ interspecific competition

16. _____ predation

17. _____ parasitism

18. _____ symbiosis

a. An interaction that directly helps one species but does not affect the other much, if at all

b. Both species can be harmed

c. A mechanism that favors individuals who minimize their cost and maximize their benefits

d. Defined as a close association between two or more species during part or all of the life cycle

e. Possess physical and chemical features, such as temperature, and chemical features, such as temperature, and an array of species; an organism's "address"

f. Kill prey

g. An organism that is killed and eaten by a predator

h. The sum of an organism's activities and interactions as it goes about acquiring and using the resources required to survive and reproduce; the "profession" a species has in the community

i. Weaken hosts

j. Some species interactions in a community where some species help others in roundabout ways; an example is Canadian lynx eating hares fattened by plants

k. Both species benefit; a two-way exploitation

Completion

Complete the following table which summarizes the effect of different types of species interactions on both species involved. You might want to consult Table 41.1 for additional information. [p.692]

Type of interaction	Effect on species 1	Effect on species 2
19.	helpful	none
mutualism	20.	helpful
Interspecific competition	harmful	21.
Predation	helpful	22.
23.	helpful	harmful

Complete the Table

24. Complete the following table by describing how each organism listed in intimately dependent on the other for survival and reproduction in a mutualistic interaction. [p.693]

Organism	Dependency
a. Yucca moth and yucca plant	
b. Various fungi and plant roots in a mycorrhizal association	
c. Anemone fish and sea anemones	
d. Endosymbiosis; phagocytic cells and aerobic bacterial cells	

Choice

Choose from the following [pp.694-695]:

    a.   interference competition
    b.   exploitative competition
    c.   competitive exclusion
    d.   resource partitioning
    e.   interspecific competition

25. ____ Bristly foxtail grasses, Indian mallow plants, and smartweed plants coexist in the same habitat.

26. ____ Two species of *Paramecium* are grown in the same culture but cannot coexist indefinitely.

27. ____ One species of chipmunk chases another species of chipmunk out of its habitat

28. ____ Nine species of chipmunks live in different habitats on the slopes of the Sierra Nevada.

29. ____ deer have no acorns due to the fact that blue jays ate most of them

30. ____ Many sages exude aromatic compounds from their leaves that taint the soil around the plant and prevent potential competitors from taking root

31. ____ Two species of *Paramecium* do not overlap much in requirements; the two continued to coexist

Matching

Match each term with its description [p. 696-697]

32. _____ The species in a predation relationship that does the hunting

33. _____ The species in a predation relationship that is hunted

34. _____ The similar appearance between bees and wasps is called

35. _____ A form, patterning, color, or behavior that allows an animal to blend into its background

36. _____ Eating plants or plant parts

a. camouflage
b. predator
c. prey
d. herbivory
e. mimicry

Choice

Choose from the following: (Note: the same letter may be used more than once, but use only one letter per blank.) Also some of these are not given as examples in the book but you should be able to apply the concepts to these examples. [p. 696-697]

a. camouflaging
b. warning coloration
c. mimicry
d. cornered animals and plants under attack
e. little or no attempt at concealment
f. adaptive responses of predators

37. _____ Grasshopper mice plunging the noxious chemical-spraying tail end of their beetle prey into the ground to feast on the head end

38. _____ Aggressively stinging yellow jackets are the likely model for nonstinging edible wasps, beetles, and flies

39. _____ Polar bears stalking seals over ice, striped tigers crouched in tall-stalked, golden grasses, and scorpionfish well hidden on the seafloor

40. _____ Dangerous or repugnant species such as skunks who spray repellants and poisonous frogs of the genus *Dendrobates*

41. _____ Resemblance of *Lithops*, a desert plant, to a small rock

42. _____ Yellow-banded wasps or an orange-patterned monarch butterfly; predators learn to avoid

43. _____ Least bittern with coloration similar to that of surrounding withered reeds

44. _____ Peach, apricot, and rose seeds loaded with cyanide; lethal ricin in castor bean plants

45. _____ Animals may startle the predator, by hissing, puffing up, showing teeth, or flashing big-eye-shaped spots; opossum and hognose snakes pretend to be dead; many animals secrete of squirt irritating chemical repellents or toxins

41.6 PARASITES, BROOD PARASITES, AND PARASITOIDS [pp.698-699]

Selected Term:
endoparasites [p.698], ectoparasite [p.698], evolutionary fitness [p.698], sickle-cell anemia [p.698], European cuckoos & North American cowbirds [p.698], roundworms [p.698], tick [p.698], biological control agents [p.699], Hawaiian butterfly [p. 699]

Boldfaced Terms
[p.698] parasitism _____

_____

[p.698] brood parasitism _____

_____

[p.699] parasitoids _____

_____

True-False [pp.698-699]
If the statement is true, write a "T" in the blank. If the statement is false correct it by writing the correct word(s) for the underlined word(s) in the answer blank.

1. _____ All viruses, some bacteria, protists, fungi, tapeworms, flukes, and some roundworms are notorious invertebrate <u>parasitoids</u>.

2. _____ Dodder is a type of <u>predatory</u> plants.

3. _____ Social parasites are animals that take advantage of the <u>social</u> behavior of a host species in order to carry out their life cycle.

4. _____ North American cowbirds alter the <u>digestive</u> behavior of another species to complete their life cycle.

5. _____ Because parasites and parasitoids can help control the population growth of other species, many are raised commercially and then selectively released as <u>biological controls</u>.

Short Answer
6. List the advantage displayed by effective biological controls. [p.699]

_____

_____

_____

_____

7. List two examples of brood parasites, one example of an endoparasite parasite, and one example of an ectoparasite.

_____

_____

_____

_____

_____

41.7 ECOLOGICAL SUCCESSION [pp.700-701]

Selected Terms
colonize [p.700], "climax community" [p. 700], species richness [p.701], disturbance [p.701]

Boldfaced Terms
[p.700] pioneer species _____

_____

[p.700] primary succession _____

_____

[p.700] secondary succession _____

_____

[p.719] intermediate disturbance hypothesis _____

_____

Choice
For questions 1-10, choose from the following: [pp. 700-701]
     a. primary succession
     b. secondary succession
     c. intermediate disturbance hypothesis

1. ____ A process that begins when pioneer species colonize a barren habitat, such as a new volcanic island and land exposed when a glacier retreats

2. ____ When trees fall in tropical forests, small gaps open in the dense tree canopy, and more light reaches patches on the forest floor; then conditions favor growth of previously suppressed small trees and germination of pioneers or shade-intolerant species

3. ____ Included are lichens and mosses that are small, have short life cycles, and can survive intense sunlight, extreme temperature changes, and nutrient-poor soil

4. ____ A disturbed area within a community recovers; if improved soil is still there, secondary succession can be quite rapid

5. ____ This pattern is common in abandoned fields, burned forests, and areas cleared by volcanic eruptions

6. ____ Many types are mutualists with nitrogen-fixing bacteria, so they can grow in nitrogen-poor habitats

7. ____ Once established, the pioneers improve conditions for other species and often set the stage for their own replacement.

8. ____ In time, organic wastes and remains accumulate, adding volume and nutrients to soil that help more species take hold.

9. ____ The type that would occur when land is exposed by the retreat of a glacier.

10. ____ Many small-scale changes recur in patches of habitats; they contribute to the community's internal dynamics.

Listing
11. The modern view of succession cites three factors that affect the species composition of a community. List these three factors. [p.701]

_____

_____

_____

41.8 SPECIES INTERACTIONS AND COMMUNITY INSTABILITY [pp.702-703]

Selected Terms
Pisaster ochraceus [p.702], chitons, limpets, barnacles, & mussels [p.702], periwinkles [p.702], intertidal zone jump [p.702], kudzu [p.703], *Pueraria lobata* [p.703], Gypsy moth [p.703] *Lymantria dispar* [p.703], nutrias [p.703], *Myocastor coypus* [p.703]

Boldfaced Terms
[p.702] keystone species _____

_____

[p.702] indicator species _____

_____

[p.702] exotic species _____

_____

Matching
Match each term with its description. [p.702-703]

1. ____ keystone species

2. ____ exotic species

a. Residents of established communities disperse from their home range and become established elsewhere; permanently insinuates itself into a new community

b. A species that has a disproportionately large effect on a community relative to its abundance; the periwinkle is an example

Community Ecology **437**

Choice
Choose from the following [p. 722]

    a. kudzu

    b. Gypsy moth

    c. nutrias

3. ____ An exotic species that is native to Europe and Asia and feeds on oaks

4. ____ In 1876, this vine was introduced to the United States; grows faster in the southeast where it now blankets streambanks, trees, telephone poles, houses and almost everything else

5. ____ Arrived in the United States in the mid 1700s

6. ____ large semi-aquatic rodent

7. ____ Asians use an extracted starch that is used in drinks, herbal medicines, and candy; it may also help save trees as it is an alternative source of paper

8. ____ Its voracious appetite and burrowing activities contribute to marsh erosion and levee damage

41.9 BIOGEOGRAPHIC PATTERNS IN COMMUNITY STRUCTURE [pp.704-705]

Selected Terms
biogeography [p.704], Surtsey [p.704], island patterns [p.704], latitude [p.704], island [p.705], "sea" of habitat [p.705], MacArthur & Wilson [p.705]

Boldfaced Terms
[p.704] species richness _____

_____

[p.704] equilibrium model of island biogeography _____

_____

[p.705] distance effect _____

_____

[p.705] area effect _____

_____

Fill in the Blanks [p. 724]

The most striking pattern of biodiversity corresponds to distance from the (1) _____. For most groups of plants and animals, on land and in the seas, the number of coexisting species is greatest in the (2) _____ and it systematically declines toward the poles. Tropical latitudes intercept more incoming intense (3) _____ and receive more rainfall, and their growing season is longer. As one outcome, (4) _____ availability tends to be greater and more reliable in the tropics than elsewhere.

Tropical communities have been (5) _____ for a longer time than temperate ones, some of which did not start forming until the end of the last ice age.

Species diversity might be self-reinforcing. Diversity of tree species is much (6) _____ than in comparable forests at higher latitudes. When more plant species compete and coexist, more species of (7) _____ similarly compete and coexist, partly because no single (7) _____ can overcome all of the chemical defenses of all the different plants. More (8) _____ and (9) _____ [p.722] species tend to evolve in response to a diversity of prey and hosts. The same applies to diversity on tropical (10) _____.

Species Diversity Problem

11. Study of the distance effect, the area effect, and species diversity patterns as they relate to the equator, and answer the following question. There are two islands (b and c) of the same size and topography that are equidistant from the African coast (a), as shown in the diagram. Which will have the higher species diversity values? [p.724]

_____

_____

_____

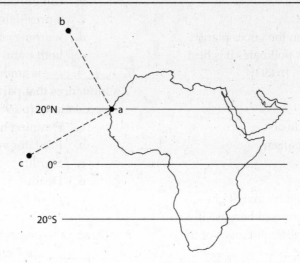

SELF QUIZ

1. _____ have physical and chemical features, such as temperature, and an array of species. [p.692]
   - a. Habitats
   - b. Niches
   - c. Interactions
   - d. Ecosystems
   - e. Climates

2. A lopsided interaction that directly benefits one species but does not affect the other much, if at all, is _____. [p.692]
   - a. commensalism
   - b. competitive exclusion
   - c. predation
   - d. mutualism
   - e. parasitism

3. The relationship between the yucca plant and the yucca moth that pollinates it is best described as _____. [p.693]
   - a. camouflage
   - b. commensalism
   - c. competitive exclusion
   - d. obligatory mutualism
   - e. mimicry

4. _____ is represented by foxtail grass, mallow plants, and smartweed because their root systems subdivide different areas of the soil in a field. [p.695]
   - a. Succession
   - b. Resource partitioning
   - c. A climax community
   - d. A disturbance
   - e. Competitive exclusion

5. G. Gause grew two species of *Paramecium* separately and then together. They required identical resources and could not coexist indefinitely. This is an example of _____. [pp.694-695]
   - a. warning coloration
   - b. mimicry
   - c. camouflage
   - d. competitive exclusion
   - e. obligatory mutualism

6. Edible insect species often resemble toxic or unpalatable species that are not at all closely related. This is an example of _____. [p.696]
   - a. mimicry
   - b. camouflage
   - c. a prey defense
   - d. warning coloration
   - e. both c and d

7. _____ is among the pervasive influences that parasites have on populations. [p.698]
   - a. Draining host nutrients
   - b. Predator vulnerability
   - c. Sterility
   - d. Death
   - e. all of the above

8. During the process of primary succession, _____. [p.700]
   - a. pioneer populations colonize a barren habitat
   - b. pioneers include lichens and mosses
   - c. pioneers improve conditions for other species
   - d. pioneers set the stage for their own replacement
   - e. all of the above

9. The _____ is an example of a keystone species. [p.702]
   a. kudzu vine
   b. sea star
   c. chiton
   d. nutrias
   e. Gypsy moth

10. The most striking patterns of biodiversity corresponds to _____ [p.704]
   a. distance effect
   b. area effect
   c. immigration rate for new species
   d. distance from the equator
   e. resource availability

CHAPTER OBJECTIVES/REVIEW QUESTIONS

1. Discuss possible controls for invading Argentine fire ants. [p.691]

2. Be able to define the following terms and cite examples: *habitat, community, ecological niche, resource partitioning, character displacement, commensalism, mutualism, interspecific competition, predation, parasitism,* and *symbiosis.* [pp.693-695]

3. Lichens and mycorrhizae are cases of _____ mutualism. [p.693]

4. Describe and cite examples of interference competition and competitive exclusion. [p.694-695]

5. Describe the interactions of predator-prey, warning coloration, mimicry, camouflage, when an animal is cornered, and predator responses to prey; cite examples. [pp.696-697]

6. Be able to list the adaptations of prey as well as the adaptive responses of predators [p. 697]

7. List pervasive influences that parasites have on populations, the kinds of parasites and tell how parasites may be potential biological control agents. [pp.698-699]

8. _____ are insects that develop inside another species of insect, which they devour from the inside out as they mature. [p.699]

9. Describe the patterns of ecological succession; include the following terms: pioneer species, climax community, primary and secondary succession. [pp.701-702]

10. List and explain examples of forces leading to community instability. [pp.702-703]

11. Explain why kudzu, *Gypsy moths*, and nutrias are exotic invaders that lead to community instability. [pp.702-703]

12. Describe mainland, marine, and island patterns of diversity. [pp.704-705]

CHAPTER SUMMARY

A community consists of all (1) _____ in a habitat. Each species has a (2) _____ which is the sum of its activities and relationships. A habitat's history, its biological and physical characteristics, and interactions among species in the habitat affect (3) _____ structure. Interspecific interactions include (4) _____, mutualism, competition, (5) _____ and parasitism. They influence the population (6) _____ of participating species, which in turn influences the community's structure. A community changes over time. Factors that affect this change include: physical characteristics of the habitat, (7) _____, (8) _____ and chance events.

(9) _____ identify regional patterns in species distribution. The (10) _____ regions hold the greatest number of species. The characteristics of islands can be used to predict how many (11) _____ an island will hold. (12) _____ biologists are working to preserve species richness by resource management that is not in conflict with the human need for survival.

INTEGRATING AND APPLYING KEY CONCEPTS

1. When a U. S. citizen re-enters the USA after traveling in another country, they are asked if they have any plants or animals in their baggage or on their person. Why?

2. An "island" doesn't have to be an island in the middle of an ocean. Where else could "islands" exist? One hint – a lake could be an island surrounded by land.

42

ECOSYSTEMS

INTRODUCTION

Ecosystems represent an interaction between the biotic (living) and abiotic (non-living) factors in a given area. In this chapter you will explore how energy moves through an ecosystem. Energy flow is one way. This will be followed by an introduction to the cycling of biologically important elements in ecosystems. The cycling of various chemicals between the atmosphere, hydrosphere, and lithosphere is shown with numerous diagrams. Food chains and food webs are discussed as well as the different types of pyramids that ecologist use to describe ecosystems.

FOCAL POINTS

- Figure 42.2 [p.710] illustrates the difference between nutrient and energy cycling in an ecosystem.
- Figure 42.4 [p.712] shows some organisms in an Arctic food web.
- Figure 42.6 [p.713] illustrates an ecological pyramid.
- Figure 42.7 [p.714] provides an overview of biogeochemical cycles.
- Table 42.1 [p.715] provides information about environmental water reservoirs.
- Figure 42.8 [p.715] illustrates the water cycle.
- Figure 42.9 [p715] shows a map of the groundwater troubles in the United States.
- Figure 42.10 [p.716] illustrates the carbon cycle.
- Figure 42.11 [p.717] diagrams the greenhouse effect on the Earth.
- Figure 42.12 [p.718] illustrates the nitrogen cycle.
- Figure 42.13 [p.719] illustrates the phosphorus cycle.

INTERACTIVE EXERCISES

42.1 TOO MUCH OF A GOOD THING [p.709]

42.2 THE NATURE OF ECOSYSTEMS [p.710]

42.3 FOOD CHAINS [p.711]

42.4 FOOD WEBS [pp.712-713]

42.5 ECOLOGICAL PYRAMIDS [p.713]

42.6 BIOGEOCHEMICAL CYCLES [p.714]

Selected Terms

elements [p.709], essential nutrients [p.709], ecosystem [p.709], laundry detergent [p.709], dish detergents [p.709], lawn fertilizers [p.709], algal bloom [p.709], nutrient cycle [p.709], organism [p.710], herbivores [p.710], carnivores [p.710], omnivores [p.710], parasite [p.710], physical environment [p.710], photoautotrophs [p.710], photosynthesis [p.710], organic molecules [p.710], decomposition [p.710], herbivores [p.711], primary producer, primary consumer, secondary consumer, & third level consumer [p.711], first, second, third, & fourth trophic levels [p.711], detritus [p.713], troph [p.711], biomass pyramid [p.713], ecological pyramid [p.713], energy flow pyramid [p.713], sediments, water, & atmosphere [p.714], chemical & geologic processes [p.714], erosion & uplifting [p.714]

Boldfaced Terms

[p.709] eutrophication _____

_____

[p.710] primary producers _____

_____

[p.710] primary production _____

_____

[p.710] consumers _____

_____

[p.710] detritivores _____

_____

[p710] decomposers _____

_____

[p.711] trophic levels _____

_____

[p.711] food chain _____

_____

[p.712] food webs _____

_____

[p.712] grazing food web _____

_____

[p.712] detrital food web _____

_____

[p.714] biogeochemical cycle _____

_____

Matching

Match each of the following terms with the correct description. [pp.710-712]

1. _____ herbivores
2. _____ carnivores
3. _____ omnivores
4. _____ parasites
5. _____ consumers
6. _____ ecosystem
7. _____ primary producers
8. _____ detritivores
9. _____ trophic levels
10. _____ food chain
11. _____ food web
12. _____ decomposers

a. autotrophic organisms that obtain energy from the sun
b. consumers that eat plants
c. consumers that live inside a host and eat its tissues
d. consumers that eat the flesh of animals
e. consume small organic matter called detritus
f. hierarchy of feeding relationships
g. a food chain that has cross-connected levels
h. consumers that eat both plants and animals
i. feed on organic wastes and break them down to inorganic building blocks
j. sequence of steps that transfers energy from producer to organisms in higher trophic levels
k. the general term for organisms that eat other organisms
l. an array of organisms and the physical environment

Choice

The figure below represents a theoretical food web. The numbers in the circles represent different species. Organisms 1, 2 and 3 are producers. The arrows indicate the direction of energy flow. For each statement, list the correct numbers. [pp.710-713]

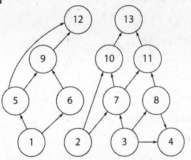

13. Which organisms are the decomposers? _____

14. Which organisms represent only the first trophic level? _____

15. Which organisms represent only the second trophic level? _____

16. Which organisms represent only the third trophic level? _____

17. Which organism(s) represent(s) only the fourth trophic level? _____

18. Which organisms represent more than one trophic level? _____

19. Which organism(s) could be classified as a herbivore? _____

20. Which organism(s) could be classified as an omnivore? _____

Short Answer

21. Distinguish between a grazing and detrital food web. [p.712]

_____

_____

_____

_____

_____

Matching

Match each of the following statements to the correct term. [pp.709-714]

22. _____ rate at which producers capture and store energy in a given interval

23. _____ dry weight of all organisms at each trophic level of an ecosystem

24. _____ the movement of elements from reservoirs to ecosystems and back to reservoirs

25. _____ biogeochemical cycle that involves the movement of carbon and hydrogen

26. _____ biogeochemical cycle that involves elements that do not have a gaseous form

27. _____ enrichment of lakes and rivers by additions of nitrogen and phosphorus, usually by activities associated with human activities

28. _____ biogeochemical cycle that involves the gaseous form of a nutrient

29. _____ illustrates how usable energy is reduced through an ecosystem

a. atmospheric cycles
b. sediment cycles
c. water cycles
d. energy pyramid
e. biogeochemical cycles
f. primary productivity
g. eutrophication
h. biomass pyramid

Ordering

Place the following organisms in the correct order of a typical food chain from the producers forward. [p.712]

hawk, grass, spider, grasshopper, sparrow

30. _____   33. _____

31. _____   34. _____

32. _____

Matching [p.714]
Match the letters in the diagram of a generalized biogeochemical cycle with the terms that describe the place of the letter in the diagram below.

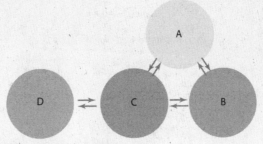

35. ____ rocks and sediments

36. ____ atmosphere

37. ____ living organisms

38. ____ sea water and fresh water

42.7 THE WATER CYCLE [pp.714-715]

Selected Terms
evaporation [p.715], precipitation [p.715], transpiration [p.715], hydrologic cycle [p.715], water vapor [p.715], ice [p.715], irrigation [p.715], global water supply [p.715]

Boldfaced Terms
[p.714] water cycle _____

_____

[p.714] watershed _____

_____

[p.714] soil water _____

_____

[p.714] aquifers _____

_____

[p.714] groundwater_____

_____

[p.714] runoff_____

_____

Matching

Match the terms below to the letters presented on the water cycle. [p.715]

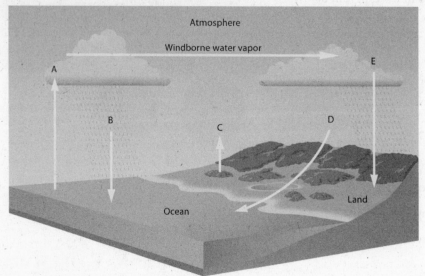

1. ____ Evaporation from land plants (transpiration)

2. ____ Evaporation from ocean

3. ____ Precipitation onto the land

4. ____ Precipitation into ocean

5. ____ Surface and groundwater flow

42.8 THE CARBON CYCLE [pp.716-717]

Selected Terms

aerobic respiration [p.716], bicarbonate & carbonate ions [p.716], aquatic producers [p.716], carbon-rich shells & rocks [p.716], "greenhouse gases" [p.717], cutting forests [p.717], burning fossil fuels [p.717], Intergovernmental Panel on Climate Change [p.717], light & heat energy [p.717], carbon dioxide [p.717], Earth's tilt [p.717]

Boldfaced Terms

[p.716] carbon cycle _____

_____

[p.716] atmospheric cycle_____

_____

[p.717] greenhouse effect _____

_____

[p.717] global climate change _____

_____

Matching

Match the number in the figure with the correct choice below (a – f) of the carbon cycle. You may consult Figure 42.10 for assistance. [p.716]

1. ____   photosynthesis

2. ____   aerobic respiration

3. ____   diffusion between the ocean and the atmosphere

4. ____   marine organisms are involved

5. ____   sedimentation

6. ____   burning fossil fuels

Short Answer

7. Describe the relationship between greenhouse gases and global warming. [pp.716-717].

_____

_____

_____

_____

_____

42.9 THE NITROGEN CYCLE [p.718]
42.10 THE PHOSPHOROUS CYCLE [p.719]

Selected Terms

nitrogen gas [p.718], ammonia [p.718], ammonium [p.718], hydrogen ions [p.718], acid rain [p.718], nitrogen-fixing [p.718], cyanobacteria [p.718], nodules [p.718], legumes [p.718], fungal decomposition [p.718], ammonia-oxidizing bacteria [p.718], archaean [p.718], nitrites & nitrates [p.718], ammonia fertilizer [p.718], nitrous oxide [p.718], phosphorus [p.719], weathering [p.719], erosion [p.719], phosphate [p.719], limiting factor [p.719], phosphate-rich dropping [p.719]

Boldfaced Terms

[p.718] nitrogen cycle _____

_____

[p.718] nitrogen fixation_____

_____

[p.718] nitrification_____

_____

[p.718] denitrification _____

_____

[p.719] phosphorous cycle _____

_____

[p.719] sedimentary cycle _____

_____

Choice

For each of the following terms or statements, choose the appropriate cycle from the list below.

    a. nitrogen cycle [pp.748-749]
    b. phosphorous cycle [pp.750-751]
    c. both the nitrogen and phosphorous cycles

1. ____ Earth's crust is the largest reservoir

2. ____ Eutrophication may change this nutrient's concentration in an ecosystem

3. ____ The atmosphere is the largest reservoir

4. ____ *Rhizobium* bacteria are involved in fixing this for use by organisms

5. ____ Deforestation and conversion to grassland cause losses

6. ____ Use of fertilizers disrupts normal cycles

7: ____ Ammonium is one form

8. ____ Moved by hydrologic cycle through ecosystems

SELF QUIZ

1. Which of the following breaks down organic material into inorganic building blocks? [p.710]
   a. herbivores
   b. detritivores
   c. decomposers
   d. omnivores
   e. none of the above

2. In an ecosystem, energy _____ the ecosystem while nutrients _____ the ecosystem. [p.710]
   a. flows through; flow through
   b. flows through; cycle within
   c. cycle within; flow through
   d. cycle within; cycle within

3. The members of feeding relationships are structured in a hierarchy, the steps of which are called _____. [p.711]
   a. predator-prey relationships
   b. trophic levels
   c. feeding groups
   d. functional groups

4. Eutrophication is caused by? [p.709]
   a. to much nitrogen being added to aquatic ecosystems
   b. to much phosphorus being added to aquatic ecosystems
   c. to much carbon dioxide being added to the atmosphere
   d. both a and b choices
   e. choice a, b, and c

5. In a _____ food web, energy flows from producers to herbivores, then carnivores and decomposers. [p.712]
   a. primary
   b. secondary
   c. detrital
   d. grazing
   e. atmospheric

6. Which of the following does not typically have a major reservoir in the atmosphere? [p.719]
   a. nitrogen
   b. carbon
   c. water
   d. phosphorous

7. The amount of organic materials in the bodies of all of the organisms in an ecosystem is called the _____. [p.713]
   a. biomass
   b. biototal
   c. gross productivity
   d. net energy

8. Global warming is directly associated with which of the following cycles? [pp.716-717]
   a. phosphorous
   b. water
   c. carbon
   d. nitrogen

9. Which of the following makes up the majority of the atmosphere? [p.718]
   a. carbon dioxide
   b. water
   c. nitrogen
   d. oxygen

10. Which of the following captures solar energy for an ecosystem? [p.710]
    a. detritivores
    b. decomposers
    c. primary producers
    d. herbivores
    e. parasites

CHAPTER OBJECTIVES/REVIEW QUESTIONS

1. Define eutrophication. [p.709]

2. Be able to define the term ecosystem. [p.710]

3. Understand the role of producers, consumers, detritivores and decomposers in an ecosystem. [p.710]

4. Understand the concept of a trophic level. [p.711]

5. Distinguish between a food chain and a food web. [pp.711-712]

6. Distinguish between a detrital and grazing food web. [p.712]

7. Understand the relationship between productivity and biomass with respect pyramid models. [p.713]

8. Understand the principles of the biogeochemical cycles. [p.714]

9. Be able to provide recognize terms associated with the water cycle. [pp.714-715]

10. Understand the major components of the carbon cycle. [pp.716-717]

11. Be able to define a greenhouse gas. [p.717]

12. Recognize the relationship between greenhouse gases and global warming. [pp.716-717]

13. Understand the terminology and major steps of the nitrogen cycle. [p.718]

14. Understand the major steps of the phosphorous cycle. [p.719]

CHAPTER SUMMARY

An (1) _____ consists of a community and its physical environment. A one-way flow of (2) _____ and a cycling of raw materials among its interacting participants maintain it. It is an open system, with inputs and outputs of energy and nutrients.

Food (3) _____ are linear sequences of feeding relationships, from producers through consumers, decomposers and (4) _____. The chains cross-connect as food (5) _____. Most of the energy that enters a food web returns to the environment, mainly as metabolic heat. Most the nutrients are cycled, some reenter the environment.

(6) _____ is the the enrichment of aquatic ecosystems by the addition of nitrogen and phosphorus, mainly by human activities.

Primary (7) _____ is the rate at which an ecosystem's producers capture and store energy in their tissues during a given interval. The number of producers and the balance between (8) _____ and aerobic respiration influence the amount stored.

The availability of water, (9) _____, (10) _____, phosphorous and other substances influences primary productivity. Ions or molecules of these susbstances move slowly in global cycles, from environmental (11) _____, into food webs, then back to reservoirs. (12) _____

activities can disrupt these cycles.

INTEGRATING AND APPLYING KEY CONCEPTS

1. In 1971, *Diet for a Small Planet* was published. Frances Moore Lappé, the author, felt that people in the United States of America wasted protein and ate too much meat. She said, "We have created a national consumption pattern in which the majority, who can pay, overconsume the most inefficient livestock products [cattle] well beyond their biological needs (even to the point of jeopardizing their health), while the minority, who cannot pay, are inadequately fed, even to the point of malnutrition." Cases of marasmus (a nutritional disease caused by prolonged lack of food calories) and kwashiorkor (caused by severe, long-term protein deficiency) have been found in Nashville, Tennessee, and on an Indian reservation in Arizona, respectively. Lappé's partial solution to the problem was to encourage people to get as much of their protein as possible directly from plants and to supplement that with less meat from the more efficient converters of grain to protein (chickens, turkeys, and hogs) and with seafood and dairy products. Most of us realize that feeding the hungry people of the world is not just a matter of distributing the abundance that exists—it is also a matter of political, economic, and cultural factors. Yet, it is still valuable to consider applying Lappé's idea to our everyday living. Devise two full days of breakfasts, lunches, and dinners that would enable you to exploit the lowest acceptable trophic levels to sustain yourself healthfully.

2. What would be the easiest solution to reducing climate change? Is this idea practical? Why? Why not.

3. If you were on a governmental advisory panel and had to make recommendations about reducing carbon emissions, what would be on your list?

43

THE BIOSPHERE

INTRODUCTION

This chapter takes a look at the biosphere, or Earth. The chapter begins with a discussion of climate and then introduces how climate relates to the major ecosystems, both terrestrial and aquatic, on the planet. The introduction of this chapter is about El Nino and how it affects the climate across the world. The biosphere concerns all of us as this is where our species lives with all the other species on Earth. Anything that we do that affects another species, ultimately affects us.

FOCAL POINTS

- Figure 43.3 [p.724] illustrates the intensity of the solar radiation on the Earth.
- Figure 43.4 [p.725] illustrates global air circulation patterns.
- Figure 43.5 [p.726] shows major climate zones correlated with ocean currents.
- Figure 43.6 [p.727] illustrates the rain shadow effect.
- Figure 43.8 [p.728] shows the global distribution of major categories of biomes and marine ecoregions.
- Figure 43.24 [p.738] shows lake zonation.
- Figure 43.26 [p.739] shows a lake in all four seasons.
- Figure 43.32 [p.742] diagrams the major oceanic zones.

INTERACTIVE EXERCISES

43.1 EL NINO EFFECTS [p.723]
43.2 AIR CIRCULATION PATTERNS [pp.724-725]
42.3 THE OCEAN, LANDFORMS, AND CLIMATES [pp.726-727]

Selected Terms
climate event [p.723], lithosphere [p.726], atmosphere [p.723], equatorial waters [p.723], anchovies [p.723], torrential rains [p.723], flooding [p.723], landslides [p.723], hurricanes [p.723], temperature, humidity, & windspeed [p.724], elliptical path [p.724], Northern Hemisphere [p.724], Southern Hemisphere [p.724], equinoxes [p.724], greenhouse gases [p.724], global air circulation [p.724], poles [p.725], latitude [p.725], latitudinal variation [p.726], topography [p.726], leeward [p.726], Gulf Stream

Boldfaced Terms
[p.723] biosphere _____

_____

[p.723] El Nino _____

_____

[p.723] El Nina _____

_____

[p.725] climate _____

_____

[p.726] rain shadow _____

_____

[p.727] monsoons _____

_____

Matching

Match each of the following terms to its correct definition. [pp.723-727]

1. ____ biosphere
2. ____ climate
3. ____ atmosphere
4. ____ leeward
5. ____ ozone layer
6. ____ hydrosphere
7. ____ topography
8. ____ El Nino
9. ____ La Nina
10. ____ rain shadow
11. ____ monsoons

a. gases and airborne particles that envelop the Earth
b. average humidity, cloud cover, temperature, and wind speed, over time
c. periodic warming of the equatorial Pacific waters
d. the influence of mountains on precipitation patterns
e. a thin layer of the atmosphere that contains high levels of O_3
f. periodic cooling of equatorial Pacific waters
g. the ocean, ice caps, and other bodies of water, liquid or frozen
h. an alternation of dry and wet seasons, producing periods of heavy rainfall
i. the lay of the lay, valleys, mountains, etc
j. the sum of all places on Earth where life is found
k. the side of a mountain facing away from the wind

Labeling

Label each of the indicated portions of the diagram below. [pp.724-725]

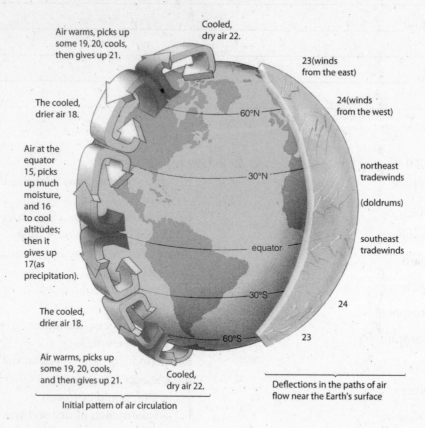

Air warms, picks up some 19, 20, cools, then gives up 21.

Cooled, dry air 22.

23(winds from the east)

24(winds from the west)

The cooled, drier air 18.

60°N

Air at the equator 15, picks up much moisture, and 16 to cool altitudes; then it gives up 17(as precipitation).

30°N

northeast tradewinds

(doldrums)

equator

southeast tradewinds

30°S

The cooled, drier air 18.

24

Air warms, picks up some 19, 20, cools, and then gives up 21.

60°S

Cooled, dry air 22.

23

Initial pattern of air circulation

Deflections in the paths of air flow near the Earth's surface

15. _____	20. _____
16. _____	21. _____
17. _____	22. _____
18. _____	23. _____
19. _____	24. _____

43.4 BIOMES [pp.728-729]
43.5 DESERTS [pp.730-731
43.6 GRASSLANDS [p.732]
43.7 DRY SHRUBLANDS AND WOODLANDS [p.733]

Selected Terms

North America prairie [p.728], African veld [p.728], South American pampa [p.728], Eurasian steppe [p.728[, deserts [p.728], tundra [p.728], cactus [p.729], euphorbs [p.729], C4 photosynethesis [p.729], C3 photosyntheisis [p.729], Chile's Atacama desert [p.730], China's Gobi [p.730], CAM plant [p.731], crust community [p.731], cyanobacteria [p.731], lichen [p.731], mosses [p.731], fungi [p.731], periodic fires [p.732], shortgrass & tallgrass prairie [p.732], savannas [p.732], shrublands [p.733], dry woodlands [p.733]

Boldfaced Terms

 [p.728] biomes _____

[p.730] desert _____

_____

[p.732} grasslands _____

_____

[p.733] chaparral _____

_____

Matching

Match each of the following terms with the correct definition. [pp.728-731]

1. _____ deserts

2. _____ tundra

3. _____ crust community

4. _____ biomes

a. occur where temperatures soar to the highest on the planet.

b. occur where temperatures drop to the lowest on the planet.

c. includes cyanobacteria, lichens, mosses, and fungi.

d. a discontinuous region characterized by its climate and dominant vegetation.

Choice

For each statement, choose the correct biome. Some answers may be used more than once. [pp.732-733]

a. savannahs

b. shortgrass and tallgrass prairies

c. dry shrublands

d. dry woodlands

e. grasslands

5. _____ precipitation is between 25-60 cm per year.

6. _____ precipitation is between 40 – 100 cm per year.

7. _____ tall trees, but not a continuous canopy

8. _____ warm summers and cold winters, not enough precipitation to support forests

9. _____ broad belts of grasslands with few shrubs and trees.

10. _____ vulnerable to lightening-sparked, wind-driven firestorms

11. _____ a form of grassland in North America, now largely destroyed

43.8 BROADLEAF FORESTS [pp.734-735]
43.9 CONIFEROUS FORESTS [p.736]
43.10 TUNDRA [p.737]

Selected Terms

semi-evergreen forests [p.734], broadleaf trees [p.734], angiosperm [p.734], tropical deciduous forests [p.734], deciduous[p.734], O horizon, A horizon, B horizon & C horizon [p.734], deforestation [p.735], chemotherapy [p.735], vincristine & vinblastine [p.735], rosy periwinkle [p.735], coniferous forests [p.736], evergreen trees [p.736], seed-bearing cones [p.736], periodic fires [p.736], pole ice cap [p.737]

Boldfaced Terms

[p.734] temperate deciduous forests _____

_____

[p.734] topical rain forests _____

_____

[p.736] boreal forests _____

_____

[p.737] arctic tundra _____

_____

[p.737] permafrost _____

_____

[p.737] alpine tundra _____

_____

Matching

Match each of the following to its correct definition. [pp.734-737]

1. ____ evergreen broadleaf forests

2. ____ semi-evergreen forests

3. ____ coniferous forests

4. ____ A-E horizons

5. ____ arctic tundra

6. ____ alpine tundra

7. ____ permafrost

8. ____ O horizon

9. ____ tropical deciduous forests

10. ____ temperate deciduous forests

a. trees dormant in winter; rainfall between 50 to 150 cm per year

b. the top layer of soil in a temperate deciduous forest.

c. rainfall between 130 to 200 cm per year, average temperature of 25ºC

d. longer dry seasons than evergreen broadleaf forests; slow decomposition

e. a frozen layer with anaerobic growing conditions

f. dry and cold, but occurs in the high mountains

g. dry and cold for most of year, rainfall less than 25 cm per year

h. found just below the sparse litter layer in tropical rain forests

i. dominated by evergreen trees with needle-shaped cones

j. trees that shed leaves at the start of the dry season once a year

43.11 FRESHWATER ECOSYSTEMS [pp.738-739]
43.12 COASTAL ECOSYSTEMS [p.740]
43.13 CORAL REEFS [p.741]
43.14 THE OPEN OCEAN [pp.742-743]

Selected Terms

fresh water [p.738], lake [p.738], phytoplankton [p.738], detritivores [p.738], decomposers [p.738], primary productivity [p.738], aquatic plants [p.738], profundal zone [p.738], primary producers [p.738], limnetic zone [p.738], littoral zone [p.738], zooplankton [p.738], producers [p.738], diatoms [p.738], green algae [p.738], succession [p.738], eutrophic [p.738] eutrophication [p.738], oligotrophic [p.738], spring overturn [p.738], thermocline [p.738], fall overturn [p.738], ecosystems [p.739], solute concentration [p.738], *Tubifex* [p.739], dissolved oxygen [p.739], leeches [p.739], hemoglobin [p.739], detrital food chains [p.740], marine nurseries [p.740], mangroves [p40], mangrove wetland [p.740], *upper littoral zone* [p.740], *midlittoral zone* [p.740], *lower littoral zone* [p.740], rocky shores [p.740], grazing food chains [p.740], tidal flats [p.740], sandy shores [p.740], marsh grass [p.740], *Spartina* [p.740], intertidal zone [p.740], seawater [p.740], calcium carbonate [p.741], Australia's Great barrier ref [p.741], biological architecture [p.741], photosynthetic dinoflagellates [p.741], sodium cyanide [p.741], exotic algae [p.741], staghorn coral [p.741], *Acropora* [p.741], continental shelve [p.742], species richness [p.742], trawling [p.742], flytrap anemone [p.742], superheated water [p.743], tectonic plates [p.743], chemoautotrophic bacteria [p.743]

Boldfaced Terms

[p.740] estuary _____

_____

[p.741] coral reefs _____

_____

[p.741] coral bleaching _____

_____

[p.742] pelagic province _____

_____

[p.742] benthic province _____

_____

[p.742] seamounts _____

_____

[p.743] hydrothermal vents _____

_____

Matching

Match each of the following terms to the correct definition. [pp.738-743]

1. _____ fall overturn
2. _____ eutrophication
3. _____ phytoplankton
4. _____ zooplankton
5. _____ estuaries
6. _____ spring overturn
7. _____ seamounts

a. tiny consumers such as copepod .
b. a natural or artificial process that enriches a body of water with nutrients.
c. the action of winds move dissolved oxygen into the depths of a lake
d. a group of photosynthetic microorganisms including green algae, diatoms, and cyanobacteria.
e. undersea mountains.
f. partially enclosed areas where sea water mixes with nutrient-rich fresh water.
g. as lake water cools in the fall, it becomes more dense and sinks

Choice

For the following questions, choose from the following answers. Some answers may be used more than once. [pp.738-743]

a. open ocean
b. lake ecosystems
c. wetlands and intertidal zones
d. coral reefs
e. coastlines

8. _____ may be either oligotrophic or eutrophic

9. _____ in response to environmental stress, the symbiotic relationships may break

10. _____ divided into pelagic and benthic provinces

11. _____ consist of estuaries with complex food webs

12. _____ these consist of three zones, the upper, middle, and lower littoral zones

13. _____ nutrients may be cycled by spring and fall overturn

14. _____ represents the accumulated remains of marine organisms

15. _____ seamounts are located within this body of water

Labeling

Label the indicated areas of the ocean in the diagram below. [p.778]

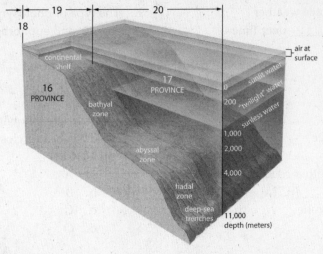

16. _____

17. _____

18. _____

19. _____

20. _____

SELF QUIZ

1. Tropical rain forests belong to the _____ biome. [pp.734-735]
   a. evergreen broadleaf
   b. deciduous broadleaf
   c. coniferous
   d. taigas

2. Permafrost would be found in which of the following? [p.737]
   a. taiga
   b. artic tundra
   c. alpine tundra
   d. all of the above

3. Which one of these has the lowest amount of annual rainfall? [pp.732-733]
   a. grassland
   b. dry woodland
   c. dry shrubland

4. A rain shadow is associated with _____ [p.726]
   a. eutrophication
   b. acid rain
   c. mountains
   d. El Nino
   e. La Nino

5. This aquatic ecosystem starts as a seep. [p.739]
   a. lakes
   b. wetlands
   c. oceans
   d. streams
   e. all of the above

6. The season of the year when a thermocline can exist in a lake is _____. [p.739]
   a. spring
   b. summer
   c. fall
   d. winter

7. Climate refers to average weather conditions, such as _____ over time. [p.724]
   a. cloud cover
   b. temperature
   c. humidity
   d. wind speed
   e. all of the above

8. The average of all weather conditions for a region is called the _____ [p.724]
   a. temperature zone
   b. rain shadow
   c. climate
   d. biome

9. The type of photosynthesis used by cacti, agaves, and euphorbs is _____. [pp.730-731]
   a. CAM
   b. C3
   c. C4

10. Which of the following soil horizons is typically located closest to the surface? [pp.732-735]
    a. A
    b. B
    c. C
    d. O

CHAPTER OBJECTIVES/REVIEW QUESTIONS

1. Explain the effects of El Nino. [p.723]

2. Explain how climate is related to the heating and movement of the atmosphere. [pp.724-725]

3. Understand the concept of an ocean current. [pp.726-727]

4. Describe the relationship between mountains and rainfall patterns. [p.727]

5. Explain biome. [pp.728-729]

6. Understand C3, C4, and CAM photosynthesis with respect of which biomes these are found within. [p.729-731

7. Recognize what is meant by a soil profile. [pp.730-735]

8. List the key characteristics of deserts, dry shrublands, dry woodlands, and grasslands. [pp.7732-733]

9. Distinguish between evergreen broadleaf forests, deciduous broadleaf forests and coniferous forests. [pp.734-736]

10. Understand the concept of permafrost and its relationship to the types of tundra. [p.737]

11. Define a lake and stream ecosystem. [pp.738-739]

12. Explain why eutrophication is dangerous for a lake ecosystem. [p.738]

13. Give the key characteristics of a wetland, intertidal zone, estuary, coastline, coral reef, and open ocean. [pp.740-743]

CHAPTER SUMMARY

Air circulation patterns start with (1) _____ differences in (2) _____ inputs from the sun. Earth's rotation and orbit, and the (3) _____ of land and seas. These factors give rise to great weather systems and regional (4) _____.

Interactions among (5) _____ currents, air (6) _____ patterns, and landforms produce regional climates, which affect the distribution and dominant features of (7) _____.

(8) _____ realms are vast regions characterized by species that evolved nowhere else. They are subdivided into (9) _____, or regions characterized mainly by the dominant vegetation. Sunlight intensity, moisture, (10) _____ type, species interactions, and evolutionary history vary within and between biomes.

Water provinces cover more than (11) _____ percent of the Earth's surface. All freshwater and marine ecosystems have gradients in (12) _____ availability, temperature, and dissolved (13) _____ that vary daily and seasonally. The variations influence primary (14) _____.

Understanding interactions among the atmosphere, (15) _____ and land can lead to discoveries about specific events.

INTEGRATING AND APPLYING KEY CONCEPTS

1. One species, *Homo sapiens,* uses about 40 percent of all of Earth's productivity, and its representatives have invaded every biome, either by living there or by dumping waste products there. Many of Earth's residents are being denied the minimal resources they need to survive, while human populations continue to increase exponentially. Can you suggest a better way of keeping Earth's biomes healthy while providing at least the minimal needs of all Earth's residents (not just humans)? If so, outline the requirements of such a system and devise a way in which it could be established.

2. Evidence is mounting that we are experiencing a period of global climate change that is frequently called "global warming". Choose the type of ecosystem that you live in and suggest what would occur if temperature were to increase? What type of ecosystem would your be turning into?

44

HUMAN EFFECTS ON THE BIOSPHERE

INTRODUCTION

The last chapter of the text examines the human influences and impacts on our planet Earth. Our species is responsible for a tremendous amount of change that has occurred to biodiversity through our life styles and practices. We have influenced the waters, fresh and marine, the air, and the soil. Since we have created these problems it is up to us to solve them. Your education through this class at the college or university you are attending will provide ideas and suggestions about a brighter future for the biodiversity on Earth. Yet you as a responsible human will make the choices to either help the planet or harm the planet. Throughout your study of this course biological connection have been made. Now you must realize that everything is in fact connected. Any organism that goes extinct because of human activity affects the other species it interacted with as these species affect the ones they interact with. Consider the line of dominoes and what happens if one is tipped into another. Consider your impacts on Earth daily and influence others to do the same.

FOCAL POINTS

- Figure 44.5 [p.751] shows a map of the United States and average precipitation acidities.
- Figure 44.6 [p.752] shows DDT residues in various organisms demonstrating biological magnification.
- Figure 44.8 [p.754] illustrates the concentration of CFCs in the upper atmosphere.
- Figure 44.9 [p.755] shows a comparison of changing global average temperatures and atmospheric carbon dioxide.
- Figure 44.11 [p.756] is a map of the world show location and conservation status of the land ecoregions deemed most important by the World Wildlife Fund.

INTERACTIVE EXERCISES

44.1 A LONG REACH [p.747]
44.2 A GLOBAL EXTINCTION CRISIS [pp.748-749]
44.3 HARMFUL LAND USE PRACTICES [p.750]

Selected Words
human explores [p.747], human influences [p.747], Earth's atmosphere [p.747], sea ice [p.747], top predators [p.747], mercury & organic pesticides [p.747], pollutants [p.747], Proterozoic [p.747], photosynthetic cells [p.747], extinction [p.748], speciation [p.748], mass extinction [p.748], asteroid impact [p.748], passenger pigeons [p.748], overharvesting species [p.748], Atlantic codfish & white abalone [p.748], captive breeding [p.748], altering their habitat [p.748], giant panda [p.748], ivory-billed woodpecker [p.748], harlequin frog [p.748], gorillas [p.748], *Rafflesia* [p.748], Texas blind salamander [p.748], aquifer [p.748], sea turtles [p.748], manatees [p.748], predators [p.748], koa bug [p.749], parasitoid fly [p.749], ammonia-oxidizing bacteria [p.749], wind erosion [p.750], positive feedback [p.750], drought [p.750], waterlogged soil [p.750], nutrient poor [p.750], deforestation [p.750]

Boldfaced Words

[p.748] endangered species _____

_____

[p.748] threatened species _____

_____

[p.748] endemic species _____

_____

[p.748] desertification_____

_____

Matching

Choose the most appropriate answer for each term. [pp.748-750]

1. ____ endangered species

2. ____ threatened species

3. ____ endemic species

a. a species that is confined to the limited area in which it evolved

b. a species that faces extinction in all or part of its range

c. a species that is likely to become endangered in the near future

Fill in the Blank [p.748-750]

(4) _____ and (5) _____ are natural processes. A species is considered (6) _____ when one or more of its populations have declined or are declining. The decline of the Atlantic codfish and the white abalone is a direct result of (7) _____ species. An example of an endemic species is the giant (8) _____ of China. Ammonia-oxidizing bacteria convert ammonia to (9) _____. (10) _____ is the conversion of grasslands or woodland into a desert.

44.4 ACID RAIN [p.751]
44.5 BIOLOGICAL EFFECTS OF CHEMICAL POLLUTANTS [p.752]
44.6 THE TROUBLE WITH TRASH [p.753]
44.7 OZONE DEPLETION AND POLLUTION [p.754]
44.8 GLOBAL CLIMATE CHANGE [p.755]

Selected Words

sulfur & nitrogen oxides [p.751], wet acid deposition [p.751], pH [p.751], aluminum [p.751], pathogen [p.751], phytoremediation [p.752], hydrophobic chemical pollutants [p.752], DDT [p.752], osprey [p.752], egg shell formation [p.752], methylmercury [p.752], point source [p.752], nonpoint source [p.752], trash [p.753], engineered landfills [p.753], recycle [p.753], ozone (O_3) [p.754], ultraviolet radiation (UVR) [p.754], Antarctica [p.754], "ozone hole" [p.754], skin cancer [p.754], chlorofluorocarbons (CFCs) [p.754], temperature [p.755], polar latitudes [p.755], equator [p.755], thermal expansion [p.755], greenhouse gas emissions [p.755]

Boldfaced Terms

[p.751] pollutants _____

_____

[p.751] acid rain _____

_____

[p.752] bioaccumulation _____

_____

[p.752] biological magnification _____

_____

[p.754] ozone layer _____

_____

Choice

Choose the appropriate type of communication cue for each statement. [pp.751-755]

    a. acid rain
    b. bioaccumulation
    c. biological magnification
    d. ozone layer
    e. pollutants

1. _____ a natural or man-made substance released into the soil, air, or water in greater than natural amounts.

2. _____ the concentration of a chemical as it moves up the food chain

3. _____ produced when water combines with sulfur and nitrogen oxides

4. _____ tissues of an organism store pollutants taken up from the environment

5. _____ a layer that exist in the upper atmosphere that protects the Earth from UV radiation

Fill in the blank [pp.754-755]

Chlorofluorocarbons are the destroyers of (6) _____ layer. The "ozone hole" exist over (7)

_____. With a thinner ozone layer, humans will be exposed to more (8) _____ which can

lead to skin cancer. Carbon dioxide is a (9) _____ gas. Warming of tropical waters as a result of

climate change is causing coral (10) _____.

44.9 CONSERVATION BIOLOGY [pp.756-757]
44.10 REDUCING NEGATIVE IMPACTS [pp.758-759]

Selected Terms

material wealth [p.756], cultural wealth [p.756], biological wealth [p.756], genetic diversity [p.756], species diversity [p.756], ecosystem diversity [p.756], detoxify [p.756], extinction [p.756], northern spotted owl [p.757], logging [p.757], ecoregion [p.757], tourism [p.757], restoration [p.757], energy flow [p.758], resource limitation [p.758], globalization [p.758], resource extraction [p.758], nonrenewable mineral resources [p.759], recycling [p.759], renewable energy resources [p.759]

Boldfaced Terms

[p.756] biodiversity _____

_____

[p.756] conservation biology _____

_____

[p.756] hot spots _____

_____

[p.756] ecological restoration _____

_____

Matching

Match each of the following terms to its correct definition. [pp 756-757]

1. ____ biodiversity
2. ____ conservation biology
3. ____ hot spots
4. ____ ecological restoration

a. the branch of biology that promotes biodiversity and its conservation
b. the biological wealth of the Earth
c. work designed to bring about the renewal of a natural ecosystem that has been degraded or destroyed.
d. places that are home to species found nowhere else and are under great threat of destruction

Fill in the Blanks [pp.756-759]

On Figure 44.11 the black square on the map legend represents (5) _____ or (6) _____ ecoregions. Looking on the same figure, the large island (just east of Africa) of (7) _____ is entirely black. (8) _____ are areas of land or water characterized by climate, geography, and the species found within them. The health of our planet depends upon our abilities to recognize that the principles of energy (9) _____ and resource (10) _____ apply to all and do not change. Promoting sustainability begins with the recognition of the environmental consequences of one's own (11) _____ style. (12) _____ makes it difficult to know the source of the raw materials in products you purchase. If you want to make a difference, learn about the threats to (13) _____ in your own area.

SELF QUIZ

1. A(n) _____ is a species that faces extinction in all or part of its range.[p.748]
   a. endemic species
   b. threatened species
   c. endangered species
   d. overharvested species

2. A(n) _____ is a species that is likely to become endangered in the near future. [p.748]
   a. endemic species
   b. threatened species
   c. overharvested species

3. Desertification effects which of the following? [p.766]
   a. grasslands
   b. intertidal zones
   c. wetlands
   d. broadleaf deciduous forests

4. Which gases combine with water to produce acid rain? [p.751]
   a. sulfur oxides
   b. nitrogen oxides
   c. carbon oxides
   d. a and b
   e. a, b, and c

5. _____ is the process in which an organism's tissues store a pollutant taken up from the environment. [p.752]
   a. biomagnification
   b. bioaccumulation

6. _____ is the process in which the concentration of a chemical increases as the pollutant moves up a food chain. [p.752]
   a. biomagnification
   b. bioaccumulation

7. CFCs are directly related to the severity of _____ [p.754]
   a. global warming
   b. ozone thinning
   c. eutrophication
   d. thermal inversions
   e. acid rain

8. Which of the following is a greenhouse gas?
   a. sulfur oxides
   b. nitrogen oxides
   c. carbon dioxide

Fill in the blank

9. The biological wealth of the Earth is referred to as _____. [p.756]

10. Biologist have identified _____, places that are home to species found nowhere else and are under great threat of destruction.

11. Ecological _____ is work designed to bring about the renewal of a natural ecosystem that has been degraded or destroyed fully or in part.

CHAPTER OBJECTIVES/REVIEW QUESTIONS

1. Give examples of how humans have a "long reach" with respect to their impacts and influences on Earth. [p.747]

2. Define endangered, threatened, and endemic species [p.748]

3. Give examples of plants and animals that have lost habitats because of humans. [p.748]

4. Give examples of animals that have been overharvested by humans. [p.748]

5. Understand the sad history of the extinction of the passenger pigeon. [p.748]

6. Give examples of exotic species and where they were introduced. [pp.748-749]

7. Explain how the koa bug became a victim of a biological control program gone awry. [p.749]

8. Explain why other life forms (other than vertebrates and plants) need and should to be considered when we discuss loss of biodiversity. [p.749]

9. Define desertification and give an example. [p.750]

10. Explain why deforestation becomes a problem, especially in areas that receive lots of rain. [p.750]

11. What is acid rain? What causes it? Why is it harmful to plants? [p.751]

12. Explain the difference between bioaccumulation and biological magnification. [p.752]

13. Explain how DDT affected ospreys. [p.752]

14. Explain the difference in point and nonpoint sources with respect pollutants. [p.752]

15. What are the troubles caused by trash? What can we do to reduce these problems? [p.753]

16. Where is the ozone layer/ Why is it important to Earth? [p.754]

17. Where is the ozone hole? [p.754]

18. What is a CFC? [p.754]

19. What is the scientific evidence for global warming? Is there a relationship between carbon dioxide and global warming? [p.755]

20. What is a greenhouse gas? What is a fossil fuel? [p.755]

21. What is biodiversity and why is it important? [p.756]

22. What is biological conservation? [p.756]

23. What is a hot spot? Where is a hot spot on Earth? [p.756]

24. What is ecological restoration? Give an example of ecological restoration. [p.757]

25. What is the difference between renewal and nonrenewal resources? Give examples of each? [pp.758-759]

26. List ways that YOU can reduce YOUR negative impacts on the Earth. [pp.758-759]

CHAPTER SUMMARY

Extinctions, like (1) _____, are a natural process. The rate of extinction picks up dramatically during a (2) _____ extinction. The crash of the Atlantic codfish was a direct result of (3) _____ a species. Human activities that degrade, rather than destroy, habitats also endanger (4) _____. Deliberate or accidental species (5) _____ threatened species. The decline or loss of one species can (6) _____ other species.

Endangered species listings have historically focused on (7) _____. Deserts naturally expand and contract over times as (8) _____ conditions vary. Like desertification, (9)

_____ affects local weather. Forests take up and store huge amounts of (10) _____. (11) _____ are natural or human made substances released into the soil, air, or water in greater amounts than natural amounts. Sulfur and nitrogen (12) _____ are examples of common air pollutants. Acid rain that falls on or drains into water catchments affect (13) _____ organisms.

The ability for some plants to (14) _____ toxic substances makes them useful in phytoremediation of polluted soil. By the process of biological (15) _____ the concentration of a chemical increases as the pollutant moves up the food chain. Pollution from a (16) _____ source is much more challenging to deal with. Once in the ocean, (17) _____ can persist for a long time. Between 17 to 27 kilometers above the Earth, is a layer called the (18) _____ layer.

CFCs are the main (19) _____ of ozone. Global warming is more pronounced at temperate and (20) _____ latitudes than at the equator. All living species are the result of an ongoing (21) _____ process that stretches back billions of years. Promoting sustainability begins with recognizing the (22) _____ consequences of one's own lifestyle.

INTEGRATING AND APPLYING KEY CONCEPTS

1. Have you ever planted a tree? How do you feel now that you know the connection between photosynthesis and global warming and climate change? Whould you plant another?

2. Make a list of recyclables items that you can recycle if you choose.

3. Can one person make a difference? If so, what difference can you make to help Earth be a better place?

4. Make a list all the thing that you throw away in one day? What was the most common item?

5. List as many ways as you can to reduce our society's use of fossil fuels.

ANSWERS

Chapter 1: Invitation to Biology
1.1 The Secret Life of Earth [p.3]
1.2 The Science of Nature [pp.4-5]
1. g; 2. d; 3. a; 4. c; 5. i; 6. f; 7. b; 8. e; 9. k; 10. l; 11. h; 12. molecules; 13. organs; 14. multicelled organism; 15. population; 16. community; 17. ecosystem
1.3 How Living Things Are Alike [pp.6-7]
1.4 How Living Things Differ [pp.8-9]
1. f; 2. h; 3. I; 4. c; 5. j; 6. e; 7. d; 8. b; 9. a; 10. g; 11. a; 12. b; 13. a; 14. b; 15. c; 16. a; 17. f; 18. b; 19. c; 20. e; 21. d;
1.5 Organizing Information About Species [pp.11-12]
1. d; 2. a; 3. e; 4. b; 5. c; 6. b; 7. c; 8. a; 9. 3; 10. 6; 11. 1; 12. 1; 13. 4
1.6 The Nature of Science [pp.12-13]
1. h; 2. a; 3. f; 4. b; 5. g; 6. d; 7. e; 8. c; 9. m; 10. i; 11.n; 12. k; 13. j; 14. l; 15. observation; 16. hypothesis; 17. prediction; 18. experiments; 19. results; 20. scientific community
1.7 Examples of Biology Experiments [pp.14-15]
1.8 Asking Useful Questions [pp.16-17]
1.9 Philosophy of Science [pp.18-19]
1. b; 2. e; 3. d; 4. a; 5. c; 6. b; 7. c; 8. d; 9. e; 10.b; 11. d; 12. a; 13. c 14. scientific theory; 15. law of nature; 16. sampling error; 17. probability; 18. statistically significant
Self-Quiz
1. b; 2. c; 3. a; 4. c; 5. c; 6. d; 7. a; 8. b; 9. c; 10. a
Chapter Summary
1. organization; 2. cells; 3. unity; 4. cells; 5. environment; 6. DNA; 7. reproduction; 8. species; 9. species; 10. epithet; 11. observations

Chapter 2: Life's Chemical Basis
2.1 Mercury's Rising [p.23]
2.2 Start With Atoms [pp.24-25]
1. b; 2. j; 3. i; 4.a; 5. k; 6. g; 7. e; 8. m; 9. h; 10. f; 11. l; 12. c; 13. d
2.3 Why Electrons Matter [pp.26-27]
2.4 Why Atoms Interact? [pp.28-29]
1. c; 2. e; 3. b; 4. d; 5. a; 6. sodium; 7. chlorine; 8. carbon; 9. oxygen; 10. helium; 11. d; 12. c; 13. a; 14. c; 15. a; 16. b; 17. a; 18. b
2.5 Water's Life-Giving Properties [pp.30-31]
2.6 Acids and Bases [pp.32-33]
1. hydrogen; 2. hydrophobic; 3. hydrophilic; 4. sphere of hydration; 5. temperature; 6. freezing; 7. evaporation; 8. c; 9. e; 10. g; 11. j; 12. a; 13. h; 14. i; 15. f; 16. b; 17. d; 18. 8, base; 19. 2, acid; 20. 10, base; 21. 5, acid; 22. 1, acid; 23. 5, acid
Self Quiz
1. b; 2. d; 3. b; 4. d; 5. b; 6. d; 7. a; 8. d; 9. c; 10. c
Chapter Summary
1. atoms; 2. neutrons; 3. protons; 4. isotopes; 5. electrons; 6. sharing; 7. ionic; 8. hydrogen; 9. water; 10. cohesion; 11. ions

Chapter 3: Molecules of Life
3.1 Fear of Frying? [p.37]
3.2 Molecules of Life – From Structure to Function [pp.38-39]
1. methyl; 2. hydroxyl; 3. carbonyl (ketone); 4. amino; 5. phosphate; 6. carboxyl; 7. carbonyl (aldehyde); 8. Condensation 9. Hydrolysis 10. hydrocarbon; 11. functional; 12. Organic; 13. polymers; 14. metabolism 15. a. hydrolysis, b. condensation
3.3 Carbohydrates [pp.40-41]
1. lactose (oligosaccharide); 2. chitin (polysaccharide); 3. glucose (monosaccharide); 4. cellulose (polysaccharide); 5. starch (polysaccharide); 6. sucrose (oligosaccharide); 7. monosaccharide; 8. monosaccharide; 9. glycogen (polysaccharide)
3.4 Lipids [pp.42-43]
1. b; 2. d; 3. a; 4. c; 5. a; 6. c; 7. a; 8. c; 9. b; 10. c; 11. a; 12. c
3.5 Proteins – Diversity in Structure and Function [pp.44-45]
3.6 The Importance of Protein Structure [p.46]
1. h; 2. f; 3. e; 4. d; 5. b; 6. l; 7. m; 8. i; 9. c; 10. j; 11. a; 12. k; 13. g
3.7 Nucleic Acids [p.47]
1. ATP; 2. coenzymes; 3. nucleic acids; 4. DNA; 5. RNA
Self Quiz
1. b; 2. b; 3. a; 4. d; 5. c; 6. c; 7. d; 8. a; 9. a; 10. b; 11. b; 12. c; 13. a; 14. a; 15. d

Chapter Summary

1. proteins; 2. fatty acids; 3. carbon; 4. functional; 5. carbohydrates; 6. energy; 7. complex; 8. reservoirs; 9. membranes; 10. proteins; 11. enzymes; 12. nucleic; 13. DNA; 14. proteins

---

Chapter 4: Cell Structure

4.1 Food for Thought [p.51]
4.2 What, Exactly Is A Cell [pp.52-53]
4.3 Spying on Cells [pp.54-55]

1. h; 2. a; 3. i; 4.d; 5. e; 6.f; 7. c; 8. g; 9. b; 10. Surface-to-volume ratio constrains cell size due to the fact that past a certain point of growth the surface area will not be sufficient to allow the movement of nutrients and wastes. As a cell grows the surface area increases with the square of its diameter, whereas the volume increases with the cube of its diameter; 11. Every organism consists of one or more cells. The cell is the smallest unit that has all the properties of life. Each new cell arises from a preexisting cell; 12. a; 13. c; 14. b; 15. c; 16. d; 17.e

4.4 Membrane Structure and Function [pp.56-57]

4.5 Introducing Bacteria and Archaeans [pp.58-59]

1. f; 2. h; 3. g; 4. b; 5. e; 6. d; 7. c; 8. a; 9. flagellum; 10. pilus; 11. capsule; 12. cell wall; 13. plasma membrane; 14. cytoplasm; 15. polypeptide; 16. plasmid; 17. flagellum; 18. wall; 19. biofilm

4.6 Introducing Eukaryotic Cells [p.60]
4.7 The Nucleus [p.61]

4.8 The Endomembrane System [pp.62-63]

1. b; 2. d; 3. f; 4. g; 5. h; 6. e; 7. c; 8. a; 9.a. nucleolus, b. nuclear envelope, c. DNA, d. nucleoplasm, e. nuclear pore; 10. ribosomes; 11. rough; 12. smooth; 13. Golgi body; 14. lipid; 15. vesicles; 16. plasma membrane

4.9 Mitochondria and Plastids [pp.64-65]
4.10 The Dynamic Cytoskeleton [pp.66-67]

1. b; 2. a; 3. a; 4. a; 5. b; 6. a; 7. b; 8. c; 9. c; 10. b; 11. b; 12. a; 13. c; 14. a; 15. b; 16. a; 17. b

4.11 Cell Surface Specializations [pp.68-69]

1. b; 2. d; 3. c; 4. f; 5. a; 6. g; 7. e; 8. h;

4.12 Visual Summary of Eukaryotic Cell Components [p.70]

1. o; 2. i, w; 3.e, s; 4. h, v; 5. n; 6. j, x; 7. c, q; 8. g, u; 9. l, z; 10. f, t; 11. y; 12. a, aa; 13. k; 14. m; 15. b, p; 16. d, r

Self Quiz

1. c; 2. d; 3. e; 4. b; 5. a; 6. d; 7. a; 8. b; 9. d; 10. b

Chapter Summary

1. plasma membrane; 2. cytoplasm; 3. cells; 4. bilayer; 5. poteins; 6. molecules; 7. bacteria; 8. smallest; 9. eukaryotic; 10. nucleus; 11. protein; 12. chromosomes

---

Chapter 5: Ground Rules of Metabolism

5.1 A Toast to Alcohol Dehydrogenase [p.75]
5.2 Energy and the World of Life [pp.76-77]
5.3 Energy in the Molecules of Life [pp.78-79]

1. b; 2. a; 3. a; 4. b; 5. a; 6. b; 7. b; 8. g; 9. b; 10. f; 11. e; 12. d; 13. c; 14. a; 15. releases; 16. ADP; 17. phosphorylation; 18. energy

5.4 How Enzymes Work [pp.80-81]

1. b; 2. c; 3. a; 4. d; 5. c; 16. f; 7. a; 8. e; 9. b; 10. d; 11. e

5.5 Metabolism: Organized, Enzyme-Mediated Reactions [pp.82-83]

1. b; 2. g; 3. e; 4. f; 5. a; 6. d; 7. c

5.6 Movement of Ions and Molecules [pp.84-85]

5.7 Membrane-Crossing Mechanisms [pp.86-87]

1. e; 2. d; 3. b; 4. a; 5. c; 6. pressure, charge, size of the molecule, steepness of the concentration gradient, temperature; 7. passive transport does not expend energy and occurs down the concentration gradient; 8. a; 9. b; 10. a; 11. a; 12. b; 13. c; 14. a; 15. b; 16. d; 17. c; 18. a; 19. b 20. c; 21.a; 22. b 23. hypotonic; 24. isotonic; 25. hypertonic

5.8 Membrane Trafficking [pp.88-89]

1. c; 2. a; 3. b; 4. a; 5. g; 6. d; 7. f; 8. h; 9. b;; 13. T; 14. T; 15. F, hypotonic; 16. F, isotonic; 17. F, hypertonic; 18. T

Self Quiz

1. a; 2. d; 3. d; 4. c; 5. c; 6. b; 7. d; 8. a; 9. d; 10.d

Chapter Summary

1. energy; 2. organization; 3. ATP; 4. enzymes; 5. environmental; 6. metabolic pathways; 7. enzymes; 8. concentration; 9. proteins; 10. solute; 11. metabolism

Chapter 6: Where It Starts - Photosynthesis

6.1 Green Energy [p.93]

6.2 Sunlight as an Energy Source [pp.94-95]

6.3 Exploring the Rainbow [p.96]

1. f; 2. e; 3. a; 4. c; 5. d; 6. g; 7. b; 8. the peaks represent those wavelengths that the pigment absorbs the most

6.4 Overview of Photosynthesis [p.97]

6.5 Light-Dependent Reactions [pp.98-99]

1a.ADP; 1b. ATP; 2a. NADP+; 2b. NADPH; 3a. H_2O; 3b. O_2; 4. energy; 5a. ATP; 5b. ADP; 6a. NADPH; 6b. NAD+; 7a. CO_2; 7b. glucose; 8.H_2O; 9. T; 10. F, light dependent; 11. F, thylakoid membrane; 12. F, stroma; 13. T; 14. light energy; 15. electron transfer chain; 16. H2O; 17. oxygen; 18. ATP; 19. stroma; 20. thylakoid compartment; 21. NADPH

6.6 Energy Flow in Photosynthesis [p.100]

6.7 Light-Independent Reactions: The Sugar Factory [p.101]

1. c; 2. b; 3. b; 4. c; 5. a; 6. b; 7. a; 8. CO2 (e); 9. PGA (c); 10. ATP (g); 11. NADPH (a); 12. PGAL (d): 13. glucose (f): 14. PGAL (b)

6.8 Adaptations: Different Carbon-Fixing Pathways [p.102]

1. c; 2. a; 3. c; 4. c; 5. b; 6. b; 7. a; 8. a; 9. c; 10. c; 11. c; 12. a; 13. b; 14. a; 15. b; 16. The primary cause of global warming is the burning of fossil fuels which releases carbon dioxide into the atmosphere. Carbon dioxide acts as a greenhouse gas that traps heat in the atmosphere.

Self Quiz

1. c; 2. d; 3. c; 4. d; 5. d; 6. e; 7. e; 8. a; 9. d; 10. e

Chapter Summary

1. chlorophylls; 2. glucose; 3. chloroplasts; 4. light; 5. chemical; 6. carbohydrates; 7. photosynthesis; 8. ATP; 9. NADPH; 10. synthesis; 11. carbon dioxide; 12. photosynthesis; 13. electrons; 14. autotrophs; 15. metabolism; 16. global warming

Chapter 7: How Cells Release Chemical Energy

7.1 When Mitochondria Spin Their Wheels [p.107]

7.2 Extracting Energy from Carbohydrates [pp.108-109]

7.3 Glycolysis – Glucose Breakdown Starts [pp.110-111]

7.4 Second Stage of Aerobic Respiration [pp.112-113]

1. c; 2. b; 3. a; 4. b; 5..a; 6. a; 7. b; 8. a; 9. c; 10. oxygen; 11. glycolysis; 12. Krebs cycle; 13. CO_2; 14. ATP; 15. NADH; 16. $FADH_2$; 17. 32;18. ATP; 19. PGAL; 20. NADH; 21. ATP; 22. phosphorylation; 23. ATP; 24. 2; 25. 2; 26. pyruvate; 27. NADH; 28. CO_2; 29. NAD+; 30. ATP; 31. $FADH_2$; 32. oxaloacetate; 33. a. 6, b. 8, c. 2, d. 2

7.5 Aerobic Respiration's Big Energy Payoff [p.114-115]

7.6 Fermentation Pathways [pp.116-117]

1. NADH; 2. pyruvate; 3. cytoplasm; 4. mitochondrial; 5. CO_2; 6. $FADH_2$; 7. ATP; 8. Krebs cycle; 9. oxygen; 10. a. 6, b. 10, c. 2, d. 36; 11. it serves as an electron acceptor; 12.c; 13. a; 14. c; 15. c; 16. b; 17. c; 18. a; 19. c; 20. b; 21. a

7.7 Alternative Energy Sources in the Body [pp.118-119]

1. a; 2. b; 3. c; 4. b; 5. a; 6. a; 7. c; 8. a;

Self Quiz

1. b; 2. c; 3. d; 4. a; 5. a; 6. e; 7. d; 8. c; 9. e; 10. d

Chapter Summary

1. degradative; 2. chemical; 3. aerobic; 4. mitochondria; 5. glycolysis; 6. NAD+; 7. electrons; 8. two; 9. Krebs cycle; 10. carbon dioxide; 11. phosphorylation; 12. oxygen; 13. glycolysis; 14. electron; 15. small; 16. glucose; 17. lipids; 18. cellular

Chapter 8: DNA Structure and Function

8.1 A Hero Dog's Golden Clones [p.123]

8.2 Eukaryotic Chromosomes [pp.124-125]

1. f; 2. i; 3. a; 4. d; 5. b; 6.h; 7. g; 8. j; 9. c; 10.e

8.3 The Discovery of DNA's Function [pp.126-127]

1. DNA from a dead pathogenic strain of bacteria was taken up by a live nonpathogenic strain of the same bacteria. The live bacteria expressed the "new" DNA and became pathogenic, causing disease in animals.; 2.a. In the 1800s, Miescher discovered what came to be known as deoxyribonucleic acid, or DNA; b. Griffith; c. Avery; d. Hershey and Chase;

8.4 The Discovery of DNA's Structure [pp.128-129]

1. A five-carbon sugar called deoxyribose, a phosphate group, and one of the four nitrogen-containing bases (A, T, G, or C); 2. guanine (pu);

3. cytosine (py); 4. adenine (pu); 5. thymine (py);
6. deoxyribose; 7. phosphate; 8. nucleotide; 9.
Erwin Chargaff; 10. x-ray diffraction; 11. Watson;
12. T

8.5 Fame and Glory [p.130]
8.6 Replication and Repair [pp.130-131]
1. d; 2. c; 3. b; 4. replication; 5. Enzymes; 6.
strand; 7. nucleotides; 8. T; 9. C; 10. double helix;
11. conserved; 12. semiconservative; 13.
Replication; 14. rewinding; 15. Enzyme; 16.
Hydrogen; 17. polymerases; 18. phosphate; 19.
energy; 20. ligases; 21. double helix; 22. ligases;
23. polymerases; 24. survival; 25. mutation

25.	T-	A	T	-A
	G-	C	G	-C

A-	T	A	-T
C-	G	C	-G
C-	G	C	-G
C-	G	C	-G

Self-Quiz
1. d; 2. a; 3. b; 4. a; 5. a; 6. c; 7. d; 8. e; 9. d; 10. a
Chapter Summary
1. DNA; 2. nucleotides; 3. hydrogen; 4. double; 5.
cytosine; 6. order; 7. species; 8. structure; 9. copy;
10. errors; 11. mutations; 12. function; 13. cloning
Integrating and Applying Key Concepts
If A = 23 %, then T would be 23% leaving 54% to
be divided by C and G, and the % of those two
would be 27% each.

Chapter 9: From DNA to Proteins
9.1 Ricin and Your Ribosomes [p.137]
9.2 The Nature of Genetic Information [p.138-139]
9.3 Transcription [pp.140-141]
1. d; 2. d; 3. b; 4. c; 5. b; 6.a. mRNA, the only kind
of RNA that carries protein-building instructions;
b. rRNA, a component of ribosomes, where
polypeptide chains are built; c. tRNA, delivers
amino acids one at a time to a ribosome; 7. RNA
is single-stranded whereas most DNA is double-
stranded. RNA contains four types of
ribonucleotides. Each consists of a five-carbon
sugar, ribose (not DNA's deoxyribose), a
phosphate group, and a base. Three bases—
adenine, cytosine, and guanine—are the same in
both RNA and DNA. In RNA, however, the
fourth base is uracil, not thymine; 8. DNA; 9.
RNA polymerase; 10. polymerase; 11. RNA; 12.
one; 13. promoter; 14. transcription; 15. RNA
polymerase; 16. polymerase; 17. RNA nucleotide;
18. RNA;
19. d; 20. f; 21. g; 22. b; 23. c; 24. a; 25. e
9.4 RNA and the Genetic Code [pp.142-143]

9.5 Translating the Code: RNA to Protein [pp.
144-145]
1.i; 2. b; 3. j; 4. k; 5. a; 6. g; 7. e; 8. f; 9. h; 10. c; 11.
mRNA transcript; AUG UUC UAU UGU AAU
AAA GGA UGG CAG UAG; 12. tRNA
anticodons: UAC AAG AUA ACA UUA UUU
CCU ACC GUC AUC; 13. amino acids: (start)
met phe tyr cys asn lys gly try gln stop; 14. AUG,
TAC, Met; 15. UGA, UAA or UAG; 16.a.
termination; b. elongation; c. initiation
9.6 Mutated Genes and Their Protein Products [pp.146-147]
1. b; 2. c; 3. d; 4. b; 5. c
Self Quiz
1. b; 2. b; 3. d; 4. c; 5. d; 6. e; 7. a; 8. b; 9. e; 10. d;
11. d
Chapter Summary
1. enzymes; 2. polypeptide; 3. nucleotide; 4.
genes; 5. transcription; 6. translation; 7. DNA; 8.
template; 9. RNA; 10. messenger; 11. three; 12.
code; 13. conserved; 14. translation; 15.
messenger; 16. transfer; 17. ribosomal; 18.
peptide; 19. mutations; 20. variation.

Chapter 10: Control Over Genes
10.1 Between You and Eternity [p.151]
10.2 Gene Expression in Eukaryotic Cells [pp.152-153]
10.3 There's a Fly in my Research [pp.154-155]
10.4 A Few Outcomes of Gene Controls [pp.156-157]
10.5 Gene Control in Bacteria [pp.158-159]
1. b; 2. c; 3. a; 4. genes; 5. structure; 6. specialized;
7. function; 8. differentiation; 9. genes; 10. b; 11. b;
12. a; 13. a; 14. a; 15. a; 16. b; 17. b; 18. a; 19. c; 20.
d
Self Quiz
1. c; 2. d; 3. a; 4. b; 5. c; 6. d; 7. d; 8. a; 9. c; 10. b

Chapter Summary

1. control; 2. expressed; 3. conditions; 4. outside; 5. transcription; 6. product; 7. master; 8. selective; 9. differential; 10. function; 11. localized; 12. multicellular; 13. inactivated; 14. complex; 15. activated; 16. suppress; 17. nutrient; 18. environment; 19. transcription

---

Chapter 11: How Cells Reproduce

11.1 Henrietta's Immortal Cells [p.163]
11.2 Multiplication by Division [pp.164-165]
11.3. Mitosis [pp.166-167]

1. HeLa cells are tumor cells taken from and named for a cancer patient, Henrietta Lacks, in 1951. Henrietta Lacks died (at age thirty-one) two months following her diagnosis of cancer. HeLa cells have continued to divide in culture and are used for cancer research in laboratories all over the world. The legacy of Henrietta Lacks continues to benefit humans everywhere. 2. h; 3. j; 4. g; 5. b; 6. k; 7. d; 8. a; 9. i; 10. c; 11. e; 12. f; 13. G1 interval; 14. S interval; 15. G2 interval; 16. prophase; 17. metaphase; 18. anaphase; 19. telophase; 20. cytoplasmic division; 21. interphase; 22. mitosis; 23. daughter cells; 24. 15; 25. 22; 26. 14; 27. 13; 28. 21; 29. 20; 30. 21; 31. interphase-daughter cells (f); 32. anaphase (a); 33. late prophase (g); 34. metaphase (d); 35. cells at interphase (e); 36. early prophase (c); 37. transition to metaphase (b); 38. telophase (h)

11.4 Cytokinesis: Division of the Cytoplasm [p.168]

1. a; 2. b; 3. a; 4. a; 4. b; 5. b; 6. b; 7. a; 8. a. 9. a

11.5 Control Over Cell Division [p.169
11.6 Cancer: When Control Is Lost [p.170}

1. First, cancer cells grow and divide abnormally. Second, cancer cells have profoundly altered cytoplasm and plasma membranes. Third, cancer cells have a weakened capacity for adhesion. Four, cancer cells usually have a lethal effect 2. d; 3. f; 4. h; 5. a; 6. f; 7. g; 8. c; 9. b

Self-Quiz

1. a; 2. d; 3. b; 4. e; 5. c; 6. d; 7. b; 8. a; 9. b; 10. b; 11. a ; 12. c

Chapter Summary

1. species; 2. chromosomes; 3. shape; 4. hereditary; 5. daughter; 6. cytoplasm; 7. cycle; 8. mitosis; 9. mass; 10. DNA; 11. nucleus; 12. anaphase; 13. microtubular; 14. nuclei; 15. nuclear; 16. cytoplasm; 17. animal; 18. plant; 19. timing; 20. rate; 21. tumor; 22. cancer

---

Chapter 12: Meiosis and Sexual Reproduction

12.1 Why Sex? [p.175]
12.2 Sexual Reproduction and Meiosis [pp.176-177]
12.3 The Process of Meiosis [pp.178-179]

1. b; 2. a; 3. a; 4. b; 5. a; 6. b; 7. a; 8. b; 9. a; 10. b; 11. Meiosis; 12. gamete; 13. Diploid; 14. Diploid; 15. Meiosis; 16. sister chromatids; 17. sister chromatids; 18. one; 19. four; 20. two; 21. interphase preceding meiosis I; 22. chromosome; 23. haploid; 24. meiosis II; 25. twenty-three; 26. anaphase II (H); 27. metaphase II (F); 28. metaphase I (A); 29. prophase II (B); 30. telophase II (C); 31. telophase I (G); 32. prophase I (E); 33. anaphase I (D); 34. E ($2n = 2$); 35. D ($2n = 2$); 36. B ($2n = 2$); 37 A ($n = 1$); 38. C ($n = 1$)

12.4 How Meiosis Introduces Variation in Traits [pp.180-181]
12.5 From Gametes to Offspring [pp.182-183]
12.6 Mitosis and Meiosis – An Ancestral Connection [pp184-185]

1. f; 2. g; 3. b; 4. d; 5. h; 6. c; 7. e; 8. a; 9. b; 10. a; 11. a; 12. a; 13. b; 14. b; 15. a; 16. b; 17. b; 18. a; 19. 2 (2n); 20. 5 (n); 21. 4 (n); 22. 3 (n); 23. A; 24. B; 25. E; 26. D; 27. C; 28. Fertilization also adds to variation among offspring. During prophase I, an average of two or three crossovers take place in each human chromosome. Random positioning of pairs of paternal and maternal chromosomes at metaphase I results in one of millions of possible chromosome combinations in each gamete. Of all male and female gametes produced, which two get together is a matter of chance. The sheer numbers of combinations that can exist is staggering.

Self Quiz

1. d; 2. a; 3. c; 4. e; 5. b; 6. a; 7. d; 8. d; 9. b; 10. c

Chapter Summary

1. asexual; 2. genetic; 3. sexual; 4. alleles; 5. gene; 6. trait; 7. diploid; 8. meiosis; 9. sexual; 10. haploid; 11. cytoplasmic; 12. chromosomes; 13. alleles; 14. shuffled; 15. chance; 16. variation; 17. gametes; 18. spore; 19. molecular; 20. mitosis; 21. DNA

Chapter 13: Observing Patterns in Inherited Traits

13.1 Menacing Mucus [p.189]

13.2 Mendel, Pea Plants, and Inheritance Patterns [pp.190-191]

13.3 Mendel's Law of Segregation [pp.192-193]

13.4 Mendel's Theory of Independent Assortment [pp.194-195]

1. c; 2. d; 3. b; 4. a; 5. genotype: 1/2 Tt; 1/2 tt, phenotype: 1/2 tall; 1/2 short; 6.a. 1 tall : 1 short; 1 heterozygous tall : 1 homozygous short; b. all tall; 1 homozygous tall : 1 heterozygous tall; c. all short; all homozygous short; d. 3 tall : 1 short; 1 homozygous tall : 2 heterozygous tall : 1 homozygous short; e. 1 tall : 1 short; 1 homozygous short : 1 heterozygous tall; f. all tall; all heterozygous tall; g. all tall; all homozygous tall; h. all tall; 1 heterozygous tall : 1 homozygous tall; 7. a. 9/16; pigmented eyes, right-handed; b. 3/16 pigmented eyes, left-handed; c. 3/16 blue-eyed, right-handed; d. 1/16 blue-eyed, left-handed; 8.a. F_1: black trotter; F_2: nine black trotters, three black pacers, three chestnut trotters, one chestnut pacer; b. black pacer; c. BbTt; d. bbtt, chestnut pacers and BBTT, black trotters.

13.5 Beyond Simple Dominance [pp.196-197]

13.6 Complex Variation in Traits [pp.198-199]

1. Genotypes: 1/4 AA; 1/4 AB; 1/4 AO; 1/4 BO, phenotypes: 1/2 A; 1/4 AB; 1/4 B; 2. Genotypes: 1/4 AB; 1/4 BO; 1/4 AO; 1/4 OO, phenotypes: 1/4 AB; 1/4 B; 1/4 A; 1/4 O; 3. Genotypes: 1/4 AA; 1/2 AB; 1/4 BB, phenotypes: 1/4 A; 1/2 AB; 1/4 B; 4.a. phenotype: all pink; genotype: all RR; b. phenotypes: 1/4 red; 1/2 pink; 1/4 white; genotypes: 1/4 RR; 1/2 RR'; 1/4 R'R'; 5. The genotype of the male parent is RrPp and the genotype of the female parent is rrpp. The offspring are 1/4 walnut comb, RrPp; 1/4 rose comb, Rrpp; 1/4 pea comb, rrPp; 1/4 single comb, rrpp; the process is called epistasis 6. 3/8 black; 1/2 yellow; 1/8 brown; 7.a. Incomplete dominance; b. Codominance; c. Multiple alleles; d. Gene interactions; e. Pleiotropy; 8. b; 9. b; 10. a; 11. b; 12. a; 13. a; 14. b; 15. An individual's phenotype is an outcome of complex interactions among its genes, enzymes and other gene products, and environmental factors (the phenotype is a product of genetic and environmental factors).

Self Quiz

1. d; 2. b; 3. a; 4. c; 5. d; 6. b; 7. e; 8. a; 9. a; 10. c; 11. d

Chapter Review

1. experimental; 2. inheritance; 3. heritable; 4. units; 5. genes; 6. segregation; 7. homologous; 8. gametes; 9. independent; 10. distributed; 11. recessive; 12. codominant; 13. traits; 14. environment

Chapter 14: Human Inheritance

14.1 Shades of Skin [p.203]

14.2 Human Genetic Analysis [pp.204-205]

14.3 Autosomal Inheritance Patterns [p.206-207]

1. In autosomal dominant inheritance, if one parent does not have a mutated allele, but the other parent is heterozygous, every child has a 50% chance of getting the allele. The trait usually appears every generation. In autosomal recessive inheritance, if both parents are heterozygous, each child will have 50% chance of being heterozygous, and 25% will be homozygous. The heterozygous individuals often show no characteristics.; 2. Assuming the father is heterozygous with Huntington disorder and the mother normal, the chances are 1/2 that the son will develop the disease; 3. The woman (the children's mother) parents were father nn because he is albino; The woman's mother was Nn; The woman is heterozygous normal, Nn. The albino man, nn, has two heterozygous normal parents, Nn. The two normal children are heterozygous normal, Nn; the albino child is nn;

14.4 X-Linked Inheritance Patterns [pp.208-209]

14.5 Heritable Changes in Chromosome Structure [p.210-211]

14.6 Heritable Changes in Chromosome Number [p.212-213]

1. sons; 2. mothers; 3. daughters; 4.duplication; 5. inversion; 6. deletion; 7. translocation; 8. deletion; 9. Structural alterations in DNA may have been the origin of species differences among closely related organisms, such as apes and humans. Eighteen of the twenty-three pairs of human chromosomes are almost identical with chromosomes of chimpanzees and gorillas. The other five differ at inverted and translocated regions; 10. If only male offspring are considered, the probability is 1/2 that the couple will have a color-blind son; 11. The probability is that 1/2 of the sons will have hemophilia; the probability is 0

that a daughter will express hemophilia; the probability is that 1/2 of the daughters will be carriers; 12. If the woman marries a normal male, the chance that her son would be color-blind is 1/2. If she marries a color-blind male, the chance that her son would be color blind is also 1/2; 13.a. With aneuploidy, individuals have one extra or one less chromosome; a major cause of human reproductive failure; b. With polyploidy, individuals have three or more of each type of chromosome; common in flowering plants and some animals but lethal in humans; c. Nondisjunction is due to a failure of one or more pairs of chromosomes to separate in mitosis or meiosis; some or all forthcoming cells will have too many or too few chromosomes; 14. All gametes will be abnormal; 15. One-half of the gametes will be abnormal; 16. Half of all species of flowering plants, some insects, fishes, and other animals are polyploid; 17. If a normal gamete fuses by chance with an n + 1 gamete (one extra chromosome), the new individual will be trisomic (2n + 1), with three of one type of chromosome and two of every other type. Should a n – 1 gamete fuse with a normal n gamete, the new individual will be monosomic (2n – 1). Mitotic divisions perpetuate the mistake as an early embryo forms; 18. d; 19. b; 20. c; 21. e; 22. a; 23. f; 24. d;

14.7 Genetic Screening [pp.214-215]
1. a; 2. e; 3. f; 4. b; 5. c; 6. g; 7. d; 8. A genetic abnormality is nothing more than a rare, uncommon version of a trait that society may judge as abnormal or merely interesting; a genetic disorder is an inherited condition that sooner or later causes mild to severe medical problems; a syndrome is a recognized set of symptoms that characterize a given disorder; a genetic disease is illness caused by a person's genes increasing susceptibility to infection or weakening the response to it; 9a. Autosomal recessive; b. Autosomal dominant; c. X-linked recessive; d. Autosomal dominant; e. Changes in chromosome number; f. Changes in chromosome structure; g. Changes in chromosome number; h. X-linked recessive; i. Changes in chromosome number; j. X-linked recessive; k. Autosomal dominant; l. Changes in chromosome number; m. Autosomal recessive;

Self-Quiz
1. d; 2. d; 3. b; 4. c; 5. d; 6. a; 7. c; 8. b; 9. b; 10. c

Chapter Summary
1. autosomes; 2. genes; 3. sex; 4. gene; 5. karyotyping; 6. chromosomes; 7. Mendelian; 8. dominance; 9. traits; 10. males; 11. females; 12. structure; 13. duplicated; 14. translocated; 15. number; 16. genetic; 17. diagnostic

---

Chapter 15: Biotechnology
15.1 Personal DNA Testing [p.219]
15.2 Cloning DNA [pp.220-221]
15.3 From Haystacks to Needles [pp.222-223]
15.4 DNA Sequencing [pp.224-225]
15.5 Genomics [pp.226-227]
1. c; 2. f; 3. a; 4. d; 5. b; 6. b; 7. f; 8. e; 9. a; 10. d; 11. c; 12. separates DNA fragments based upon molecular size; 13. used to replicate genes, or part of genes, into multiple copies 14; uses probes to distinguish one DNA sequence from others in a library; 15. provides a rapid method of obtaining the DNA sequence of a gene; 16. d; 17. f; 18. b;1 9. c; 20. a; 21. e.
15.6 Genetic Engineering [p.228]

15.7 Designer Plants [p.228-229]
15.8 Biotech Barnyards [pp.230-231]
15.9 Safety Issues [p. 231]
15.10 Genetically Modified Humans? [p.232-233]
1. F, transgenics; 2. F, genetic engineering; 3. T; 4. T; 5. b; 6. a; 7. d; 8. a; 9. c; 10. d;
Self Quiz
1. c; 2. b; 3. d; 4. a; 5. a; 6. d; 7. a; 8. c; 9. e; 10.a
Chapter Review
1. recombinant; 2. species; 3. vectors; 4. PCR; 5. fragments; 6. sequencing; 7. fingerprint; 8. genomes; 9. species; 10. genes; 11. ethical

Chapter 16: Evidence of Evolution

16.1 Reflections of a Distant Past [p.237]

16.2 Early Beliefs, Confounding Discoveries [pp.238-239]

16.3 A Flurry of New Theories [pp.240-241]

16.4 Darwin, Wallace, and Natural Selection [pp.242-243]

1. f; 2. c; 3. e; 4. a; 5. b; 6. d; 7. c; 8. b; 9. a; 10. d; 11. c; 12. b; 13. b; 14. d; 15. b; 16. b; 17. b; 18. d; 19. b; 20. c; 21. a

16.5 Fossils – Evidence of Ancient Life [pp.244-245]

16.6 Putting Time in Perspective [pp.246-247]

1. g; 2. c; 3. a; 4. b; 5. f; 6. e; 7. d; 8. 0.5 grams, 16,110 years; 9. d; 10. e; 11. e; 12. d; 13. d; 14. d; 15. a; 16. c; 17. e

16.7 Drifting Continents, Changing Seas [pp.248-249]

16.8 Similarities in Body Form and Function [pp.250-251]

16.9 Similarities in Patterns of Development [pp.252-253]

1. d; 2. b; 3. a; 4. c; 5. d; 6. d; 7. b; 8. a; 9. d; 10. c; 11. a; 12. a

Self Quiz

1. d; 2. d; 3. a; 4. b; 5. d; 6. c; 7. c; 8. c; 9. a; 10. d

Chapter Summary

1. global; 2. fossils; 3. evolution; 4. Charles Darwin; 5. natural selection; 6. heritable; 7. physical; 8. geologic; 9. species; 10. lineages; 11. ancestor; 12. Gene

Chapter 17: Processes of Evolution

17.1 Rise of the Super Rats [p.257]

17.2 Individuals Don't Evolve, Populations Do [pp.258-259]

1. a; 2. f; 3. d; 4. c; 5.g; 6. b; 7. h; 8. e; 9. i

17.3 A Closer Look at Genetic Equilibrium [pp.260-261]

17.4 Natural Selection Revisited [p.261]

1. 0.16; 2. 0.4; 3. 0.6; 4. 0.36; 5. 0.48; 6. 0.4; 7. no mutation, large population size, isolated population, random mating, equal reproductive success; 8. directional; 9. disruptive; 10. stabilizing

17.5 Directional Selection [pp.262-263]

17.6 Stabilizing and Disruptive Selection [pp.264-265]

1. a; 2. c; 3. c; 4. b; 5. a; 6. a; 7. c; 8. b; 9. c

17.7 Fostering Diversity [p.266-267]

17.8 Genetic Drift [pp.268-269]

17.9 Gene Flow [p.269]

1. a; 2. f; 3. g; 4. h; 5. c; 6. d; 7. b; 8. e; 9. F, loss; 10. F, balanced polymorphism; 11.T; 12. T; 13. F, inbreeding; 14. F, decreases

17.10 Reproductive Isolation [pp.270-271]

17.11 Allopatric Speciation [pp.272-273]

17.12 Sympatric and Parapatric Speciation [pp.274-275]

1. prezygotic; 2. temporal isolation; 3. mechanical isolation; 4. behavioral isolation; 5. ecological isolation; 6. gamete incompatibility; 7. postzygotic; 8. hybrid inviability; 9. hybrid sterility; 10. c; 11. b; 12. a; 13. b; 14. a; 15. c; 16. b

17.13 Macroevolution [pp.276-277]

17.14 Phylogeny [pp.278-279]

1. a; 2. i; 3. g; 4. h; 5. c; 6. d; 7. e; 8. f; 9. c; 9. b

Self Quiz

1. a; 2. d; 3. b; 4. b; 5. c; 6. d; 7. c; 8. a; 9.b; 10. b

Chapter Summary

1. populations; 2. alleles; 3. phenotypes; 4. frequency; 5. mutation; 6. genetic; 7. flow; 8. interbreed; 9. natural; 10. fertile; 11. isolated; 12. flow; 13. independently; 14. divergence; 15. reproductive; 16. macroevolution; 17. adaptive; 18. extinctions; 19. adaptation; 20. behavior; 21. survive; 22. environment

Chapter 18: Life's Origin and Early Evolution

18.1 Looking for Life [p.283]

18.2 Earth's Origin and Early Conditions [pp.284-285]

18.3 From Polymers to Cells [pp.286-287]

1. d; 2. e; 3. a; 4. b; 5. e; 6. a; 7. c; 8. b; 9. c; 10. e; 11. abiotic; 12. lipid; 13. DNA; 14. RNA; 15. selectively-permeable; 16. proto-cells; 17. lightning-fueled atmospheric reactions, reactions at hydrothermal vents, delivery from space;18. The scientific study of life's origin and

distribution in the universe. The possibility that life forms may exist in on other planets in our solar system.

18.4 Life's Early Evolution [pp.288-289]

18.5 Evolution of Organelles [pp.290-291]

18.6 Time Line for Life's Origin and Evolution [pp.292-293]

1. a; 2. d; 3. e; 4. b; 5. c; 6. life could not longer arise spontaneously, aerobic respiration evolved, ozone layer formed; 7. a; 8. c; 9. b; 10. a; 11. c; 12.

a; 13. d; 14. I; 15. g; 16. b; 17. c; 18. a; 19. f; 20. j; 21.
e; 22. h
Self Quiz
1. a; 2. d; 3. a; 4. a; 5. d; 6. b; 7. d; 8. d; 9. b; 10. d

Chapter 19: Viruses, Bacteria, and Archaeans
19.1 Evolution of a Disease [p.297]
19.2 Viral Structure and Function [pp.298-299]
19.3 Viral Effects on Human Health [pp.300-301]
19.4 Viroids: Tiny Plant Pathogens [p.301]
1. infectious; 2. RNA; 3. protein; 4. host; 5.
Bacteriophages; 6. tail; 7. head; 8. proteins;
9. f; 10. j; 11. a; 12. l; 13. b; 14. i; 15. c; 16. h; 17. g;
18. d; 19. e; 20. a; 21. c; 22. a; 23. b; 24. b; 25. a.; 26.
f; 27. b; 28. a; 29. e; 30. d; 31. c
19.5 Bacterial Structure and Function [p.302]
19.6 Bacterial Reproduction and Gene Exchange
    [p.303]
1. DNA (b); 2. pilus (c); 3. flagellum (f); 4. capsule
(a); 5. cell wall (d); 6. plasma membrane (g); 7.
cytoplasm (e); 8. b; 9. a; 10. c; 11. d; 12. b; 13.
binary fission; 14. DNA; 15. membrane; 16. cell
wall; 17. cells; 18. coccus; 19. bacillus; 20.
spirillum

Chapter 20: The Protists
20.1 Harmful Algal Blooms [p.311]
20.2 A Collection of Lineages [p.312]
20.3 Flagellated Protozoans [p.313]
1. lineages; 2. group; 3. animals; 4. unicellular; 5.
colonial; 6. multicellular; 7. sexually;
8. f; 9. h; 10. a; 11. e; 12. g; 13. d; 14. b; 15. c; 16.
autotroph;
20.4 Mineral-Shelled Protozoans [p.314]
20.5 The Alveolates [p.314]
20.6 Malaria and the Night Feeding Mosquitos
    [p.316]
1. g; 2. b; 3. d; 4. a; 5. e; 6. f; 7. c; 8. f; 9. c; 10. d; 11.
e; 12. a; 13. e; 14. e; 15. c; 16. e; 17. e; 18. e; 19. c;
20. a
20.7 Single Celled Stramenophiles [p.317]
20.8 Red Algae and Green Algae [p.318-319]

Chapter Summary
1. 4 billion; 2. seas; 3. organic; 4. meteorites; 5.
experiments; 6. chemical; 7. membrane; 8.
anaerobic; 9. bacteria; 10. eukaryotic; 11.
photosynthetic; 12. atmosphere; 13. selection; 14.
membrane-bound; 15. plasma membrane; 16.
chloroplasts; 17. time-line; 18. organisms

19.7 Bacterial Diversity [pp.304-305]
19.8 The Archaeans [pp.306-307]
1. a; 2. e; 3. b; 4. c; 5. d; 6. a; 7. e; 8. cyanobacteria;
9. proteobacteria; 10. proteobacteria; 11. gram
staining bacteria; 12. archaea; 13. proteobacteria;
14. proteobacteria; 15. proteobacteria; 16.
proteobacteria; 17. proteobacteria; 18. spirochete;
19. gram staining; 20. gram staining; 21.b and c ;
22. b; 23. c; 24. c; 25. a
Self Quiz
1. a; 2. a; 3. b; 4. d; 5. c; 6. d; 7. d; 8. c; 9. c; 10. e
Chapter Summary
1. bacteria; 2. archaea; 3. nucleus; 4. organelles; 5.
metabolic; 6. fission; 7. plasmid; 8. bacteria; 9.
extreme; 10. replicate; 11. protein; 12.viroids 13.
RNA

1. a, e; 2. c; 3. b; 4. b; 5. e; 6. e; 7. b; 8. b; 9. c; 10. b;
11. green algae; 12. red algae; 13. green algae; 14.
green algae; 15. water molds;
20.9 Amoebozoans [p.320-321]
1. ciliate; 2. apicomplexa; 3. green algae; 4.
flagellate; 5. red algae; 6. flagellate; 7. green algae;
8. flagellate; 9. amoeba; 10. flagellate; 11. NP; 12.
malaria; 13. NP; 14. dysentery; 15. NP; 16. African
sleeping sickness; 17. NP; 18. vaginosis; 19. NP;
20. plant destroyers.
Self Quiz
1. c; 2. d; 3. d; 4. e; 5. d; 6. a; 7. d; 8. a; 9. c; 10. e
Chapter Summary
1. eukaryotic; 2. multicelled; 3. fungi; 4. animals;
5. protozoans; 6. radiolarians; 7. dinoflagellates;
8. photoautotrophs; 9. sacs; 10. stramenophiles; 11.
a; 12. a; 13. b; 14. plants; 15. amoebozoans; 16.
fungi

Chapter 21: Plant Evolution

21.1 Speaking for the Trees [p.325]

21.2 Adaptative Trends Among Plants [pp.326-327]

1. g; 2. i; 3. h; 4. c; 5. a; 6. b; 7. d; 8. e; 9. j; 10. f

21.3 The Bryophytes [pp.336-337]

1. T; 2. lack; 3. rhizoids; 4. T; 5. sporophyte; 6. T; 7. mosses (bryophytes); 8. T; 9. gametangium; 10. T; 11. sporophyte, diploid, gametophyte, haploid; 12. Meiosis, haploid spores; 13. gametophyte, haploid; 14. gametophyte; 15. gametophyte; 16. haploid sperms; 17. haploid eggs

21.4 Seedless Vascular Plants [pp.330-331]

1. b; 2. a; 3. d; 4. c; 5. c; 6. d; 7. a; 8. c; 9. b; 10. b; 11. c; 12. d; 13. a; 14. c; 15. b; 16. d; 17. c; 18. gametophyte, haploid; 19. sporophyte, diploid; 20. haploid, mitosis; 21. gametophyte, haploid; 22. diploid, gametophyte; 23. spore production, diploid, sporophyte

21.5 History of the Vascular Plants [pp.332-333]

1.h; 2. e; 3. j; 4. f; 5. b; 6. k; 7. a; 8. g; 9. c; 10. d; 11. i

21.6 Gymnosperms—Plants With Naked Seeds [pp.334-335]

1. b; 2. d; 3. b; 4. c; 5. a; 6. k; 7. e; 8. b; 9. e; 10. c; 11. d; 12. e; 13. sporophyte, diploid; 14. seed; 15. sporophyte, diploid; 16. pollen sac, cone; 17. ovule, diploid; 18. female gametophyte, haploid; 19. eggs, haploid, mitosis; 20. male gametophyte, haploid; 21. megaspore, meiosis, haploid; 22. microspores, haploid, pollen grain; 23. pollen grain, female cone

21.7 Angiosperms—The Flowering Plants [pp.336-337]

21.8 Ecological and Economic Importance of Angiosperms [pp. 338-339]

1. f; 2. j; 3. e; 4. i; 5. c; 6. a; 7. h; 8. b; 9. g; 10. d; 11. sporophyte, diploid; 12. female gametophyte, haploid; 13. egg, haploid, mitosis; 14. ovule, diploid, seed; 15. male gametophyte, haploid, mitosis; 16. released pollen grain, haploid; 17. embryo sporophyte plant, diploid, mitosis, diploid zygote; 18. seed, ovule

Self Quiz

1. c; 2. c; 3. a; 4. b; 5. e; 6. e; 7. c; 8. c; 9. d; 10. c

Chapter Summary

1. 475; 2. divergences; 3. radiations; 4. functional; 5. nonvascular; 6. gamete; 7. water; 8. vascular; 9. seeds; 10. internal; 11. sperm; 12. drier; 13. pollen; 14. seeds; 15. flowers; 16. diversity; 17. angiosperms; 18. secondary metabolites

---

Chapter 22: Fungi

22.1 High – Flying Fungi [p.343]

22.2 Fungal Traits and Diversity [p.344]

1. Dominican Republic; 2. Ug99; 3. decomposer; 4. heterotroph; 5. wastes; 6. hypha; 7. mycelium; 8. different

22.3 Chytrids, Zygote Fungi, and Relatives [p.345]

22.4 Sac Fungi [p.346]

22.5 Club Fungi [p.347]

1. zygote; 2. sac; 3. sac; 4. sac; 5. sac; 6. club; 7.bread mold; 8. bread, beer, wine; 9. antibiotics; 10. roundworm control; 11. eating; 12. decomposer

22.6 Fungi as Partners [p.348]

22.7 Fungi as Pathogens [p.349]

1. b; 2. d; 3. e; 4. f; 5. g; 6. c; 7. a; 8. c; 9. a; 10. c; 11. b; 12. a; 13. c; 14. d

Self Quiz

1. d; 2. a; 3. d; 4. c; 5. b; 6. b; 7. a; 8. a; 9. c; 10.c

Chapter Summary

1. heterotrophs; 2. animals; 3. organic; 4. filaments; 5. zygote; 6. club; 7. sac; 8. intracellular; 9. spore; 10. sexual; 11. spore; 12. sexual; 13. cell; 14. species; 15. lichens; 16. parasites; 17. humans

---

Chapter 23: Animals I: Major Invertebrates Groups

23.1 Old Genes, New Drugs [p.353]

23.2 Animal Traits and Trends [pp.354-355]

23.3 Animal Origins and Early Radiations [p.356]

1. T; 2. F, heterotrophs; 3. T; 4. T; 5. e; 6. f; 7. m; 8. h; 9. I; 10. o; 11. a; 12. n; 13. b; 14. j; 15. g; 16.c; 17.d; 18.l; 19. colonial protists; 20. flagellum; 21. division; 22. feeding; 23. reproduction; 24. gene; 25.animal;26. Sea; 27. Cambrian; 28.c; 29. e; 30. a; 31. b; 32. f; 33. d; 34. g

23.4 Sponges [p.357]

23.5 Cnidarians [pp.358-359]

23.6 Flatworms [pp.360-361]

1. b; 2. b; 3. c; 4. a; 5. c; 6. a; 7. c; 8. b; 9. b; 10. c; 11. d; 12. a; 13. f; 14. c; 15. e; 16. d; 17. b

23.7 Annelids [pp.362-363]

23.8 Mollusks [pp.364-365]

23.9 Roundworms [p.366]

1. a; 2. c; 3. a; 4. a; 5. c; 6. b; 7. a; 8. b; 9. c; 10. b; 11.
d; 12. e; 13. c; 14. f; 15. a; 16. b; 17. g; 18. radula;
19. mantle; 20. foot; 21. stomach; 22. digestive
gland; 23. heart; 24. excretory gland; 25. gill; 26.
trichinosis; 27. stomach pain, vomiting; 28.
elephantiasis

23.10 Keys to Arthropod Diversity [p.367]
23.11 Spiders and Their Relatives [p.368]
23.12 Crustaceans [p.369]
23.13 Insect Traits and Diversity [pp.370-371]
23.14 The Importance of Insects [p.372]
23.15 Echinoderms [p.373]

1. F, open; 2. T; 3. F, exoskeleton; 4. T; 5. F,
coelum; 6. a; 7. b; 8. c; 9. d; 10. c; 11. d; 12. b; 13. a;
14. c; 15. b; 16. c; 17. segmented; 18. thorax; 19.
abdomen; 20. antennae; 21. thorax; 22. two; 23.
invertebrates; 24. complete; 25. nymphs; 26.
adults; 27.larvae; 28. pupa; 29. T; 30. F, radial; 31.
T; 32. F, decentralized 33. arthropod -insect; 34.

arthropod-chelicerate; 35. arthropod-insect; 36.
arthropod-insect; 37. arthropod-insect; 38.
agricultural affects; 39.Lyme disease; 40. African
sleeping sickness; 41. bubonic plague; 42.
malaria; 43. hardened exoskeleton; 44. jointed
appendages; 45. highly modified segments; 46.
respiratory structures; 47. sensory specializations;
48. specialized stages of development

Self Quiz

1. a; 2. e; 3. a; 4. b; 5. c; 6. d; 7.a; 8. c; 9. b; 10. a; 12.
d; 13. b; 14. d; 15. b; 16. c

Chapter Summary

1. heterotrophs; 2. develop; 3. life; 4. tissues; 5.
simple; 6. complex; 7. integration; 8. symmetry; 9.
body; 10. segments; 11. deuterostomes; 12.
symmetry; 13. tissues; 14. radially; 15. bilateral;
16. systems; 17. embryos; 18. arthropods; 19.
crustaceans; 20. insects; 21. chordates; 22.
invertebrates; 23. radial

Chapter 24: Animals II: The Chordates

24.1 Windows on the Past [p.377]
24.2 The Chordate Heritage [pp.378-379]

1. a; 2 f; 3. d; 4. b; 5. e; 6. c; 7. b,c,d,e; 8. g; 9. j; 10. i:
11. f; 12. c; 13. d; 14. a; 15. e; 16. h; 17; e; 18.
braincase; 19. backbone; 20. jaws; 21. four limbs;
22. amniote eggs

24.3 The Fishes [pp.380-381]
24.4 Amphibians – The First Tetrapods [pp.382-
383]

1. e; 2. b; 3. d; 4. c; 5. a; 6. g; 7. h; 8. d; 9. f; 10. b;
11. e; 12. a; 13. c; 14. I 15. e; 16. a; 17. f; 18. c; 19. d;
20. b

24.5 Evolution of the Amniotes [p.384]
24.6 Nonbird Reptiles [p.385]
24.7 Birds- Reptiles with Feathers [p.386]

1. f; 2. c; 3. d; 4. e; 5. a; 6. b; 7. crocodilians; 8.
iguana; 9. Komodo dragon; 10. saltwater and Nile
crocodile; 11. shell, no teeth; 12. c; 13. b; 14. a

24.8 Mammals-The Milk Makers [p.387]

24.9 Primates Traits and Evolutionary Trends
[p.388-389]
24.10 Emergence of Early Humans [pp.390-391]
24.11 Emergence of Modern Humans [pp.392-
393]

1. T; 2. placental ; 3.T; 4.mice; 5. T; 6. back; 7. tree-
dwellers; 8. daytime; 9. bipedalism; 10. hands; 11.
hand; 12. teeth; 13. brain; 14. culture; 15.
behavioral; 16. learning; 17. traits; 18. a; 19. b; 20.
e; 21. d; 22. c

Self Quiz

1. c; 2. d; 3. b; 4. a; 5. c; 6. b; 7. d; 8. e; 9. b; 10. d

Chapter Summary

1. chordates; 2. nerve; 3. gill slits; 4. tail; 5.
vertebrates; 6. backbone; 7. muscles; 8. sensory;
9. brain; 10. gills; 11. gas; 12. limbs; 13. seas; 14.
fishes; 15. diversity; 16. tetrapods; 17. amniotes;
18. mammals; 19. land; 20. human; 21. climate; 22.
resources; 23. Africa

Chapter 25: Plant Tissues

25.1 Sequestering Carbon in Forests [p.397]
25.2 Organization of the Plant Body [pp.398-399]

1. Sequestering carbon is extremely important
since the abundance of carbon dioxide is
increasing in the atmosphere. Plants can take in
carbon dioxide and store the carbon, therefore it
is taken out of the atmosphere.2. b (c); 3. a (e); 4. e
(d); 5. c (a); 6. d (b); 7. primary growth occurs at
apical meristems, secondary growth occurs at

lateral meristems; 8. a. one, three (or multiples of
three), parallel, one pore or furrow, scattered
throughout stem ground tissue, b. two, four or
five (or multiples of four or five), netlike, three
pores or furrows, ring in the ground tissue

25.3 Components of Plant Tissues [pp.400-401]

1. parenchyma; 2. xylem, 3. phloem; 4.
sclerechyma; 5. companion cell; 6. sieve plate; 7.
d; 8. b; 9. a; 10. c; 11. a; 12. d; 13. c; 14. b; 15. b

25.4 Primary Shoots [pp.402-403]

25.5 A Closer Look at Leaves [pp.404-405]

1. leaf; 2. apical meristem; 3. lateral bud; 4. cortex; 5. vascular; 6. pith; 7. leaf; 8. apical meristem; 9. meristems; 10. cortex; 11. phloem; 12. xylem; 13. pith; 14. monocot; 15. eudicot; 16. petiole; 17. blade; 18. bud; 19. node; 20. stem; 21. blade; 22. sheath; 23. node

25.6 Primary Roots [pp.406-407]

1. endodermis; 2. pericycle; 3. xylem; 4. phloem; 5. cortex; 6. epidermis; 7. root hair; 8. matured; 9. elongate; 10. dividing; 11. root cap; 12. taproot; 13. fibrous; 14. lateral root

25.7 Secondary Growth [pp.408-409]

1. periderm; 2. phloem; 3. heartwood; 4. sapwood; 5. bark; 6. vascular cambium; 7. early; 9. late

25.8 Variation on a Stem [p.410]

25.9 Tree Rings and Old Secrets [p.411]

1. f; 2. d; 3. c; 4. e; 5. b; 6. a; 7. growth rings

Self Quiz

1. d; 2. c; 3. a; 4. b; 5. b; 6. e; 7. c; 8. d; 9. b; 10.e

Chapter Summary

1. vascular; 2. shoot; 3. root; 4. water; 5. photosynthesis; 6. environment; 7. meristems; 8. primary; 9. dermal; 10. organization; 11. sunlight; 12. water; 13. gas; 14. eudicot; 15. mineral; 16. anchor; 17. secondary; 18. wood

---

Chapter 26: Plant Nutrition and Transport

26.1 Mean Green Cleaning Machines [p.415]

26.2 Plant Nutrients and Soil [pp.416-417]

1. c; 2. j; 3. i; 4. b; 5. d; 6. e; 7. f; 8. g; 9. a; 10. h

26.3 How Do Roots Absorb Water and Minerals? [pp.418-419]

1. mutualism; 2. gaseous nitrogen; 3. root nodules; 4. sugars; 5. minerals; 6. exodermis; 7. cortex; 8. endodermal cells; 9. vascular cylinder; 10. xylem; 11. phloem; 12. endodermis; 13. Casparian strip; 14. b; 15. c; 16. d; 17. a

26.4 Water Movement Inside Plants? [pp.420-421]

26.5 Water-Conserving Adaptations of Stems and Leaves [pp.422-423]

1. stomata; 2. xylem; 3. tension; 4. hydrogen; 5. roots; 6. e; 7. a; 8. b; 9. d; 10.c

26.6 Movement of Organic Compounds in Plants? [pp.424-425]

1. b; 2. a; 3. f; 4. e; 5. d; 6. c; 7. source; 8. active; 9. solutes; 10. water; 11. turgor; 12. sink

Self Quiz

1. b; 2. a; 3. d; 4. a; 5. c; 6. c; 7. b; 8. d; 9. b; 10. d

Chapter Summary

1. water; 2. mineral; 3. root; 4. soil; 5. erosion; 6. solutes; 7. evaporation; 8. stomata; 9. gas; 10. photosynthesis; 11. phloem; 12. organic

---

Chapter 27: Plant Reproduction and Development

27.1 Plight of the Honeybee [p.429]

27.2 Reproductive Structures of Flowering Plants [pp.456-457]

1. sepal; 2. petal; 3. stamen; 4. filament; 5. anther; 6. carpel; 7. stigma; 8. style; 9. ovary; 10. meiosis; 11. microspores; 12. megaspores; 13. gametophyte; 14. fertilization; 15. zygote; 16. sporophyte

27.3 A New Generation Begins [pp.432-433]

27.4 From Zygotes to Seeds and Fruits [pp.434-435]

1. sporophyte; 2. ovule; 3. ovary; 4. meiosis; 5. megaspores; 6. mitosis; 7. eight; 8. embryo; 9. ovule; 10. gametophyte; 11. anther; 12. filament; 13. pollen sac; 14. meiosis; 15. microspores; 16. pollen grain; 17. anther; 18. pollination; 19. gametophyte; 20. pollen; 21. sperm; 22. endosperm; 23. egg; 24. ovule; 25. egg; 26. endosperm; 27. pollen; 28. endosperm; 29. embryo; 30. coat; 31. g; 32. b; 33. c; 34. d; 35. f; 36. e; 37. a

27.5 Asexual Reproduction of Flowering Plants [p.435]

1. b; 2. a

27.6 Patterns of Development in Plants [pp.436-437]

1. root; 2. branch; 3. coleoptile; 4. foliage; 5. adventitious; 6. branch; 7. primary; 8. prop; 9. coat; 10. root; 11. cotyledons; 12. hypocotyls; 13. leaf; 14. foliage; 15. primary; 16. branch; 17. primary; 18. nodule

27.7 Plant Hormones and Other Signaling Molecules [pp.438-439]

27.8 Adjusting the Direction and Rates of Growth [pp.440-441]

1. a; 2. c; 3. c; 4. d; 5. a; 6. e; 7. d; 8. e; 9.e; 10. b; 11. e; 12. b; 13. d

27.9 Sensing Recurring Environmental Changes [pp.442-443]

27.10 Plant Defenses [pp.444-445]
1. long-day; 2. short-day; 3. day-neutral; 4.
longer; 5. shorter; 6. red; 7. far-red; 8. inactivated;
9. b; 10. a; 11. c; 12. d
Self Quiz
1. d; 2. d; 3. b; 4. b; 5. d; 6. c; 7. c; 8. d; 9. d; 10.a

Chapter Summary
1. sexual; 2. pollinators; 3. flowers; 4.
gametophytes; 5. fertilization; 6. sporophyte; 7.
ovary; 8. dispersal; 9. animals; 10. asexually; 11.
hormones; 12. germination; 13. root; 14.
synthesis; 15. environmental; 16. night

Chapter 28: Animal Tissues and Organ Systems
28.1 Stem Cells [p.449]
28.2 Organization of Animal Bodies [pp.450-451]
28.3 Epithelial Tissue [pp.452-453]
28.4 Connective Tissue [pp.454-455]
1. a; 2. b; 3. m; 4. j; 5. i; 6. g; 7. h; 8. c; 9. l; 10. f; 11.
e; 12. d; 13. k; 14. b; 15. a; 16. b; 17. a; 18. a
28.5 Muscle Tissues [pp.456-457]
28.6 Nervous Tissue [p.457]
1. T; 2. F, cardiac; 3. F, muscle; 4. F. involuntary;
5. T; 6. F, smooth
28.7 Organs and Organ systems [pp.458-459]
28.8 Closer Look at a Organ-Human Skin
   [pp.460-461]
28.9 Integrated Activities [p.462-463]
1. f; 2. d; 3. a; 4. c; 5. b; 6. e; 7. d; 8. b; 9. e; 10. a;
11.c; 12. circulatory (D); 13. respiratory (E); 14.

urinary (A); 15. skeletal (C); 16. endocrine (J): 17.
reproductive (G); 18. digestive (I); 19. muscular
(H); 20. nervous (B); 21 integumentary (F); 22.
lymphatic (K); 23. hair; 24. epidermis; 25. dermis;
26. hypodermis; 27. hair follicle; 28. smooth
muscle
Self Quiz
1. a; 2. d; 3. c; 4. b; 5. b; 6. d; 7. d; 8. a; 9. a; 10. e
Chapter Summary
1. connective; 2. nervous; 3. epithelia; 4. secretory;
5. insulate; 6. cartilage; 7. adipose; 8. muscle; 9.
cardiac; 10. communication; 11. neurons; 12.
compartmentalize; 13. skin; 14. sesory; 15. water;
16. waste; 17. stimuli; 18. negative; 19. junction;
20. apoptosis

Chapter 29: Neural Control
29.1 In Pursuit of Ecstasy [p.467]
29.2 Evolution of Nervous Systems [pp.468-469]
29.3 Neurons – The Communicators [p.470]
29.4 Membrane Potentials [p.471]
29.5 A Closer Look at Action Potentials [p.472-
   473]
1. carry information toward cell body; 2. one per
cell in most neurons; 3. branch farther from cell
body; 4. no insulating sheath; 5. e; 6. f; 7. g; 8. d; 9.
a; 10. b; 11. c; 12. T; 13. T; 14. dendrite; 15. input
zone; 16. trigger zone; 17. axon; 18. conducting
zone; 19. output zone; 20. a; 21. f; 22. e; 23. d; 24.
c; 25. g; 26. b; 27. a; 28. T; 29. F, intensifies; 30. F,
sodium-potassium; 31. T; 32. T
29.6 Chemical Communication at Synapses
   [pp.474-475]
29.7 Disrupting Signaling – Disorders and
   Drugs [pp.476-477]
1. c; 2.b; 3.a; 4.d; 5. f; 6. e; 7.a; 8. b; 9. c; 10. d; 11. c;
12. e, 13. a; 14. e; 15. a; 16. b; 17. c; 18. d; 19.
dopamine; 20. low level of dopamine; 21.
Alzheimer's; 22. Parkinson's; 23. tolerance; 24.
habituation; 25. inability to stop drug use, even if
desire to do so persists; 26. concealment; 27.
dangerous actions taken to obtain drug; 28.
deterioration of professional and personal

relationships; 29. anger or defensiveness is
someone suggest a problem; 30. drug use
preferred over customary actions
29.8 The Peripheral Nervous Systems [pp.478-
479]
29.9 The Spinal Cord [pp.480-481]
1. a; 2. c; 3. e; 4. g; 5. f; 6. d; 7. b; 8. T; 9. F, white;
10. F, central; 11. T; 12. T; 13. g; 14. a; 15. d; 16. e;
17. b; 18. c; 19. f; 20. optic; 21. vagus; 22. pelvic;
23. cervical; 24. thoracic; 25. lumbar; 26. sacral; 27.
e; 28. b; 29. a; 30. d; 31. c; 31. d; 32. d; 33. d; 34. a;
35. b; 36. c; 37. a; 38. c; 39. a; 40. a; 41. a,b,c
29.10 The Vertebrate Brain [pp.482-483]
29.11 The Human Cerebrum [pp.484-485]
1. a; 2. d; 3. g; 4. f; 5. e; 6. c; 7. b; 8. spinal cord; 9.
cerebrospinal; 10. blood capillaries; 11. tight; 12.
proteins; 13. glucose; 14. ions; 15. urea; 16.
alcohol; 17. neural; 18. corpus callosum; 19.
hypothalamus; 20. thalamus; 21. pineal gland; 22.
midbrain; 23. cerebellum; 24. pons; 25. medulla
oblongata; 26. M; 27. S; 28. A; 29. M; 30. A; 31. S;
32. T; 33. F, olfactory; 34. F, long-term; 35. F,
hippocampus; 36. F, separately
Self Quiz
1. d; 2. d; 3. b; 4. a; 5. d; 6. e; 7. c; 8. b; 9. b; 10. b

Chapter Summary

1. neurons; 2. radially; 3. bilaterally; 4. plasma membrane; 5. self; 6. electric charge; 7. chemical; 8. inhibit; 9. psychoactive; 10. reflex arc; 11. brain; 12. spinal cord; 13. peripheral; 14. cerebral cortex; 15. neuroglia

---

Chapter 30: Sensory Perception

30.1 A Whale of a Dilemma [p.489]

30.2 Detecting Stimuli and Forming Perceptions [p.490]

30.3 Somatic and Visceral Sensations [p.491]

1. d; 2. c; 3. a; 4. f; 5. e; 6. b; 7. F, somatic; 8. T; 9. F, frequency; 10. T

30.4 Do You See What I See? [pp.492-493]

30.5 The Human Retina [p.494]

30.6 Visual Disorders [p.495]

1. choroid (f); 2. iris (g); 3. lens (h); 4. pupil (a); 5. cornea (e); 6. aqueous humor (b); 7. vitreous humor (i); 8. optic nerve (c); 9. retina (d); 10. F, left; 11. F, rod; 12. F, three; 13. F, epithelium; 14. F, rod; 15. e; 16. a; 17. f; 18. c; 19. b; 20. d; 21. g

30.7 The Chemical Senses [p.496]

30.8 Keeping the Body Balanced [p.497]

30.9 Detecting Sounds [pp.498-499]

1. c; 2. a; 3. b,d; 4. a; 5. c; 6. a,c; 7. d; 8. b; 9. F, pheromones; 10. F, olfactory; 11 T; 12. T; 13. T; 14. F, static; 15. T; 16. eardrum; 17. round window; 18. cochlea; 19. auditory nerve; 20. oval window; 21. stirrup; 22. anvil; 23. hammer; 24. b; 25. f; 26. d; 27. g; 28. e; 29. c; 30.a

Self Quiz

1. d; 2. c; 3. a; 4. a; 5. b; 6. c; 7. c; 8. b; 9. d; 10. e

Chapter Summary

1. nervous; 2. nerve; 3. process; 4. stimulus; 5. number; 6. frequency; 7. somatic; 8. mechanoreceptors; 9. chemoreceptors; 10. balance; 11. gravity; 12. hearing; 13. action potentials; 14. light-sensitive; 15. photoreceptors; 16. stimuli; 17. retina; 18. visual cortex.

Chapter 31: Endocrine Control

31.1 Hormones in the Balance [p.503]

31.2 The Vertebrate Endocrine System [pp.504-505]

1. d; 2. b; 3. c; 4. a; 5. hypothalamus (d); 6. pituitary gland (b); 7. adrenal gland (a); 8. ovaries (f); 9. testes (i); 10. pancreatic islets (c); 11. thymus (h); 12. parathyroid gland (g); 13. thyroid gland (j); 14. pineal gland (e)

31.3 Nature of Hormone Action. [pp.506-507]

1. steroid, amines, peptides, proteins; 2. must have receptors, interaction of hormones, concentration of the hormone, cell's metabolic and nutritional state, environmental cues; 3. a; 4. b; 5. b; 6. a; 7. b; 8. steroid; 9. lipid; 10. nucleus; 11. transcription; 12. response; 13. peptide; 14. enzyme; 15. cyclic AMP

31.4 The Hypothalamus and Pituitary Gland [pp.508-509]

31.5 Sources and Effects of Other Vertebrate Hormones [p.510]

31.6 Thyroid and Parathyroid Glands [p.511]

1. e; 2. d; 3. b; 4. a; 5. b; 6. a; 7. c; 8. d; 9. d; 10. c

31.7 The Adrenal Glands [pp.512-513]

31.8 Pancreatic Hormones [p.514]

31.9 Diabetes [p.515]

1. a; 2. b; 3. b; 4. b; 5. b; 6. a; 7. a; 8. T; 9. F, fight-flight; 10. F, insulin; 11. F, beta; 12. T; 13. F, pancreatic islets; 14. T; 15. T; 16. F, Type I

31.10 The Gonads, Pineal Gland, and Thymus [p.516]

31.11 Invertebrate Hormones [p.517]

1. c; 2. d; 3. f; 4. a; 5. e; 6. g; 7. b

Self Quiz

1. d; 2. b; 3. a; 4. b; 5. d; 6. d; 7. e; 8. e; 9. b; 10. a

Chapter Summary

1. hormones; 2. development; 3. endocrine; 4. receptors; 5. transduction; 6. hypothalamus; 7. glands; 8. negative; 9. environment; 10. molting; 11. invertebrates; 12. mammalian; 13. receptors

---

Chapter 32: Structural Support and Movement

32.1 Muscles and Myostatin [p.521]

32.2 Animal Skeletons [pp.522-523]

32.3 Bones and Joints [pp.524-525]

1.b; 2. d; 3. a; 4.f; 5. a; 6. b; 7. a; 8. c; 9. c; 10. b; 11. a; 12. e; 13. c; 14. d; 15. c; 16. e; 17. b; 18. b; 19. c; 20. g; 21. d; 22. f; 23. a; 24. e; 25. h; 26. compact bone; 27. spongy bone; 28. spongy bone; 29. compact bone; 30. connective tissue; 31. blood vessel; 32. hormones; 33. negative; 34. thyroid; 35. calcitonin; 36. osteoclasts; 37. parathyroid; 38. parathyroid hormone (PTH); 39. osteoclast; 40.a; 41. b; 42. e; 43. c; 44. d

32.4 Skeletal-Muscular Systems [pp.526-527]

32.5 How Does Skeletal Muscle Contract
[pp.528-529]

1. F, skeletal; 2. T; 3.T; 4. T; 5. triceps brachii (e); 6. pectoralis major (h); 7. external oblique (j); 8. rectus abdominis (k); 9. sartorius (b); 10. quadriceps femoris (i); 11. biceps brachii (c); 12. deltoid (d); 13. trapezius (f); 14. gluteus maximus (a); 15. biceps femoris (g); 16. a; 17. c; 18. d; 19. e; 20. b; 21. 5; 22. 3; 23. 1; 24. 2; 25. 4

32.6 From Signal to Responses [pp.530-531]
32.7 Muscles and Health [p.532]

1. h; 2. b; 3. c; 4. g; 5. e; 6. f; 7. d; 8. a; 9. tetanus; 10. botulism.

Self Quiz
1. d; 2. c; 3. c; 4. c; 5. d; 6. b; 7. b; 8. b; 9. a; 10. d.

Chapter Summary
1. contractile; 2. skeleton; 3. hydrostatic; 4. exoskeleton; 5. endoskeleton; 6. collagen; 7. minerals; 8. blood; 9. joints; 10. skeletal; 11. reverse; 12. tendons; 13. myofibrils; 14. sarcomeres; 15. actin; 16. ATP; 17. contraction; 18. motor units; 19. neuron; 20. tension; 21. contracts; 22. exercise

---

Chapter 33: Circulation

33.1 And Then My Heart Stood Still [p.537]
33.2 Internal Transport Systems [pp.538-539]

1. a; 2. c; 3. e; 4. d; 5. b; 6. c; 7. a; 8. b; 9. c; 10. a; 11. b; 12. b.

33.3 The Human Cardiovascular System [pp.540-541]
33.4 The Human Heart [pp.542-543]

1. jugular veins (h); 2. superior vena cava (j); 3. pulmonary veins (d); 4. hepatic portal vein (a); 5. renal vein (o); 6. inferior vena cava (q); 7. iliac veins (f); 8. femoral vein (b); 9. femoral artery (c); 10. iliac arteries (i); 11. abdominal aorta (p); 12. renal artery (g); 13. brachial artery (n); 14. coronary arteries (e); 15. pulmonary arteries (k); 16. ascending aorta (m); 17. carotid arteries (l); 18. superior vena cava; 19. semilunar valve; 20. pulmonary veins; 21. right atrium; 22. AV valve; 23. right ventricle; 24. inferior vena cava; 25. septum; 26. left ventricle; 27. AV valve; 28. left atrium; 29. pulmonary veins; 30. semilunar valve; 31. ventricles; 32. SA; 33. T; 34. T; 35. pericardium.

33.5 Characteristics and Functions of Blood [pp.544-545]

1. a; 2. c; 3. b; 4. d; 5. c; 6. b; 7. a; 8. a; 9. c; 10.a; 11. water; 12. plasma proteins; 13. 1-2%; 14. red blood cells; 15. neutrophils; 16. lymphocytes; 17. phagocytosis; 18. defense against parasitic worms; 19. basophils; 20. platelets; 21. b; 22. a

33.6 Blood Vessel Structure and Function [p.546]
33.7 Blood Pressure [p.547]
33.8 Capillary Exchange [p.548]
33.9 Vein Function [p.549]
33.10 Cardiovascular Disorders [pp. 550-551]
33.11 Interactions with the Lymphatic System [pp.576-577]

1. h; 2. j; 3. k; 4. f; 5. I; 6. g; 7. a; 8. d; 9. e; 10. b; 11. c; 12. b 13. d 14. c 15. a 16. g; 17. d; 18. e; 19. f; 20. a; 21. b; 22. c; 23. lymph vascular; 24. water; 25. interstitial; 26. lymph; 27. lymph; 28. plasma; 29. blood; 30. fats; 31. small; 32. pathogens; 33. nodes.

Self Quiz
1. e; 2. a; 3. d; 4. b; 5. c; 6. a; 7. c; 8. a; 9. d; 10.d

Chapter Summary
1. circulatory; 2. tissues; 3. connective; 4. platelets; 5. plasma; 6. gas; 7. defend; 8. four; 9. ventricles; 10. lung; 11. heart; 12. rhythmically; 13. ventricles; 14. atria; 15. arterioles; 16. capillaries; 17. vessels; 18. rhythms; 19. life-styles; 20. lymph; 21. infectious

---

Chapter 34: Immunity

34.1 Frankie's Last Wish [p.557]
34.2 Integrated Response to Threats [pp.588-559]
34.3 Surface Barriers [pp.560-561]

1. b; 2. c; 3. f; 4. e; 5. a; 6. d; 7.lysozyme enzyme; 8. low pH, *Lactobacillus* secretions; 9. dead keratin-packed cells, harmless bacteria and yeast; 10. harmless microbes, saliva lysozyme; 11. low pH.

34.4 Innate Immune Responses [pp.562-563]
34.5 Antigen Receptors in Adaptive Immunity [pp.564-565]

34.6 Overview of Adaptive Immune Response [pp.566-567]

1. c; 2. b; 3. a; 4. a; 5. c; 6. b; 7. c; 8. a; 9. e; 10. c; 11. b; 12. d; 13adaptive; 14. diversity; 15. specificity; 16. memory; 17. self vs nonself recognition; 18. molecular; 19. one; 20. receptors; 21. remember; 22.d; 23. e; 24. a; 25. c; 26. b; 27. f; 28. a; 29. e; 30. b; 31. c; 32. d; 33. b; 34. c; 35. e; 36. d; 37. a.

34.7 The Antibody-Mediated Immune Response [pp.568-569]

34.8 Blood Typing [p.569]
34.9 The Cell Mediated Response [p.570-571]
1. dendritic cell; 2. antigen presenting cell; 3.
cytokines; 4. effector cytotoxic T cell; 5. activated
cytotoxic T cell; 6. effector T helper cell; 7. naïve T
helper cell; 8. B; 9. A; 10. AB; 11. receive; 12.
foreign; 13. O; 14. Rh; 15. marker; 16. Rh +; 17.
Rh-; 18. antibodies 19 .Rh+; 20. Charles
34.10 Allergies [p.571]
34.11 Vaccines [p.572]
34.12 Antibodies Awry [p.573]

34.13 AIDS [pp.574-575]
 1. DPT; 2. HiB; 3. Pneumonococcal; 4. c; 5. a; 6. e;
7. h; 8. g; 9. f; 10. I; 11. d; 12. b
Self Quiz
1. b; 2. b; 3. c; 4. d; 5. b; 6. e; 7. c; 8. c; 9. a; 10. a
Chapter Summary
1. surface barriers, innate immunity, adaptive
immunity; 2. innate; 3. adaptive; 4. skin, mucous
membranes; 5. innate; 6. phagocytic; 7.
antibodies; 8. cytotoxic T; 9. self; 10. allergies,
autoimmune disorders

Chapter 35: Respiration
35.1 Up in Smoke [pp.579]
35.1 The Process of Respiration [p.580]
35.3 Invertebrate Respiration [p.581]
1. c; 2. a; 3. c; 4. a; 5. b; 6. b; 7. c; 8. a; 9. d
35.4 Vertebrate Respiration [pp.582-583]
35.5 Human Respiratory System [pp.584-585]
1. c; 2. a; 3. b; 4. c; 5. b,d; 6. oral cavity (b); 7.
pleural membrane (j); 8. intercostals muscles (f);
9. diaphragm (a); 10. bronchial tree (g); 11. lung
(h); 12. trachea (d); 13. larynx (k); 14. epiglottis (i);
15. pharynx (c); 16. nasal cavity (e);
35.6 How You Breath [pp.586-587]
35.7 Gas Exchange and Transport [pp.588-589]

1. b; 2. a; 3. b; 4. b; 5. a; 6. a; 7. CO_2; 8. acidity; 9.
chemoreceptors; 10. brain stem; 11. diaphragm;
12. decline; 13. tidal volume
**35.8 Common Respiratory Diseases and
    Disorders** [pp.590-591]
1. d; 2. b; 3. e; 4. a; 5. f; 6. c; 7. g
Self Quiz
1. b; 2.d; 3. e; 4. e; 5. e; 6. a; 7. b; 8. d; 9. b; 10. b
Chapter Summary
1. oxygen; 2. carbon dioxide; 3. pH; 4. respiration;
5. tissues; 6. gills; 7. internal; 8. tubes; 9. gills; 10.
lungs; 11. lungs; 12. blood; 13. respiratory; 14.
interstitial; 15. gas; 16. respiratory; 17. depth; 18.
brain; 19. infectious disease; 20. pollutants; 21.
cigarette

Chapter 36: Digestion and Human Nutrition
36.1 The Battle Against Bulge [p.595]
36.2 Animal Digestive System [pp.596-597]
36.3 The Human Digestive System [pp.598-599]
1. incomplete; 2. sac-like; 3. wastes; 4. gut; 5.
pharynx; 6. digested; 7. specialization; 8.
complete; 9. mouth; 10. anus; 11. specialized; 12.
absorption; 13. salivary glands (j); 14. liver (d); 15.
gallbladder (c); 16. pancreas (a); 17. mouth (e); 18.
pharynx (h); 19. esophagus (b); 20. stomach (i);
21. small intestine (f); 22. large intestine (k); 23.
anus (g); 24. f; 25. g; 26. e; 27. c; 28. a; 29. d; 30.b.
36.4 Digestion in the Mouth [p.599]
36.5 Food Storage and Digestion in the Stomach
    [p.600]
36.6 Structure of the Small Intestine [pp.601-
    602]
**36.7 Digestion and Absorption in the Small
    Intestine** [pp.602-603]
1. polysaccharides; 2. small intestine; 3. pepsins;
4. small intestine; 5. pancreas; 6. intestinal lining;
5. triglycerides; 6. b; 7. d; 8. e; 9. c; 10. a; 11. g; 12.

f; 13. b; 14. c; 15. d; 16. a; 17. b ; 18. c; 19. f ; 20. e ;
21. d ; 22. a
36.8 The Large Intestine [p.604]
36.9 The Fate of Absorbed Compounds [p.605]
36.10 Human Nutritional Requirements [pp.606-
    607]
36.11 Vitamins and Minerals [pp.608-609]
36.12 Maintaining a Healthy Weight [pp.610-
    611]
1. T; 2. T; 3. low; 4. T; 5. 3 cups/week; 6.2
cups/week; 7.3 cups/week; 8. 3 cups/week; 9. 2
cups/day; 10. 3 cups/day; 11. 3 ounces/day 12. 5.5
ounces/day; 13. beta carotene in yello fruits,
fortified milk, egg, liver; 14.inactive form made in
sin and activated in the liver; fatty fish, egg yolk;
15. counters free radicals, maintains cell
membranes; 16. K; 17. connective tissue
formation; 17. coenzyme action; 19. niacin; 20.
spinach, tomatoes, potatoes, meats; 21. common
in many foods; 22. folate; 23. poultry, fish, red
meat; 24. coenzyme in fat, glycogen formation;

25. fruits and vegetables; 26. a; 27. d; 28. b; 29. f;
30. e; 31. c.
Self Quiz
1. d; 2. a; 3. a; 4. c; 5. c; 6. e; 7. b; 8. a; 9. c; 10. d

Chapter Summary
1. sac; 2. tube; 3. organ systems; 4. nutrients; 5.
wastes; 6. mouth; 7. stomach; 8. salivary; 9.
pancreas; 10. liver; 11. small intestine; 12. water;
13. wastes; 14. nutrients; 15. vitamins; 16. calories;
17. calories

Chapter 37: The Internal Environment
37.1 Truth in a Test Tube [p.615]
**37.2 Maintaining the Volume and Composition
of Body Fluids** [pp.616-617]
37.3 Structure of the Urinary System [pp.618-
619]
1. freshwater; 2. marine; 3. f; 4. a; 5. h; 6. g; 7.c; 8.
d; 9. e; 10. b; 11. d; 12. a; 13. c; 14. b; 15. kidney;
16. ureter; 17. urinary bladder; 18. urethra; 19.
kidney cortex; 20. kidney medulla; 21. renal
artery; 22. renal vein;23.ureter
37.4 Urine Formation [pp.620-621]
37.5 Kidney Disease [p.622]
1. c; 2. d; 3. f; 4. b; 5. f; 6. e; 7. a; 8. g; 9. c; 10. b; 11.
d; 12. a; 13. e; 14. f

37.6 Heat Gains and Losses [p.623]
37.7 Temperature Regulation in Mammals
[pp.624-625]
1. h; 2. i; 3. g; 4. j; 5. a; 6. c; 7. k; 8. e; 9. b; 10. d; 11.
f; 12. I;
Self Quiz
1. a; 2. b; 3. c; 4. c; 5. d; 6. c; 7. b; 8. a; 8. b; 10.b
Chapter Summary
1. wastes; 2. water; 3. volume; 4. extracellular; 5.
homeostasis; 6. urinary; 7. kidney; 8. bladder; 9.
nephrons; 10. urine; 11. filtration; 12.
reabsorption; 13. secretion; 14. ADH; 15.
aldoseterone; 16. body temperature

Chapter 38: Reproduction and Development
38.1 Mind-Boggling Births [p.629]
38.2 Modes of Animal Reproduction [pp.630-
631]
1. meiosis; 2. fertilization; 3. zygote; 4. single; 5.
offspring; 6. asexual; 7. genetically; 8. uniformity;
9. environment; 10. variations; 11. packages; 12.
animals; 13. resources; 14. sexually; 15. alleles; 16.
variation; 17. survive; 18. reproduce; 19. c; 20. d;
21. e; 22. b; 23. g; 24. a; 25.f
38.3 Reproductive Function of Human Males
[pp.632-633]
1.h; 2. f; 3.a; 4. e; 5. i; 6.g; 7. d; 8. j; 9. b; 10. c; 11.
anterior; 12. hypothalamus; 13. decreased;
38.4 Reproductive Function of Human Females
[pp.634-635]
38.5 Hormones and the Menstrual Cycle
[pp.636-637]
1. ovary; 2. oviduct; 3. uterus; 4. urinary bladder;
5. myometrium; 6. endometrium; 7. vagina; 8.
labium major; 9. labium minor; 10. clitoris; 11.
urethra; 12. c; 13. d; 14. b; 15. e; 16. f; 17. g; 18. h;
19. a; 20. b; 21. d; 22. c; 23. a
38.6 When Egg and Sperm Meet [pp.638-639]
38.7 Preventing Pregnancy [p.640]
38.8 Sexually Transmitted Diseases [p.641]
1. ejaculation; 2. vagina; 3.ovulation; 4.
contractions; 5. hundred; 6. oviduct; 7. jelly coat;
8. enzymes; 9. digestive; 10. jelly coat; 11.

chemical; 12. oocyte; 13. polar body; 14. ovum;
15. nucleus; 16. nucleus; 17. diploid; 18. j; 19. f; 20.
k; 21. d; 22. h; 23. c; 24. l; 25. a; 26. e; 27. I; 28. b;
29. g; 30. m; 31. f; 32. a,c; 33. d,e; 34. f; 35. f; 36. d;
37. c; 38. f; 39. c; 40. f; 41. c; 42. e; 43. d; 44. b; 45.
a,b,c,d,e,f.
38.9 Overview of Animal Development [pp.642-
643]
1. f; 2. b; 3.c; 4. a; 5. e; 6. d 7.a. mesoderm; b.
ectoderm ; c. endoderm; d. mesoderm; e.
ectoderm; f. mesoderm; g. endoderm; h.
mesoderm; i. mesoderm
38.10 Early Marching Orders [pp.644-645]
1. i; 2. d; 3. e; 4. g; 5. h; 6. c; 7. j; 8. f; 9. a; 10. b; 11.
k
38.11 Specialized Cells, Tissues, and Organs
[pp.646-647]
1. a; 2. c; 3. f; 4. d; 5. i; 6. e; 7. h; 8. b; 9. g
38.12 Early Human Development [pp.648-649]
38.13 Emergence of Distinctly Human Features
[pp.6650-651]
38.14 Function of the Placenta [p.652]
38.15 Birth and Lactation [p.653]
1. f; 2. g; 3. h; 4. e; 5. d; 6. b; 7. a; 8. c; 9. a; 10. g; 11.
c; 12. h; 13. d; 14. e; 15. b; 16. f; 17. c; 18. a; 19. b;
20. maternal; 21. C; 22. E; 23. A; 24. F; 25. G; 26. B;
27. rubella; HIV; 28. alcohol; tobacco; caffeine;
illegal drugs; 29. Accutane, Paxil; 30. e; 31. c; 32.

d; 33. a; 34. b; 35. d; 36. b; 37. f; 38. a; 39. e; 40. c; 41. a; 42. d; 43. c; 44. a; 45. d; 46. b.

Self Quiz

1. b; 2. a; 3. b; 4. a; 5. d; 6. e; 7. c; 8. e; 9. c; 10. a; 11. b; 12. d; 13. e; 14. d; 15. b; 16. d; 17. e; 18. a; 19. b; 20. c

Chapter Summary

1. sexual; 2. asexual; 3. variable; 4. gamete formation; 5. fertilization; 6. cleavage; 7. gastrulation; 8. organ formation; 9. growth and tissue specialization; 10. sperm; 11. ovaries; 12. hypothalamus; 13. cyclic; 14. sexual intercourse; 15. pathogens; 16. fertilization; 17. blastocyst; 18. placenta; 19. genetic; 20. cells

Chapter 39: Animal Behavior

39.1 An Aggressive Defense [p.657]
39.2 Behavior's Genetic Basis [pp.658-659]
39.3 Instinct and Learning [pp.660-661]

1. c; 2. e; 3. b; 4. f; 5. a; 6. d; 7. the banana slug; 8. ate; 9. inland; 10. were not; 11. genetic; 12.a. cukoo birds are social parasites in that adult females lay eggs in the nests of other bird species, b. young toads instinctively capture edible insects with sticky tongues; if a bumblebee is captured and the stings the tongue, the toad learns to leave bumblebees alone.

39.4 Adaptive Behavior [p.662]
39.5 Communication Signals [pp.662-663]

39.6 Mates, Offspring, and Reproductive Success [pp.664-665]

1. a; 2. c; 3. b; 4. d; 5. a; 6. hangingflies; 7. sage grouse; 8. lions, sheep, elephant seals, bison

39.7 Living in Groups [pp.666-667]
39.8 Why Sacrifice Yourself? [pp.668-669]

1. d; 2. b; 3. c; 4. a

Self Quiz

1. c; 2. d; 3. ; 4. d; 5. b; 6. b; 7. a; 8. b; 9. c; 10. e

Chapter Summary

1. behavior; 2. pheromones; 3. environmental; 4. heritable; 5. natural selection; 6. social; 7. communication; 8. benefits; 9. evolution; 10. human; 11. animals; 12. moral

Chapter 40: Population Ecology

40.1 A Honking Mess [p.673]
40.2 Population Demographics [pp.674-675]

1. population crashes occur when population size exceeds that allowed by resources; 2. I; 3. f; 4. b; 5. e; 6. g; 7. a; 8. d; 9. k; 10. j; 11. l; 12. c; 13. h; 14. m; 15. 2,500

40.3 Population Size and Exponential Growth [pp.676-677]
40.4 Limits on Population Growth [pp.678-679]

1. a. 0.4 births per rat per month, b. 0.1 deaths per rat per month, c. 0.3 per rat per month; 2. e; 3. j; 4. I; 5. f; 6. b; 7. a; 8. h; 9. c; 10. g; 11. d; 12. b; 13. a; 14. a; 15. b; 16. a; 17. b

40.5 Life History Patterns [pp.680-681]
40.6 Evidence of Evolving Life History Patterns [pp.682-683]

1. e; 2. c; 3. a; 4. b; 5. f; 6. d; 7. F, can; 8. T; 9. T; 10. F, guppy; 11. F, rapidly; 12. T; 13. F, a genetic; 14. F, pike-cichlids; 15. F, had; 16. T

40.7 Human Population Growth [pp.684-685]
40.8 Population Growth and Economic Effects [pp.686-687]

1. geographic expansion, increased carrying capacity, sidestepped limiting factors; 2. limited natural resources (carry capacity); 3. the average number of children born to a woman during her reproductive years 4; d; 5. c; 6. a; 7. b; 8. c; 9. b; 10. c; 11. b; 12. d; 13. a; 14. c

Self Quiz

1. d; 2. a; 3. d; 4. b; 5. b; 6.a; 7. c; 8. a; 9. a; 10.b

Chapter Summary

1. growth; 2. size; 3. distribution; 4. age; 5. density; 6. reproductive; 7. exponential; 8. carrying capacity; 9. environmental; 10. stabilize; 11. disease; 12. restrict; 13. limiting; 14. history; 15. sidestepped; 16. global; 17. cultural; 18. technological; 19. resources

Chapters 41: Community Ecology

41.1 Fighting Foreign Fire Ants [p.691]
41.2 Community Structure [p.692]
41.3 Mutualism [p.693]
41.4 Competitive Interactions [pp.694-695]
41.5 Predation and Herbivory [pp.696-697]

1. *Solenopsis invicta,* one of the Argentine fire ants, can be attacked and killed in its native habitat by parts of the life cycle of two flies. Both are parasitoids. Other options include importation of fungal or protistan pathogens that will infect *S. invicta* but not native ants; 2. climate; 3. food; 4. adaptive; 5. survive; 6. numbers; 7. community; 8. e; 9. h; 10. g; 11. c; 12. j; 13. a; 14. k; 15. b; 16. f; 17. i; 18. d; 19. commensalisms; 20. helpful; 21. harmful; 22. harmful; 23. parasitism; 24. a. The larval stages of the moth grow only in the yucca plant; they eat only yucca seeds; every plant species of the genus *Yucca* can be pollinated only by one species of the yucca moth genus; the yucca moth is the plant's only pollinator; b. Fungal hyphae obtain sugar molecules from plant roots; plant roots obtain mineral ions obtained by mycorrhiza; c. Anemone fish are sheltered and protected by nematocyst-laden tentacles of cnidarians; aggressive anemone fish chase away predatory butterflyfishes than can bite off the sea anemone's tentacles; d. Long ago, phagocytic cells engulfed aerobic bacterial cells that tapped host nutrients; host cells obtain ATP produced by the guests (origin of mitochondria and chloroplasts); 25. d; 26. c; 27. a; 28. e; 29. b; 30. a; 31. e; 32. b; 33. c; 34. e; 35. a; 36. d; 37. f; 38. c; 39. f; 40. e; 41. a; 42. b; 43. a; 44. d; 45. d

41.6 Parasites, Brood Parasites, and Parasitoids [pp.698-699]

1. parasites; 2. parasitic; 3. T; 4. reproductive; 5. T; 6. The biological control agent only targets one species rather than killing many species as some pesticides do. 7. European cuckoo & North American cowbird; roundworm; tick

41.7 Ecological Succession [pp.700-701]

1. a; 2. c; 3. a; 4. b; 5. b; 6. a; 7. a; 8. a; 9. a; 10. c 11. physical factors like soil and climate, chance events, the extent of disturbances

41.8 Species Interactions and Community Instability [pp.702-703]

1. b; 2. a; 3. c; 4. a; 5. e; 6. b; 7. a; 8. b; 9. c; 10. a; 11. c

41.9 Biogeographic Patterns in Community Structure [pp.704-705]

1. equator; 2. tropics; 3. sunlight; 4. resource; 5. evolving; 6. greater; 7. herbivores; 8. predatory; 9. parasitic; 10. reefs; 11. Island C

Self –Quiz

1. a; 2. a; 3. d; 4. b; 5. d; 6. a; 7. e; 8. e; 9. b; 10. d

Chapter Summary

1. species; 2. niche; 3. community; 4. commensalism; 5. predation; 6. size; 7. species interactions; 8. disturbances; 9. biogeographers; 10. tropical; 11. species; 12. conservation.

Chapter 42: Ecosystems

42.1 Too Much of a Good Thing [p.709]
42.2 The Nature of Ecosystems [p.710]
42.3 Food Chains [p.711]
42.4 Food Webs [pp.712-713]
42.5 Ecological Pyramids [p.713]
42.6 Biogeochemical Cycles [p.714]

1. b; 2. d; 3. h; 4. c; 5. k; 6. l; 7. a; 8. e; 9. f; 10. j; 11. g; 12. i;13. 4; 14. 1,2,3; 15. 5,6,7,8; 16. 9,11; 17. none; 18. 10,12,13; 19. 5,6,7,8; 20. 10; 21. In a grazing food web energy flows from producers to herbivores and then to carnivores and decomposers. In a detrital food web energy flows from producers to detritivores and then decomposers; 22. f; 23. h; 24. e; 25. c; 26. b; 27. g; 28. a; 29. d; 30.grass; 31. grasshopper; 32. spider; 33. sparrow; 34. hawk

42.7 The Water Cycle [pp.714-715]

1. c; 2.a ; 3.e ; 4.b ; 5.d

42.8 The Carbon Cycle [pp.716-717]

1. ; 2. ; 3. ; 4. ; 5. ; 6.; 7. The greenhouse gases trap heat radiated from the earth, thus increasing the temperature of the planet. An increase in greenhouse gases equates to an increase in global temperatures.

42.9 The Nitrogen Cycle [p.718]
42.10 The Phosphorous Cycle [p.719]

1. b; 2. c; 3. a; 4. a; 5. a; 6. c; 7. a; 8. b

Self Quiz

1. c; 2. b; 3. b; 4. d; 5. d; 6. d; 7. a; 8. c; 9. c; 10. c

Chapter Summary

1. ecosystem; 2. energy; 3. chains; 4. detritivores; 5. webs; 6. eutrophication; 7. productivity; 8. photosynthesis; 9. carbon; 10. nitrogen; 11. reservoirs; 12. human

Chapter 43: The Biosphere

43.1 El Nino Effects [p.723]
43.2 Air Circulation Patterns [pp.724-725]
42.3 The Ocean, Landforms, and Climates [pp.726-727]
1. j; 2. b; 3. g; 4. k; 5. e; 6. a; 7. i; 8. c; 9. f; 10. d; 11. h; 12. warms; 13. ascends; 14. moisture; 15. descends; 16. moisture; 17. ascends; 18. moisture; 19. descends; 20. easterlies; 21. westerlies; 22. March; 23. December; 24. September
43.4 Biomes [pp.728-729]
43.5 Deserts [pp.730-731]
43.6 Grasslands [p.732]
43.7 Dry Shrublands and Woodlands [p.733]
1. a; 2. b; 3. c; 4. d; 5. c; 6. d; 7. d; 8. e; 9. a; 10. c; 11. b;
43.8 Broadleaf Forests [pp.734-735]
43.9 Coniferous Forests [p.736]

43.10 Tundra [p.737]
1. c; 2. d; 3. I; 4. h; 5. g; 6. f; 7. e; 8. b; 9. j; 10. a
43.11 Freshwater Ecosystems [pp.738-739]
43.12 Coastal Ecosystems [p.740]
43.13 Coral Reefs [p.741]
43.14 The Open Ocean [pp.742-743]
1. g; 2. b; 3. d; 4. a; 5. f; 6. c; 7. e; 8. b; 9. d; 10. a; 11. c; 12. e; 13. b; 14. d; 15. a; 16. benthic; 17. pelagic; 18. intertidal zone; 19. neritic zone; 20. oceanic zone.
Self Quiz
1. a; 2. b; 3. c; 4. c; 5. d; 6. b; 7. e; 8. c; 9. a; 10. d
Chapter Summary
1. regional; 2. energy; 3. distribution; 4. climates; 5. ocean; 6. circulation; 7. ecosystems; 8. biogeographic; 9. biomes; 10. soil type; 11. 71; 12. light; 13. gases; 14. productivity; 15. ocean

Chapter 44: Human Effects on the Biosphere

44.1 A Long Reach [p.747]
44.2 A Global Extinction Crisis [pp.748-749]
44.3 Harmful Land Use Practices [p.750]
1. c; 2. b; 3. a; 4. speciation; 5. extinction; 6. endangered; 7. overharvesting; 8. panda; 9. nitrite; 10. desertification
44.4 Acid Rain [p.751]
44.5 Biological Effects of Chemical Pollutants [p.752]
44.6 The Trouble with Trash [p.753]
44.7 Ozone Depletion and Pollution [p.754]
44.8 Global Climate Change [p.755]
1.e; 2. b; 3. a; 4. c; 5. d; 6. ozone; 7. Antarctica; 8. UV; 9. greenhouse; 10. bleaching
44.9 Conservation Biology [pp.756-757]
44.10 Reducing Negative Impacts [pp.758-759]

1. b; 2. A; 3. D; 4. C; 5. critical; 6. endangered; 7. Madagascar; 8. ecoregions; 9. flow; 10. limitations; 11. life; 12. globalization; 13. ecosystems
Self Quiz
1. c; 2. b; 3.a ; 4. d; 5. b; 6. a; 7.b; 8. c; 9. biodiversity; 10. hot spots; 11. restoration
Chapter Summary
1. speciations; 2. mass; 3. overharvesting; 4. species; 5. introductions; 6. endanger; 7. vertebrates; 8. climate; 9. deforestation; 10. carbon; 11. Pollutants; 12. oxides; 13. aquatic; 14. bioaccumulate; 15. magnification; 16. nonpoint; 17. trash; 18. ozone; 19. destroyer; 20. polar; 21. evolutionary; 22. environmental